D0081388

SPACE TECHNOLOGY

Sourcebooks in Modern Technology

Space Technology

Joseph A. Angelo, Jr.

GREENWOOD PRESS
Westport, Connecticut • London

Library of Congress Cataloging-in-Publication Data

Angelo, Joseph A.
Space technology / Joseph A. Angelo, Jr.
p. cm.—(Sourcebooks in modern technology)
Includes bibliographical references and index.
ISBN 1–57356–335–8 (alk. paper)
1. Astronautics. 2. Space sciences. 3. Outer space—Exploration. I. Title. II. Series.
TL790.A54 2003
629.4—dc21 2002075310

British Library Cataloguing in Publication Data is available.

Library of Congress Catalog Card Number: 2002075310
ISBN: 1–57356–335–8

First published in 2003

Greenwood Press, 88 Post Road West, Westport, CT 06881
An imprint of Greenwood Publishing Group, Inc.
www.greenwood.com

Printed in the United States of America

The paper used in this book complies with the Permanent Paper Standard issued by the National Information Standards Organization (Z39.48–1984).

10 9 8 7 6 5 4 3 2 1

To my beloved daughter, Jennifer April Angelo (1975–1993),
whose unfortunately brief, but very special, presence here on
Earth continues to inspire me that a higher destiny
awaits us all among the stars.

Contents

Preface

The arrival of space technology in the middle of the twentieth century changed the course of human history. Modern military rockets with their nuclear warheads redefined the nature of strategic warfare. Because there would be no victors under a strategy of mutual assured destruction, the avoidance of global nuclear war became the prime national security objective for both the United States and the former Soviet Union during the Cold War. However, stimulated by the politics of that same period, powerful space launch vehicles, initially derived from military rockets, allowed us to escape from the relentless pull of our home planet's gravity—an uncompromising, embracing force within which all previous human history occurred. Through space technology, smart robot exploring machines visited all the major planets in our solar system (except tiny, frigid Pluto), making these distant celestial objects almost as familiar as the surface of our own Moon. Then, as part of history's greatest exploration adventure, space technology enabled human beings to walk for the first time on another world. For many, the Apollo lunar landing missions conducted between 1969 and 1972 by the U.S. National Aeronautics and Space Administration (NASA) represent humanity's most magnificent technical accomplishment.

Through space technology, orbiting instruments and human eyes began to observe the universe directly, rising above the blurring influence of Earth's protective atmosphere and discovering its long-hidden immensity, variety, beauty, and dramatically violent processes. Today, the almost daily discoveries made by sophisticated orbiting astronomical observatories remind us that the universe is not only a strange place, but a much stranger place than anyone dares to imagine. Such sometimes startling, but always exciting, fresh insights about the physical universe also challenge us to revisit age-old philosophical questions about our cosmic origins and our ultimate destiny among the stars.

Closer to home, space technology helped fan the flames of the information revolution, especially through the arrival of the communications satellite and its important contribution in the creation of a global electronic village. The magnificent long-distance views of Earth captured from space during NASA's Apollo Project inspired millions of people to a heightened level of environmental awareness. Weather satellites transformed the practice of meteorological forecasting and severe-weather warning. Today, an armada of sophisticated Earth-observing satellites provides scientists, strategic planners, and government decision makers with a unique opportunity to study our home planet as an integrated, complex, dynamic system. Space technology is directly responsible for the rise of an exciting new multidisciplinary field called earth system science. Unobstructed by physical or political boundaries, the current family of modern Earth-observing spacecraft is creating a transparent globe. People from all around the world can now access information-rich, high-resolution satellite data in their efforts to achieve environmental security and to plan for responsible, sustainable development.

Space technology uniquely provides those very special scientific tools and research opportunities needed to discover whether life (including possibly intelligent life) exists elsewhere beyond Earth and is perhaps even a common phase of cosmic evolution. Over billions of years, matter and energy progressed through a long series of alterations following the ancient big-bang explosion allowing galaxies, stars, and planets to slowly emerge. Encouraged by space-exploration discoveries, we boldly ask: "Is Earth the only place where conscious intelligence emerged in this vast universe?" The German space visionary Hermann Oberth (1894–1989) provided us with a glimpse of the true long-range significance of space technology when he gave the following reason for pursuing space travel: "To make available for life every place where life is possible. To make inhabitable all worlds as yet uninhabitable, and all life purposeful." Within this perspective, space technology offers human beings the universe as both a destination and a destiny.

This book is part of a special series of comprehensive reference volumes that deal with the scientific principles, technical applications, and societal impacts of modern technologies. The present volume serves as your initial, one-stop guide to the very exciting field of space technology. The contents are carefully chosen to meet the information needs of high-school students, lower-level college students, and members of the general public who want to understand the nature of space technology, the basic scientific principles upon which it is based, how space technology has influenced history, and how it is now impacting society. This book serves as both a comprehensive, stand-alone introduction to space technology and an excellent starting point and companion for more detailed personal investigations. Specialized technical books and highly focused electronic (Internet) resources often fail to place an important scientific event, technical discovery, or applications breakthrough within its societal significance. This volume overcomes such serious omissions and makes it easy for you to understand and appreciate the significance and societal consequences of major space-technology developments and the historic circumstances that brought them about. As a well-indexed, comprehensive information resource designed for independent scholarship, this book will also make your electronic (Internet) searches for additional information more meaningful and efficient.

I wish to thank the public information specialists at NASA Headquarters, the NASA Goddard Space Flight Center (GSFC), the NASA Johnson Space Center (JSC), the National Reconnaissance Office (NRO), U.S. Space Command Headquarters, U.S. Air Force Headquarters, and the Washington, D.C., Regional Office of the European Space Agency (ESA) who generously provided much of the special material used in developing this volume. The high-quality support of the staff at the Evans Library of Florida Tech is also most gratefully acknowledged, as well as the patient and sustained encouragement from my editors within the Greenwood Publishing Group, who envisioned this series and its important role in relating science, technology, and society. Finally, without the unwavering support of my wife, Joan, the contents of this book would still be scattered around our home in myriad chaotic clumps.

Chapter 1

History of Space Technology and Exploration

THE SIGNIFICANCE OF SPACE TECHNOLOGY

On July 20, 1969, an American astronaut, Neil Armstrong, became the first human being to walk on another world. He (and eventually eleven other American astronauts) accomplished this amazing feat while participating in the National Aeronautics and Space Administration's (NASA's) Apollo Project. As Armstrong descended from the last step of the lunar excursion module's ladder and made contact with the Moon's surface, he spoke these famous words: "That's one small step for a man, one giant leap for mankind."

Minutes later, astronaut Edwin (Buzz) Aldrin joined Armstrong on the lunar surface. While they explored the Moon and collected rock and soil specimens, their companion (astronaut Michael Collins) orbited overhead in the Apollo command module. Back home on Earth, more than 500 million people around the world watched this event through live television broadcasts. Many people now regard the Apollo lunar landing missions as the most significant technical accomplishment of the twentieth century—if not all human history. Certainly these important missions will be viewed by future historians as the beginning of humankind's extraterrestrial civilization. In the second half of the twentieth century, space technology liberated us from the planetary cradle of Earth and helped us come of age in a vast and beautiful universe. Through space technology, the destiny of our species now lies among the stars, should future generations choose to accept and pursue that destiny.

Before we examine the evolution of space technology and its incredible impact on human destiny, a few general remarks concerning the na-

Astronaut Edwin (Buzz) Aldrin descends to the lunar surface during the first human expedition to the Moon to join fellow "Moon walker" Neil Armstrong on July 20, 1969. Many people regard the Apollo lunar landing missions (1969–1972) as the most significant technical accomplishment in all human history. Photograph courtesy of NASA.

ture and role of technology are appropriate. Science enables us to understand nature and the interactions occurring in the physical universe. One important concept (initially suggested by the brilliant English scientist Sir Isaac Newton) is the universality of the physical laws of science. For example, the action-reaction principle works in the same way on Earth as it does on Mars.

Mathematics enables us to model and predict natural processes and physical events. Mathematical principles are also universal. For example, $4 \times 5 = 20$ is true not only in the Milky Way galaxy (our home galaxy), but in the Andromeda galaxy (our neighboring galaxy) and in the billions of other galaxies that make up the universe.

Why is this so important? If the physical laws and mathematical principles that human beings discovered over the centuries applied only here on Earth and did not work elsewhere, then space travel would prove extremely difficult, if not totally impossible. In general, *technology* is the innovative application of these universal scientific and mathematical principles. Using science and mathematics, we can predictably manipulate the physical universe (matter, energy, and time) to satisfy our basic needs. Specifically, *space technology* involves the manipulation of matter and energy to send objects (including human crews) into outer space on precise, carefully navigated journeys. (A more thorough discussion of space technology occurs a little later in this chapter.)

Throughout human history, the *engineer* has been the person who creatively applied different forms of technology to satisfy human needs and to protect people from hazards. Some of these fundamental human needs include a stable food supply, fresh water, clean air, security, clothing, light, shelter, mobility, information storage, communications, and even entertainment. In addition to numerous important defense, commercial, and scientific applications, space technology also helps us respond to another very fundamental human need: the need to explore.

From birth, curiosity drives us to explore the world around us. As a child in a cradle, we want to see the unseen and to discover what lies beyond our immediate view. However, as we grow, this basic instinct is all too often repressed. We quickly get channeled into socially acceptable patterns of conformity. For example, a young child entering preschool soon learns to "color within the lines." Similarly, blue bananas and purple-faced puppies seldom become "winners" in a tiny-tot art contest.

However, despite such subtle (but relentless) social pressure to conform, a few hardy spirits in each human generation manage to keep this natural childhood sense of curiosity alive. They become the explorers and pioneers who dare to take risks and to boldly venture into unexplored territory. Over the centuries, humanity has greatly benefited from the restless urges of these curious and brave few. Step by painful step, such pioneers have carried the human spirit into unknown and previously unreachable places—to the New World, to the Antarctic, into the microscopic world of subatomic particles, and now (with space technology) to the Moon and the planets. From an evolutionary perspective, the exploration of space represents a major unfolding of human consciousness beyond the boundaries of one beautiful, but small, planet.

By exploring space, we are solving some of the most intriguing questions that have puzzled human beings throughout history. Who are we (as a species)? Where did we come from? Where are we going? Are we alone in this vast universe?

WHAT IS SPACE TECHNOLOGY AND WHERE DID IT COME FROM?

Space technology involves the launch vehicles that harness the principles of rocket propulsion and provide access to outer space, the spacecraft that operate in space or on the surface of another world, and many different types of payloads (including human crews) that accomplish many special functions and objectives. This multidisciplinary technology is not solely a twentieth-century accomplishment. On the contrary, it represents the culmination of the dreams and creative contributions of many dedicated people throughout history. These space visionaries and pioneers labored through the centuries, often at great personal sacrifice, to establish the scientific understanding that supported the astonishing technical developments that occurred in the last four decades of the twentieth century. Global politics also stimulated rapid progress in space technology. A tense Cold War environment made the dreams of spaceflight take second place to the use of rockets as weapons of war. In this tense, competitive environment, scientific space achievements also represented a graphic demonstration of national power.

Politics aside, by harnessing the force of the rocket, engineers and scientists were able to send sophisticated robot explorers to the far reaches of the solar system. For example, NASA's mighty Saturn V launch vehicle allowed human explorers to leave footprints on the Moon during the Apollo Project. Through space technology, we now boldly search for life beyond Earth and dare to explore the farthest reaches of the universe seeking our cosmic origins.

This chapter presents some of the key events, scientific concepts, and political developments that led to the age of space. The earliest events are obscured in antiquity. Other significant milestones are better documented because they occurred in the late Renaissance period when modern science started in Western Europe. Delightful nineteenth-century science-fiction tales from Jules Verne and others proved to be a strong stimulus for each of the three great space-travel visionaries (Konstantin Tsiolkovsky, Robert Goddard, and Hermann Oberth) of the early twentieth century. In the mid-twentieth century, other visionaries, including Wernher von Braun, Arthur C. Clarke, Walt Disney, John F. Kennedy, and Sergei Korolev, transformed dreams of space travel into a reality.

However, the mid-twentieth-century pathway into space was neither simple nor free from conflict. For example, the liquid-propellant rocket, invented by Robert Goddard in 1926, emerged from World War II as a novel, but deadly, weapon—the German V-2 rocket. A powerful new payload, the nuclear weapon warhead, also emerged from World War II. The combination (a long-range, nuclear-armed rocket) became the major instrument of superpower confrontation during the Cold War. Throughout this period of intense political tension between the United States and the former Soviet Union, space-exploration achievements directly reflected national prestige and power. Nowhere was this competition more intensely portrayed than during the Cold War race to land human explorers on the Moon, the so-called Moon race. Fortunately, the extensive military and politically motivated investments in space technology also produced an unanticipated bonus. The period from about 1960 to 1989 is often called the first golden age of space exploration. This period corresponds to that special moment in history when all the major bodies of the solar system except tiny, distant Pluto were initially visited and explored by scientific spacecraft.

The twentieth century closed with the post–Cold War era, a time when superpower competition in space gave way to more peaceful international cooperation. Together as a planetary society and united by common space-exploration goals, people now embarked on an expanded scientific and commercial use of space. They are also driven by an even greater quest, the answer to the fundamental question: Does life exist elsewhere in the solar system (or in the universe)? To help answer this important question at the dawn of a new millennium, an armada of robotic spacecraft is visiting Mars, and other sophisticated robot explorers scrutinize Jupiter and Saturn and their intriguing systems of moons.

Will future space-technology histories mention the discovery of microbial life (or at least its fossil remnants) on Mars? Does the mysterious Jovian moon Europa possess a life-bearing liquid-water ocean? What really lies beneath the murky, nitrogen-rich atmosphere of the Saturnian moon Titan? In the early decades of the twenty-first century, space technology will help answer these intriguing questions and will also produce many exciting surprises. These surprises will have a profound impact on our current model of the universe and our place in it. As you trace the origins of modern space technology as it emerged through the ages, enjoy being part of one of the most exciting periods in all human history—a time when we are discovering more clearly than ever before what our ultimate role and destiny are in the evolving universe.

PREHISTORY TO MEDIEVAL EUROPE

The story of space technology is interwoven with the history of astronomy and humankind's interest in flight and space travel. Most ancient societies developed myths and stories of human flight, even spaceflight, and were curious about the lights in the night sky. The Babylonians, Mayas, Chinese, and Egyptians all studied the sky and the motions of the planets. Nevertheless, it was the Greeks who had the largest impact on early astronomy, developing models and theories that shaped the discipline for centuries. Beginning in the fourth century B.C.E., Greek scientists and philosophers articulated a geocentric model of the universe with Earth at the center and the planets and stars embedded on a series of transparent, concentric spheres revolving about it. In the second century C.E., the Hellenistic astronomer Ptolemy carefully documented this model in the *Almagest*, a work that remained the basis for Arab and Western astronomy until the sixteenth century.

Space technology centers on the rocket, which was first developed by the Chinese. When it emerged is unclear. Historians believe that the Chinese formulated gunpowder by the first century C.E. and used it primarily to make fireworks for festivals. The Chinese filled bamboo tubes with gunpowder and tossed the tubes into a fire. Perhaps one of these tubes did not explode but shot out of the fire, propelled by hot gases from the burning gunpowder mixture. An unknown ancient Chinese "rocket pioneer" observed this event. Curious, he (or she) then began experimenting with a variety of gunpowder-filled tubes. At some point, the early rocketeer probably attached a bamboo tube to an arrow and launched the combination with a bow. The famous Chinese "flaming arrow" was born. The first reported use of rockets in warfare was the Battle of K'ai-fung-fu in 1232, when the Chinese repelled Mongolian invaders with a barrage of "flaming arrows." The Mongolians quickly adopted the weapon and spread rocket technology as they invaded the Middle East and Europe.

By the fourteenth century, gunpowder-rocket technology had dispersed throughout Western Europe, where engineers attempted to develop its role in warfare. During the fourteenth century, French writer Jean Froissart proposed the use of tubes to launch rockets on more accurate trajectories. His idea was the forerunner of the modern bazooka. In 1420, an Italian engineer, Joanes de Fontana, wrote *Book of War Machines*, in which he suggested a rocket-propelled battering ram and a torpedo delivered to its target by a rocket. Throughout the fifteenth and sixteenth centuries, European armies utilized rockets, but the military emphasized the increasingly accurate cannon.

ROCKETRY IN THE AGE OF SCIENCE AND REASON

During the sixteenth and seventeenth centuries, Europe experienced profound changes in intellectual thought that ushered in the early modern age. The changes, loosely called the scientific revolution, laid the foundation for understanding the workings of the universe that ultimately led to space flight. Nicolaus Copernicus began a revolution in astronomy with the publication in 1543 of *On the Revolutions of Celestial Spheres*. The book suggested a heliocentric model of the universe with Earth, like the other known planets, revolving around the Sun. Copernicus's ideas were ignored by most of his scientific contemporaries, but during the seventeenth century, Galileo Galilei made telescopic observations of Jupiter and the Moon that substantiated his theory. Galileo's visionary work set the stage for Sir Isaac Newton's research, which tied these new astronomical observations together in the late seventeenth century. Newton's universal law of gravitation and his three laws of motion, published in his great work *Mathematical Principles of Natural Philosophy* (1687), allowed scientists to explain in precise mathematical terms the motion of almost every object observed in the universe, from an apple falling to the ground to planets orbiting the Sun.

Stimulated by the scientific revolution, seventeenth-century military strategists and engineers promoted the role of the rocket in warfare. Both the Dutch and the Germans conducted experiments with military rockets, and in 1680, Russian czar Peter the Great established a facility in Moscow to manufacture the weapons. Nevertheless, in much of Western Europe, rockets were used primarily for elaborate fireworks displays that became a favorite pastime for royalty. In contrast, they remained important weapons in Asia, where Indian rulers successfully employed them in battles against the British during the late eighteenth century.

Spurred on by the British experience in India, artillerist Sir William Congreve promoted the rocket as a weapon during the early nineteenth century. Congreve developed a wide variety of military rockets, ranging in size from a mass of about 150 kilograms down to 8-kilogram devices. There were two basic types of assault rockets: the shrapnel (case-shot) rocket and the incendiary rocket. The British often employed the shrapnel rocket as a substitute for artillery. When it flew over enemy troops, its exploding warhead showered the battlefield with rifle balls and sharp pieces of metal. The warhead of the incendiary rocket was filled with sticky, flammable materials that quickly started fires when it impacted in an enemy city or in the rigging of an enemy sailing ship. The British used both types effectively during the Napoleonic Wars and the War of 1812.

Despite Congreve's extensive efforts, early gunpowder rockets remained inaccurate. The devastating nature of these military rockets was not due to their accuracy or the power of an individual warhead but their numbers. In the 1840s, English inventor William Hale improved rocket accuracy by means of spin stabilization. The rocket's exhaust impinged on angled surfaces, causing the device to spin about its main (longitudinal) axis as it flew through the air. But as advances in artillery outpaced those in rocketry, rockets became used primarily for civilian purposes. In the mid-nineteenth century, they were employed in marine rescue for throwing lines to sinking ships and in whaling to propel harpoons.

THE DREAMS AND THEORY OF SPACE TRAVEL (1865–1939)

While scientists made few advances in rocketry during the last half of the nineteenth century, writers began exploring the theoretical applications of rockets not as weapons but as vehicles for space travel. In 1865, the science-fiction writer Jules Verne published his classic novel *From the Earth to the Moon*, a tale of a human flight to the Moon via a 10-ton bulletlike capsule fired by a huge cannon. Verne was the first writer to make space travel appear possible, although the technology he chose (a large cannon firing a crew-capsule projectile) was not correct. His three passengers definitely would have been crushed to death by the acceleration. In 1869, Edward Everett Hale published "The Brick Moon," describing a human-crewed space station. Another gifted writer, H.G. Wells, also stimulated the minds of future space pioneers with such exciting tales as *The War of the Worlds* (1898) and *The First Men in the Moon* (1901). In the former tale, Earth is almost conquered by technically advanced Martians but is saved when the invaders are stopped by terrestrial microorganisms that prove fatal to them. In the second story, the English writer gets his travelers to the Moon with "cavorite," a fictitious antigravity substance that repels their spacecraft to the Moon. None of these nineteenth-century science-fiction writers handled the technical problem of space travel properly. Their exciting stories, however, planted the dream of space travel in the minds of those who would develop the theory of large rockets.

Three persons were especially significant in accomplishing the transition from the small, gunpowder-fueled rockets of the nineteenth century to the giant, liquid-fueled multistage rockets of the space age. The founding fathers of astronautics were Robert H. Goddard in the United States, Konstantin E. Tsiolkovsky in Russia, and Hermann J. Oberth in Germany.

These men, working independently during the early twentieth century, developed the theoretical foundations of rocket propulsion and identified the multistage liquid-propellant rocket as the enabling technology for space travel. Tsiolkovsky and Oberth were primarily theoreticians; Goddard went beyond theoretical work and developed the fundamental technology associated with modern rockets. On March 16, 1926, he successfully launched the world's first liquid-fueled rocket. The simple gasoline- and liquid-oxygen–fueled device burned for just two and one-half seconds, climbed 12.5 meters in height, and then unceremoniously landed in a frozen cabbage patch 56 meters away. Unnoticed by the world and witnessed by only four persons, this event is now considered the space-history equivalent to the Wright brothers' first powered-aircraft flight. All modern liquid-propellant rockets are the technical descendants of this modest device. Yet Goddard kept his research secret and so had little direct influence on later rocket development. (See Figure 1.1.)

Inspired by Verne and others, these men also articulated ideas of human spaceflight. Tsiolkovsky wrote science-fiction works on spaceflight and envisioned the development of a space station. Goddard postulated that an "atomic-propelled" space ark, possibly constructed using a small moon or asteroid, might someday carry human civilization away from a dying Sun to the safety of a new star system. Oberth published *The Rocket into Planetary Space* (1923), a detailed technical study of how a spaceship could be built, launched, and recovered. Science writers popularized his work, and the idea of spaceflight became a fad in the 1920s.

This wave of space enthusiasm prompted the formation of many small rocket societies. In 1924, the Soviets created a Central Bureau for the Study of the Problems of Rockets and an All-Union Society for the Study of Interplanetary Flight. Three years later, Oberth and the young Wernher von Braun formed the German Society for Space Travel (Verein für Raumschiffahrt [VfR]), which pursued pioneering studies in the areas of propulsion, supersonic aerodynamics, and guidance. Unable to conduct rocketry experiments in the United Kingdom because of legal restrictions on the use of explosives, space enthusiasts, led by Philip E. Cleator, founded the British Interplanetary Society (BIS) in October 1933. This organization performed a detailed investigation of landing humans on the Moon that helped establish the credibility of lunar travel and served as the intellectual precursor to NASA's Apollo lunar-surface-expedition missions. Yet researchers had difficulty solving the technical problems even of short flights within the atmosphere, and with the onset of the Great Depression and the growth of fascism in Europe, the civilian movement to explore the possibilities of spaceflight lost momentum.

Figure 1.1 Dr. Robert H. Goddard and the world's first liquid-fueled rocket, which he successfully launched in a frozen New England field on March 16, 1926. Drawing courtesy of NASA.

During the 1930s, literature and the media firmly entrenched the rocket ship and interplanetary travel in the minds of the public. In 1932, for example, Philip Wylie and Edwin Balmer published their classic science-fiction story *When Worlds Collide*. In this exciting tale, a giant rogue planet is on a collision course with Earth. Small groups of people, chosen by lottery, use rapidly constructed atomic-powered rockets to take them to safety. Movies, such as the 13-episode *Flash Gordon* series, released in 1936, also entertained the public with tales of space travel and intergalactic warfare.

MODERN ROCKETS AS WEAPONS OF WAR (1939–1956)

As war loomed in the 1930s, the military provided the impetus for further advances in rocketry. During World War II, the United States, the Soviet Union, Japan, the United Kingdom, and other combatant nations produced rockets and guided missiles of all sizes and shapes. However, it was Germany that made the most significant advances in rocket research.

Under the direction of Braun and his colleagues from the VfR, who reluctantly had gone to work for the military in 1932, Germany developed a liquid-propellant rocket that had a design range of 275 kilometers. The rocket, ultimately known as the V-2, was 14 meters long, with an engine that burned liquid oxygen and alcohol. The first successful test of the weapon on October 3, 1942, marked the birth of the modern military ballistic missile. In September 1944, the German army started launching V-2 rockets armed with high-explosive (one-metric-ton) warheads against London, Antwerp, and other Allied cities.

At the end of World War II, another new technology emerged that also had a major impact on the evolution of rocketry. On July 16, 1945, the United States successfully detonated the world's first nuclear explosion. While the use of the atom bomb against Japan helped bring World War II in the Pacific to a dramatic conclusion, its creation and military use also plunged a rapidly polarizing world into a nuclear arms race that dominated geopolitics for the entire Cold War era.

Technical expertise moved to the conquering nations of the United States and the Soviet Union at the end of the war. Many German rocketeers, including Braun, settled in the United States, where they spearheaded the development of rocketry and continued their dream of spaceflight. The Soviets captured the German rocket facilities and took V-2 rockets back to the USSR to use in their own program. Through the postwar emigration of German rocket scientists and the use of captured German rockets, the V-2 became the common technical ancestor to the major military missiles and space launch vehicles developed by the United States and the Soviet Union.

Following World War II, American and Soviet leaders ignored the inspirational space-travel visions of Goddard, Tsiolkovsky, and Oberth and chose instead to encourage a politically forced marriage between the atom and the rocket as the Cold War deepened. The U.S. military moved slowly to embrace rocket technology because of America's monopoly on nuclear power and its strategic focus on overwhelming air power for defense. Only after the Soviet detonation of its first nuclear device in 1949 and the outbreak of the

Korean War the following year did defense officials turn to rockets. The United States stressed the development of small intercontinental ballistic missiles (ICBMs), staged rockets capable of carrying the compact nuclear devices in the American arsenal. In contrast, Russian scientists, who knew that they were clearly behind the United States in sophisticated nuclear weapons technology, made a strategic decision to pursue the development of the very large, high-thrust booster rockets needed to carry their primitive, heavy nuclear weapons. This difference in early strategic nuclear-system planning inadvertently gave the Soviet Union a significant edge in rocket booster technology. This booster advantage became a major benefit for the Soviets during the initial, highly competitive decade of space exploration.

By the mid-1950s, the great ballistic missile war between the Soviet Union and the United States was on. In 1955, under the direction of Sergei Korolev, the Soviets developed the R-7, a huge missile that had 20 engines grouped around a core vehicle and four boosters that dropped off in flight. That same year, the United States, fearing a "strategic nuclear missile gap" with the Soviets, gave ballistic missile development the highest national priority. Cape Canaveral Air Force Station in Florida became the busy testing ground for such important long-range ballistic missiles as the Atlas and the Titan. The arrival of the modern digital computer in the mid-1950s and its innovative application within the emerging field of modern systems engineering made possible the development of giant missiles in a very short period of time. These powerful weapons served both nations as guardians of an uneasy nuclear strategic balance and as launch vehicles during the superpower space race.

While the military was using space technology for weapons development during the postwar years, visionaries kept alive the dream of spaceflight. In September 1948, the *Journal of the British Interplanetary Society* (JBIS) began publishing a four-part series of papers written by L.R. Shepherd and A.V. Cleaver exploring the technical feasibility of applying nuclear energy to space travel. Two years later, the British technical visionary Arthur C. Clarke published an article in the *JBIS* suggesting mining the Moon and launching the mined lunar material into space with an electromagnetic catapult. In 1952, *Collier's* magazine created a surge of American interest in space travel with a series of well-illustrated articles that introduced millions of Americans to the space station, a mission to the Moon, and a human expedition to Mars. Notable rocket scientists, including Braun and Willy Ley, served as technical consultants. That same year, Braun published *The Mars Project*, the first serious technical study of a human expedition to Mars.

In the mid-1950s, entertainment-industry visionary Walt Disney (left) collaborated with rocket scientist Wernher von Braun (right) in the production of a well-animated three-part television series that popularized the dream of space travel for millions of Americans. Photograph courtesy of NASA.

Entertainment-industry visionary Walt Disney also promoted space travel in the mid-1950s. Enlisting the technical services of many of the same rocket scientists who helped *Collier's* magazine, Disney produced an inspiring animated series for his *Disneyland* television program that was shown from 1955 to 1957. The three-part series popularized the dream of space travel for millions of young viewers. In fact, numerous American astronauts, scientists, and aerospace engineers identified these shows as a major career stimulus.

THE EARLY SPACE AGE (1957–1959)

In 1952, the International Council of Space Unions announced an International Geophysical Year for 1957–1958 to explore Earth and its atmosphere. The United States responded with the pledge to launch an artificial earth satellite as the culminating event of the project. The Soviet Union also declared that it would launch a satellite, but few in the West thought it technologically capable of doing so. Against the advice of Braun, who recommended using a modified military missile, the United States made a political decision to develop a "civilian" rocket (the Vanguard) to emphasize the peaceful uses of space and to play down any public emphasis on military applications. The Soviet Union, on the other hand, planned to use a modified R-7 ICBM.

The Soviet Union shattered the U.S. assumption of technological superiority when it launched *Sputnik 1* on October 4, 1957. Less than a month later, the Soviets confirmed their lead in the emerging "space race" with the launch of *Sputnik 2*. Stunned by the Soviet achievement, the United States rushed the launch of the Vanguard on December 6, 1957. The widely publicized attempt ended in disaster. While the world looked on, the Vanguard blew up after rising only a few centimeters from its launch pad. Its payload, a miniature spherical satellite, wound up hopelessly "beeping" at the edge of a raging palmetto-scrub inferno. Soviet premier Nikita Khrushchev sarcastically referred to the tiny (1.5–kilogram) test satellite as the "American grapefruit satellite."

Responding to the Vanguard disaster, the United States mounted an emergency mission to save its prestige. It hastily formed a joint project involving Caltech's Jet Propulsion Laboratory (JPL) and the U.S. Army Ballistic Missile Agency headed by Braun. Braun's team supplied the Jupiter C launch vehicle (a modified intermediate-range ballistic missile), and JPL supplied the fourth-stage rocket and the satellite itself. The distinction between military and civilian rockets was quickly forgotten. On January 31, 1958, America's first satellite, *Explorer 1*, successfully achieved orbit. Having learned its lesson, the United States reverted to modified military rockets for its space exploration until the 1980s.

Sputnik precipitated a race for technological supremacy in space that gave early space exploration a contest mentality. Throughout the remainder of the Cold War, accomplishments in space technology served as vivid manifestations of national power. Superiority in space became emblematic of general technological superiority and the superiority of a nation's economic and political systems.

Determined to win the space race, the United States began strengthening its space program. On October 1, 1958, the federal government transformed the National Advisory Committee for Astronautics, which had been testing flight on the edge of space, into the National Aeronautics and Space Administration (NASA) and gave it control over the nation's space program. NASA's primary goal was the peaceful exploration of space for the benefit of all humankind. Within seven days of its birth, NASA announced the start of the Mercury Project, America's pioneering program to put human beings into orbit. The critical linkage of NASA's overall program with human spaceflight was forged. Two years later, Braun and the Army Ballistic Missile Agency were transferred to NASA to become the nucleus of the agency's space program. Yet the Eisenhower administration was reluctant to commit the nation to a massive civilian space effort, fearing budget imbalances.

During the early years of the space race, the United States lagged behind the Soviet Union. Sergei Korolev's large rockets helped the Soviet Union achieve many dramatic space-technology firsts. Its *Luna 1* spacecraft, launched on January 2, 1959, missed the Moon but became the first human-made object to escape Earth's gravitational attraction and orbit the Sun. *Luna 2* successfully impacted the Moon on September 14, 1959, and became the first space probe to crash-land on another world. Finally, the following month, *Luna 3* circumnavigated the Moon and took the first images of the lunar farside. In contrast, American attempts to send spacecraft to the Moon between 1958 and 1959 were unsuccessful, largely due to the limitations of its launch vehicles.

On April 12, 1961, the Soviets achieved a dramatic space-technology milestone by successfully launching the first human into space. Cosmonaut Yuri Gagarin rode a military rocket into space inside Korolev's *Vostok 1* spacecraft and became the first person to observe Earth from an orbiting vehicle. The United States responded on May 5, 1961, by sending astronaut Alan B. Shepard, Jr., into space using a Redstone rocket as the launch vehicle. NASA was only able to achieve a suborbital flight because the Redstone was simply not powerful enough to place a spacecraft into orbit. It was not until February 20, 1962, that astronaut John H. Glenn, Jr., became the first American to orbit Earth, launched by the powerful Atlas.

THE RACE TO PUT HUMANS ON THE MOON

In the midst of the Soviet Union's technical triumphs, on May 25, 1961, President John Kennedy boldly committed the United States to landing a man on the Moon before the end of the decade. His challenge was accepted by the highly motivated NASA staff, which set about solving the myriad problems and developing the new capabilities needed for a Moon landing. The agency devised a plan that called for a command module to orbit the Moon while a lunar lander descended to the surface. Following an astronaut walk, the lander would rejoin the command module for the return to

President John F. Kennedy during his historic message to a joint session of the U.S. Congress (May 25, 1961) in which he declared: "I believe this nation should commit itself to achieving the goal, before the decade is out, of landing a man on the Moon and returning him safely to Earth." Shown in the background are (left) Vice President Lyndon B. Johnson and (right) Speaker of the House Sam T. Rayburn. Photograph courtesy of NASA.

Earth. To fulfill this plan, NASA had to create the technology for flights of long duration, build spacecraft capable of carrying multiple astronauts, develop the capability to walk in space, and learn how to rendezvous and couple two vehicles in space. Project Gemini, America's second manned space program, accomplished these goals between 1965 and 1966.

Driven by the objectives of the Moon landing, NASA also developed a series of versatile robot spacecraft designed to probe the lunar environment. From 1961 to 1965, nine Ranger spacecraft photographed the lunar surface at close range before impacting. Five Surveyor robot landers then successfully explored the lunar surface from May 1966 to January 1968, transmitting thousands of images from the lunar surface and performing numerous soil-mechanics experiments. Five Lunar Orbiters, launched between August 1966 and August 1967, mapped more than 99 percent of the lunar surface.

Yet despite NASA's accomplishments, the Soviets appeared to maintain a technological lead in the race to put man on the Moon. On October 12, 1964, the Soviet Union launched the *Voskhod 1* into Earth orbit. The six-ton spacecraft, the first to carry three men, was at that time the largest object launched into space. The Soviets made space history again in March 1965 when cosmonaut Aleksei Leonov performed the first space walk. In October 1967, two uncrewed Soviet spacecraft, *Cosmos 186* and *188*, performed the first automated rendezvous and docking operation in space. While pursuing human spaceflight advances in Earth orbit, the Soviets also continued to explore the Moon. In February 1966, *Luna 9* transmitted the first panoramic television pictures ever received from the lunar surface. Two months later, *Luna 10*, a massive 1,500-kilogram spacecraft, became the first human-made object to achieve orbit around the Moon. Yet despite these achievements, the Soviets lost the Moon race. Government interference prevented Korolev and his engineers from developing their plans efficiently, and the program lacked the massive funding the United States threw into the effort

An entire world watched as NASA's *Apollo 11* mission left for the Moon on July 16, 1969. The energy released from the first stage of that rocket was the equivalent of a large electrical power plant. Astronauts Neil A. Armstrong, Michael Collins, and Edwin (Buzz) Aldrin soon achieved what had only been dreamed about in history. On July 20, Armstrong cautiously descended the steps of the lunar module's ladder and touched the lunar surface. He spoke these immortal words: "That's one small step for a man, one giant leap for mankind." As prophesied by the Russian space visionary Konstantin Tsiolkovsky, people did use the rocket to leave the cradle of Earth and take their first steps in the universe. Despite strife around the planet, the astronauts' pioneering accomplishment sent the human spirit soaring. Decades later, people from around the world still remember precisely where they were and what they were doing when they first heard these historic words: "Houston, Tranquility Base here. The *Eagle* has landed."

The United States won the race to the Moon and demonstrated a clear superiority in space technology. National pride soared. But in the mercurial world of government budgets and nonvisionary politics, success does not necessarily support new opportunities. While millions cheered the triumphant lunar astronauts, the nation's leaders were already planning major cutbacks in the space program.

THE AGE OF REUSABLE CRAFT

By the late 1960s, the Soviet Union realized that the United States would win the race to put a man on the Moon and so redirected its space program to emphasize long-duration missions in low Earth orbit. To support this objective, the Soviets developed the world's first space station, *Salyut 1*, supplied by Soyuz spacecraft that were the technological descendants of the capsules the Soviets had used in their early space program. Developed on a crash schedule, *Salyut* was launched in April 1971 and crewed in June. Three cosmonauts remained on the station for 22 days. But tragedy struck upon their return to Earth. The Soyuz spacecraft did not have enough room for the cosmonauts to wear pressure suits for reentry. Consequently, they suffocated when a pressure valve apparently malfunctioned and air rushed out of their spacecraft. Future flights were cancelled, and *Salyut* was taken out of orbit after only 175 days.

As the Apollo Project wound down, NASA also turned to the development of a space station, quickly producing the prefabricated *Skylab*. The design utilized the empty upper stage of a Saturn rocket, a space more than 15 meters in length and 6 meters in diameter. Redesigned as the orbital workshop, the empty rocket tank was fitted with solar panels and divided into two compartments, a work area and living quarters for the crew, with an airlock for leaving and returning. NASA launched *Skylab* in May 1973 and sent three missions to the craft during 1973–1974. The missions, which lasted 28, 59, and 84 days respectively, studied the Sun and Earth as well as gathering invaluable information about human beings' ability to work in space. But due to declining budgets and a growing emphasis on the production of the space shuttle, NASA was quickly forced to abandon *Skylab*. It remained unused in orbit for six years. Unable to maintain its own orbit (a fatal design omission), the huge facility became a derelict and eventually decayed from orbit, making a final, fiery plunge through Earth's atmosphere on July 11, 1979.

Despite the tragic beginning of the Salyut program, the Soviets pressed on with a sequence of space-station designs, culminating in the launch in September 1977 of *Salyut* 6. It contained several important design improvements, including the addition of a second port that permitted the simultaneous docking of two spacecraft. The station was supplied by automated Progress spacecraft, which also acted as tugs, maintaining the station's orbit. Over its lifetime, the *Salyut* 6 station hosted 16 cosmonaut crews, including 6 long-duration crews. It was returned to Earth in July 1982. The final space station in the series, *Salyut 7*, was launched in April 1982 and returned to Earth in February 1991 after completing more than 50,000 orbits. The Salyut series provided the Soviets with a wealth of scientific data, including information about humans living in space, data invaluable for the operation of their third-generation space station, *Mir*.

The Soviets launched this sophisticated modular facility on February 20, 1986. This new station contained more extensive automation, more spacious crew accommodations, and an important new design feature, six docking ports. These ports were able to receive supply ships and special laboratory or workshop modules for space research. With the exception of a six-month period in 1989, *Mir* was crewed from 1986 until it was abandoned in 1999, eventually intentionally plunging into a remote area of the Pacific Ocean near Fiji in March 2001. In the post–Cold War period, the American space shuttle would visit *Mir* station during the initial phase of the *International Space Station* (*ISS*) program (discussed later)

THE SPACE SHUTTLE

Although the rockets that sent humans into space worked flawlessly, they were very costly because the entire vehicle was thrown away after a single use. Consequently, during the 1970s, the United States began concentrating on the development of a reusable launch system, the space shuttle, which would supply an eventual space station. NASA envisioned the space shuttle as an inexpensive "space truck" moving humans and payloads quickly in and out of space. President Richard M. Nixon approved the plan in 1972, announcing that the shuttle would "transform the space frontier of the 1970s into familiar territory, easily accessible for human endeavor in the 1980s and '90s." This decision shaped the major portion of NASA's program for the next three decades.

Following almost a decade of development, NASA launched the Space Transportation System, as the space shuttle was formally known, on April 12, 1981. It was almost two years behind schedule and $1 billion over budget. The final system consisted of three elements: a disposable external tank (the shuttle's "gas tank"), large solid booster rockets that were jettisoned after takeoff and recovered for reuse, and a delta-winged aerospace vehicle called the orbiter. The shuttle launched into space like a rocket but returned to earth as a glider. The United States built six shuttles: *Enterprise* (the test vehicle), *Columbia*, *Challenger*, *Atlantis*, *Discovery*, and *Endeavour*.

In 1983, the shuttle's usefulness was augmented by the inclusion of Spacelab, a self-contained scientific and engineering laboratory that fitted into the shuttle's cargo bay. Spacelab was a joint venture of NASA and the European Space Agency. It consisted of three components: a pressurized working area, a pallet for mounting instruments used in experiments, and a module containing support equipment. The laboratory conducted many experiments in a wide variety of sciences and proved an early example of international cooperation upon which nations could build in developing the space station.

Despite the shuttle's success, the program did not fulfill its designers' goal of providing an inexpensive "space truck." Because it required a great deal of maintenance, the shuttle proved incapable of the fast turnaround initially envisioned. NASA hoped to fly 116 missions between 1981 and 1985, but fewer than 100 took place by the end of the century. Initial costs and expensive upkeep also made the shuttle an extremely expensive way of launching satellites. Yet despite its shortcomings, the shuttle proved a valuable instrument for understanding the challenges of living and conducting scientific experiments in space. During the 1980s and 1990s, NASA used it as a space laboratory and as a vehicle for deploying, retrieving, and repairing satellites and resupplying the Russian *Mir* space station. Shuttles deployed NASA's *Magellan* spacecraft on a flight to Venus in 1989, the Hubble Space Telescope in 1990, and the Gamma Ray Observatory in 1991. Repair missions to the Hubble telescope in 1993, 1997, 1999, and 2002 corrected design flaws and provided life-extending maintenance on the orbiting observatory.

Like many space efforts, the shuttle experienced tragedy. On January 28, 1986, the space shuttle *Challenger* exploded just minutes after takeoff, killing its seven-member crew. A second tragedy took place on February 1, 2003, when the space shuttle *Columbia* disintegrated during reentry, killing all seven crew members.

THE FIRST GOLDEN AGE OF SPACE EXPLORATION (1960–1989)

The race that put human beings on the moon was the most spectacular element of the space race. Yet while it was grabbing headlines, scientists were developing space technologies that dramatically increased our understanding of the universe. Robot spacecraft traveled to all the planets in the solar system save Pluto. Space-based astronomical observatories helped scientists see farther into the universe and further into the past. Astrophysicists and cosmologists detected signals (e.g., cosmic microwave background radiation) necessary to help them understand cosmic evolution. There has been no other equivalent period of scientific discovery in human history. In this golden age of space exploration, the pioneering visions of Tsiolkovsky, Goddard, and Oberth emerged triumphant.

In the late 1960s, the Soviet Union redirected its space program from a human Moon landing to innovative robotic missions to Venus and the lunar surface. From 1967 to 1983, the Soviets sent a series of Venera spacecraft, which included orbiters, landers, and atmospheric probes, to explore Venus. These missions collected data on atmospheric and surface conditions and sent back pictures of the inhospitable terrain. The final probes in the series analyzed Venusian soil and mapped the planet's cloud-enshrouded surface. The Soviets achieved significant space-technology advances with their *Luna 16* and *Luna 17* missions, launched on September 12 and November 10, 1970, respectively. *Luna 16* was the first successful robot mission to collect samples of lunar dust and automatically return them to Earth. *Luna 17* achieved the first successful use of a mobile, remotely controlled (teleoperated) vehicle in the exploration of another planetary body. However, the technical significance of these pioneering missions was all but ignored outside the Soviet Union in the dazzle of NASA's human Moon missions.

NASA also quietly initiated a wave of robotic exploration as the Moon missions were winding down. On March 2, 1972, (local time), it launched *Pioneer 10*, which became the first spacecraft to transit the main asteroid belt and encounter Jupiter. On its flyby, it investigated the planet's magnetosphere, observed its four satellites, and collected data revealing that Jupiter had no solid surface but was composed of liquid hydrogen. *Pioneer* then went on to cross the orbit of Neptune and eventually became the first human-made object ever to leave the planetary boundaries of the solar system. The following year, on April 5, 1973, (local time), NASA sent *Pioneer 11* on an eventual encounter with Saturn, where it observed the planet's satellites and atmosphere and demonstrated a safe flight path through Saturn's rings for future long-range space missions. The *Pioneers* also investigated magnetic fields, the solar wind, interplanetary dust concentrations, and cosmic rays on their travels.

In 1973, NASA turned its attention to the inner solar system, sending *Mariner 10* (launched on November 3) to fly by Venus (on February 5, 1974). The spacecraft then used a gravity assist from that planet to fly by Mercury. *Mariner*'s successful use of the maneuver provided engineers with a technique to change a craft's direction and velocity in space. Scientists subsequently used the technique for many outer-space missions.

The golden age's greatest exploratory mission began on August 20, 1977, when NASA launched *Voyager 2* on an encounter with Jupiter, Saturn, Uranus, and Neptune. This "grand tour," as space scientists called it, was possible only once every 176 years when the planets aligned themselves so that a spacecraft heading to Jupiter could use a gravity assist to visit the three other planets as well. *Voyager 2* and its twin *Voyager 1* carried a complement of scientific instruments designed to provide detailed images of the planets, explore their moons and ring systems, and measure properties of the interplanetary medium. Since its launch, *Voyager 2* has provided more information about the outer planets than scientists had gathered throughout previous history.

With the exception of its lunar probes in preparation for the Moon landing, NASA concentrated on planet flybys until the late 1970s. Then it conducted a series of missions to Mars and Venus using spacecraft with both orbiters and landers to explore planets in depth. *Viking 1* and *2* made successful landings on Mars in July and September 1976, respectively, carrying out experiments intended to answer the question: Is there life on the planet? Unfortunately, the results were inconclusive. Nevertheless, the Viking probes provided a space-technology milestone: the first successful soft landing of a robot spacecraft on another planet. (The only previous soft landings had been on the Moon, an Earth satellite.) The Viking missions were planned to last for 90 days after landing on the planet's surface, but the spacecraft continued far beyond expectations—in the case of *Viking 1*, four years—providing scientists with massive amounts of data on the Red Planet.

NASA began exploring Venus with the Pioneer Venus mission at the end of 1978. The mission consisted of two spacecraft, both of which reached Venus that December. *Pioneer 12*, the *Pioneer Venus Orbiter*, circled the planet, mapping the surface and gathering a wealth of information about the planet's atmosphere and ionosphere. It continued transmitting valuable data until 1992, when NASA intentionally crashed it into the planet's surface. The *Pioneer Venus Multiprobe (Pioneer 13)* released four probes that entered into the Venusian atmosphere and collected in situ scientific data as they plunged toward the infernolike surface. One hardy probe even survived surface impact and continued to transmit data for about an hour.

During the 1980s, NASA turned its attention to understanding the beginnings of the universe, launching probes such as the *Infrared Astronomy Satellite* (*IRAS*) in January 1983. Unhindered by the absorbing effects of Earth's atmosphere, *IRAS* completed the first "all-sky" scientific survey of the universe in the infrared portion of the electromagnetic spectrum. In November 1989, NASA launched perhaps one of the most important scientific missions of all, the *Cosmic Background Explorer* (*COBE*). This scientific spacecraft carefully measured the cosmic-wave background spectrum and helped scientists answer some of their most important questions about the primeval "big-bang" explosion that started the expanding universe. Results from *COBE* impacted modern cosmology almost in the same way that Galileo's first telescopic observations of the Jovian satellites helped confirm the Copernican hypothesis.

In March 1986, an international armada of spacecraft encountered Comet Halley. (This famous periodic comet carries the name of the English astronomer Edmund Halley, who financed the publication of New-

A montage of planetary images taken by NASA spacecraft during the first great epoch of planetary exploration, which symbolically concluded when the far-traveling *Voyager 2* spacecraft encountered Neptune (August 25, 1989) and began a journey into the interstellar void. Included are (from top to bottom) images of Mercury, Venus, Earth (and the Moon), Mars, Jupiter, Saturn, Uranus, and Neptune. (Only tiny Pluto does not appear here because spacecraft have not yet visited this frigid world.) In this montage, the inner planets (Mercury, Venus, Earth, the Moon, and Mars) are roughly to scale with each other. Similarly, the gaseous giant outer planets (Jupiter, Saturn, Uranus, and Neptune) are also roughly to scale with each other. Image courtesy of NASA.

ton's *Principia.*) The European Space Agency's *Giotto* spacecraft made and survived the most hazardous flyby, as it streaked within 610 kilometers of the comet's nucleus on March 14 at a relative velocity of 68 kilometers per second. Two Japanese spacecraft (*Suisei* and *Sakigake*) and two Soviet spacecraft (*Vega 1* and *Vega 2*) also studied this famous comet, but at greater encounter distances than *Giotto*.

On August 25, 1989, NASA's far-traveling *Voyager 2* spacecraft encountered Neptune and continued its journey to beyond the solar system. Space scientists use this date as the symbolic end of a truly extraordinary epoch in planetary exploration. The first great period of planetary exploration successfully ended, and another, characterized by more intensive scientific missions, was about to begin.

SPACE TECHNOLOGY AND THE INFORMATION AGE

Between 1960 and 1989, space technology started a revolution that changed modern life. Communications satellites connected the globe; navigation satellites guided travelers on land, on sea, and in the air; and weather satellites supported increasingly sophisticated forecasting. Perhaps most important of all, incredible views of our planet from space raised the level of environmental consciousness for millions of its inhabitants. Earth-observing satellites allowed scientists to simultaneously examine the complex interrelationships of major planetary systems (e.g., hydrosphere, atmosphere, and biosphere) and to detect subtle environmental changes.

A year after *Sputnik 1*, the United States presaged the era of modern communications when it launched Project SCORE (Signal Communications Orbit Relay Experiment) on December 18, 1958. For 13 days a transmitter in the missile's payload compartment broadcast a prerecorded Christmas-season message from President Eisenhower. This modest precursor to the communications satellite was primarily a public-relations effort, but it was the first time the human voice came to Earth from space.

The era of satellite communications began in 1960 when NASA successfully launched the *Echo 1* experimental spacecraft on August 12. This large (30.5-meter-diameter), inflatable, metallized balloon served as the world's first passive communications satellite, reflecting radio signals sent from the United States to the United Kingdom. Passive satellites, however, were extremely limited because their reflected signals were too weak for wide-scale use. To fully exploit the potential of satellite communications, scientists had to solve two problems: they had to develop an active satellite that amplified the signal it transmitted; and they had to develop satellites capable of geostationary orbits to simplify tracking and antenna pointing. *Relay 1*, launched in December 1962, solved the first problem. It used a traveling-wave tube to boost the signal it received from the ground and retransmit it back to another location. This tube became a basic component of modern communications satellites.

NASA solved the second problem with the launch in July 1963 of *Syncom 2*, the first communications satellite to operate in a synchronous (figure-eight) orbit. A little more than a year later, on August 19, 1964, the space agency put *Syncom 3* in a true geosynchronous orbit (i.e., stationary above a point on Earth's equator). *Syncom 3* immediately went to work transmitting live coverage of the 1964 Olympics across the Pacific Ocean from Japan to the United States. The age of instantaneous global communications was born.

Communications satellites quickly moved into the commercial sector during the last half of the 1960s. At the urging of the United States, the telecommunications agencies of 18 nations formed INTELSAT in 1964 to coordinate satellite communications. The following year, INTELSAT used NASA facilities to launch *Early Bird 1*, the first commercial communications satellite, in geosynchronous orbit. By the end of the decade, the INTELSAT series of communications satellites was providing telecommunications services across the globe. Individual nations developed satellites specifically for domestic communications throughout the 1970s. Canada Telsat launched *Anik* in 1972, and Western Union launched *WESTAR I* in the United States two years later. In 1976 a new type of communications satellite, MARISAT, provided vastly improved service for maritime customers.

The Soviet Union, too, developed communications-satellite technology early in the space age. It launched its first communications satellite, *Molniya 1A*, into a special, highly elliptical (typically 500-kilometer-by-12,000-kilometer), 12-hour orbit on April 23, 1965. Known as the "Molniya" orbit, it allowed a spacecraft to spend most of its time well above the horizon in the Northern Hemisphere, greatly facilitating communications across the huge Russian landmass.

During the following decades, scientists continued to improve and refine basic communications-satellite technology to meet changing needs. On May 30, 1974, NASA launched its *Applications Technology Satellite–6* (*ATS-6*) to demonstrate the feasibility of using a large antenna structure on a geostationary communications satellite to transmit good-quality television signals to small, inexpensive ground receivers. *ATS-6* helped pioneer the beaming of educational television programs to remote locations and villages. During the 1990s, demands for personal communications systems resulted in the development of new satellite communications technology. In contrast with *Early Bird 1*, which could handle only 240 voice circuits at a time, each new small communications satellite placed in interactive constellations in low Earth orbit is capable of handling 1,100 calls simultaneously.

A revolution in navigation technology accompanied the revolution in communications. On April 13, 1960, the U.S. Navy successfully launched the first experimental navigation satellite, *Transit 1B*, into Earth orbit. Dubbed a "space lighthouse," it was the first of an anticipated 44-satellite network designed to enable surface ships and submarines to precisely calculate their position in all weather. Ships charted their position by measuring the changing Doppler shift of the spacecraft's radio signal. The U.S. Navy launched approximately four satellites annually through 1973. Transit satellites inaugurated yet another advance in space technology, the use of nuclear power in space. Because of high power requirements, a radioisotope thermoelectric generator (RTG) supplemented electric power for many of these spacecraft. The United States successfully launched the first nuclear-powered satellite, *Transit* 4A, on June 29, 1961. Advanced versions of this early plutonium-238-fueled RTG eventually powered many sophisticated NASA spacecraft as they explored the solar system throughout the twentieth century and beyond.

The Transit system had a large margin of error and could not be used by airplanes. Consequently, the satellites were eventually replaced by a new technology, the Global Positioning System. This system, first tested in 1967, ultimately expanded into a constellation of more than 20 satellites and became fully operational in March 1994. It used synchronized clocks on both the satellite and the vessel or other Earth receiver to measure the travel time of a signal from the satellite to the receiver. Users charted their position from the difference in time between receiving the signal from four satellites. The system was so accurate that location could be calculated to within a few meters.

Space technology also transformed weather forecasting with the development of the meteorological satellite. On April 1, 1960, the United States successfully launched the world's first weather satellite, *TIROS 1* (Television and Infrared Observation Satellite), which imaged clouds from space. The satellite took pictures only during daylight and only for one-quarter of each orbit, but it enabled meteorologists to track storms, thus demonstrating the potential of weather satellites for forecasting. By middecade, nine more TIROS satellites were in place, each carrying improved instruments including infrared radiometers that permitted meteorologists to study Earth's heat distribution. In February 1966, TOS (TIROS Operational Satellite) satellites became operational. Placed in polar orbits, they enabled meteorologists to view cloud and temperature changes on a 24-hour basis.

Improvements in weather satellites continued in the 1970s with the development of Geostationary Operational Environmental Satellites

(GOES), first launched on October 16, 1975. These satellites monitored weather conditions on a hemispheric scale, providing both visible-light and infrared images of Earth's surface and atmosphere as well as indirect measurements of temperature and humidity throughout the atmosphere. Later satellites in the series, launched through the end of the century, contained equipment that could make simultaneous images and temperature readings, measure water vapor, and monitor ozone levels. They also carried instruments for observing solar flares and other solar phenomena that might affect the Earth's atmosphere.

Another important space-technology-created revolution occurred in 1972 when specially designed satellites began systematically monitoring Earth's land surfaces in great spatial and spectral detail, helping scientists understand global change. On July 23, NASA successfully launched *Earth Resources Technology Satellite–1* (*ERTS-1*) into a Sun-synchronous polar orbit. Later renamed *Landsat-1*, it provided scientists with high-resolution multispectral images of Earth's surface that permitted the study of geology and resources in even the most remote regions of the world. For the rest of the twentieth century (and beyond), a technically evolving family of Landsat spacecraft (*Landsats 2, 3, 4, 5*, and *7*) created a valuable multispectral record of Earth's surface. These images started a revolution in the way scientists in many disciplines (including agriculture, forestry, water-resource management, environmental monitoring, and urban planning) observed and monitored our home planet.

In 1991, NASA inaugurated a long-range program called Mission to Planet Earth (MTPE) designed to study the interaction of all environmental components—Earth's atmosphere, land surfaces, oceans, and biosphere. The first phase of the program utilized space shuttle missions and environmental satellites to collect data, but because the missions could not simultaneously measure various environmental variables, scientists could not properly look at Earth as a dynamic, integrated system. NASA achieved simultaneity during MTPE's second phase, inaugurated on December 18, 1999, with the launch of its first Earth-Observing Satellite (EOS) system. *Terra*, as it was called, carried five state-of-the-art sensors: the moderate-resolution imaging spectrometer (MODIS), the advanced spaceborne thermal-emission radiometer (ASTER), the multiangle imaging spectrometer (MISR), the measurement of pollution in the troposphere (MOPITT) system, and the clouds and Earth radiant-energy system (CERES). This data-collection approach, using the vantage point of a single sophisticated space platform, permitted effective cross-calibration of all the monitoring instruments. It also avoided the adverse consequences that rapid atmospheric changes and illumination (incident-sunlight-level) variations can have on the scientific quality of any measurements of the same target scene or event that were taken at slightly different times and from several different perspectives. *Terra* was to monitor Earth's radiation balance, atmospheric circulation, biological productivity, land-surface properties, and air-sea interaction. Science themes of the mission included land-cover and land-use-change research, seasonal to interannual climate variability, natural-hazards identification and risk reduction, long-term climate variability (including human-activity impact), and atmospheric ozone research.

Both the Soviet and the American militaries took advantage of space technology to ensure national security and help preserve global stability during the Cold War. Little is publicly known about Soviet activities, but elements of the U.S. program were declassified during the 1990s. Side by side with the open, widely publicized civilian space efforts, the United States began developing a highly secret military space program, called the "black world," for its collection of space systems whose mere existence was not publicly acknowledged. These systems utilized a variety of reconnaissance satellites to gather intelligence.

The United States inaugurated its program of military surveillance satellites on May 24, 1960, with the launch of a MIDAS (Missile Defense Alarm System) satellite, designed to detect missile launches by observing the characteristic infrared (heat) signatures of a rocket's exhaust. Two months later, the U.S. Air Force launched its first photoreconnaissance satellites, named Corona. To maintain the program's secrecy, the air force publicly called the satellites Discoverer and identified space research and technology as their mission. Instead, the satellites replaced reconnaissance aircraft, more effectively collecting data over the huge Soviet landmass. On October 17, 1963, the air force successfully launched the first pair of Vela nuclear-detonation-detection spacecraft designed to monitor compliance with the Limited Nuclear Test Ban Treaty entering into force that month. Defense officials later incorporated similar nuclear-test-detection sensors on the air force's Defense Support Program (DSP) and Global Positioning System (GPS) spacecraft.

Extensive military use of space by both the United States and the Soviet Union continued throughout the 1970s and 1980s. The Corona program, for example, ultimately launched more than 100 missions from August 1960 to May 1972. The programs continued in tight secrecy.

THE SPACE STATION

In his 1984 State of the Union address, President Ronald Reagan called for a permanent American space station so "we can follow our dreams to distant stars." He also invited U.S. allies to participate in the project. By 1985, Canada, the European Space Agency (ESA), and Japan had signed bilateral memoranda of understanding with the United States for participation in the project. However, NASA could only generate a cumbersome, bureaucratic reaction to the president's call. Unlike the space agency's timely and focused response to President Kennedy's Moon-landing directive, almost a full decade of "paper studies" followed, and Reagan's request for a space station (named *Freedom* in 1988) became a quagmire of budget cuts and redesign exercises. Part of the problem was that the continually changing space-station concept lacked an inspirational central purpose. In addition, neither the American public nor the U.S. Congress felt an overall sense of Cold War "space-race" urgency.

On the 20th anniversary of the *Apollo 11* Moon landing in 1989, President George Bush reiterated Reagan's call for a permanent U.S. presence in space. He also recommended that the nation build a lunar base and embark on a human expedition to Mars early in the twenty-first century. "Why the Moon? Why Mars?" he asked. "Because it is humanity's destiny to strive, to seek, to find. And because it is America's destiny to lead." Bush's national space vision encouraged space advocates but failed to excite a Congress determined to slash budgets and curb federal deficits. For many, the proposed U.S. space station almost seemed like a step backwards, while a human expedition to Mars appeared far too costly. Even though the NASA program received public approval throughout the 1990s, the captivating excitement of the Apollo era became a rapidly fading part of the collective American memory.

INTO THE NEXT MILLENNIUM (1990–2000)

Developments in space technology in the last decade of the twentieth century reflected the end of the Cold War and a major transition to detailed scientific investigation of space. International cooperation replaced competition, and nations joined in space ventures. Great orbiting observatories, like the Hubble Space Telescope, furnished scientists with spectacular new views of the universe, while sophisticated robot spacecraft performed detailed investigations of Venus, Mars, and Saturn. Space technology, only a dream at the start of the twentieth century, had become an indispensable tool for human development by the start of the third millennium.

As the Cold War ended around 1989, the politically charged urgency of the superpower space race all but evaporated, and Russian-American cooperation in space, which had begun tentatively with joint Apollo-Soyuz

NASA astronauts James H. Newman (left) and Jerry L. Ross (right) work between *Zarya* and *Unity* (foreground) during an extravehicular activity as part of the STS-88 mission of the space shuttle and the first assembly mission of the *International Space Station* (*ISS*) (December 1998). Newman is tethered to the *Unity* module, while Ross is anchored at the feet to a mobile foot restraint mounted on the end of the shuttle's robot arm, called the remote manipulator system (RMS). Photograph courtesy of NASA (Johnson Space Center).

docking missions in 1975, became the dominant theme. In 1993, President Bill Clinton opened the American space-station program to extensive international participation. His actions transformed the latest version of NASA's space station *Freedom* into the *International Space Station* (*ISS*). Beginning in 1995, a joint shuttle-*Mir* program inaugurated the first phase of *ISS* development. Russian cosmonauts and American astronauts carried out a series of long-duration missions on *Mir*, while the shuttle performed nine docking operations with the Russian space station. This phase of the project provided space scientists with important information on the effects of long-duration spaceflight as well as practical experience in operating an international program. Almost at the dawn of a new millennium (December 1998), the astronaut/cosmonaut crew of the STS-88 shuttle mission assembled the first two components of the *ISS* (one Russian built and one American built) in orbit.

The 1990s ushered in an era of cooperation among other nations as well. During the 1970s, the European Space Agency and nations such as Japan had developed their own independent access to space, often competing with the United States, particularly in the commercial use of space. As the century ended, they frequently collaborated with NASA on major space research projects. In 1990, NASA and the European Space Agency deployed the Hubble Space Telescope to provide detailed observational coverage of the visible and ultraviolet portions of the electromagnetic spectrum. This was the first of NASA's great observatories designed to increase our understanding of the origins and evolution of the universe. The following year, NASA deployed the Compton Gamma Ray Observatory to investigate some of the most puzzling mysteries of the universe—gamma-ray bursts, pulsars, quasars, and active galaxies. The Chandra X-Ray Telescope was deployed in 1999 to detect X-ray sources that are billions of light-years away.

The international space community also inaugurated several bold space voyages in the 1990s. On January 1, 1990, NASA began the Voyager Interstellar Mission, which used *Voyager 1* and *2* to search for the heliopause—the location in deep space that forms the boundary between the outermost extent of the solar wind and the beginning of interstellar space. On October 6, 1990, a space shuttle crew deployed the joint NASA/ESA *Ulysses* spacecraft on a mission to fly over the Sun's polar regions, the first spacecraft to do so.

The Moon, the object of intense space-technology competition in the 1960s, became an area of scientific investigation again in 1994 when the spacecraft *Clementine* provided more than 1.6 million high-resolution images of the lunar surface. Some of the spacecraft's data suggested that the Moon might actually possess significant quantities of ice in its permanently shadowed polar regions. NASA's *Lunar Prospector*, launched in 1998, provided clearer evidence for lunar ice.

As the century closed, a new sophisticated robot explorer, *Cassini*, traveled to Saturn. Launched on October 15, 1997, it was designed for an ambitious program of studying the planet, its rings, its magnetosphere, its icy satellites, and Titan, Saturn's principal moon. The joint NASA/ESA project was the biggest, most complex interplanetary spacecraft ever built by NASA. Weighing more than 12,000 pounds, it contained sophisticated cameras for high-resolution photographs, 11 other instruments for a variety of experiments, and a probe, called *Huygens*, to land on and explore Titan. Because of the distance between Saturn and Earth, the spacecraft could not rely on humans for quick directions, and so *Cassini* carried its own computer programmed to react to many problems on its own.

In the summer of 1996, a NASA research team from the Johnson Space Center announced that it had found evidence in a Martian meteorite called ALH84001 that "strongly suggested primitive life may have existed on Mars more than 3.6 billion years ago." This exciting Martian microfossil hypothesis touched off a great deal of technical debate and rekindled both scientific and public interest in a renewed search for (microbial) life on Mars. NASA responded by placing the Red Planet at the center of its space-exploration activities for the next decade. In late 1996, NASA sent *Mars Global Surveyor* (MGS) and *Mars Pathfinder* to begin intensive investigation of the planet. MGS conducted high-resolution imaging from orbit around the planet, while *Pathfinder* explored the surface. The *Mars Pathfinder* used a new technique to deliver its lander safely to the planetary surface. As the lander descended, four airbags quickly inflated, forming a protective ball around the craft as it reached the surface. Once on Mars, the lander deployed a minirover to explore the surface. The rover was the first mobile robot teleoperated on another planet. (The Russians had previously teleoperated two Lunokhod rovers on the Moon's surface.) The minirover operated for more than 80 sols (Martian days) and stimulated interest in the search for life on Mars, especially in ancient riverbeds and flood plains.

Despite its impressive achievements, NASA had difficulty marshaling sustained public support during the decade. Critics charged that the agency was not doing anything exciting, and powerful Speaker of the House Newt Gingrich lambasted NASA for making space "as boring as possible." Congress dramatically cut the agency's budgets from the heady days of the Apollo program, when it received 4 percent of the national budget, to 0.7 percent in 1998. These cuts shaped its role and missions.

Spurred by budget cuts, NASA adopted a new approach to planetary exploration in the mid-1990s. It began moving away from reliance on costly, complex spacecraft and adopted a philosophy of "faster, cheaper, better" craft exemplified in the Discovery series, which included *Near Earth Asteroid Rendezvous* (*NEAR*) and the *Lunar Prospector*. NASA built the first Discovery craft, *NEAR*, inexpensively from "off-the-shelf" hardware. Launched on February 17, 1996, it became the first spacecraft to orbit an asteroid in February 2000 when it went into orbit around Eros, the second-largest of the known near-Earth asteroids. NASA developed and built the *Lunar Prospector* in under two years at a cost of $63 million. (In contrast, the average space shuttle mission cost about $420 million.). Launched on January 7, 1998, the small, inexpensive spacecraft worked far better than anticipated, sending NASA data confirming strong evidence of ice at the poles of the Moon and indicating that the Moon had a small magnetic core.

The decade ended in disappointment for NASA. Struggling with problems of support and funding, it faced the failure of two important robot missions to Mars in 1999. The agency successfully launched the *Mars Climate Orbiter* (MCO) on December 11, 1998, but as the MCO approached Mars on September 23, 1999, and began its aerobraking maneuvers, all contact was lost. The suspected reason was a human-induced navigational error that caused the spacecraft to enter the Martian atmosphere too steeply and either burn up or crash. This disappointment quickly repeated itself when the *Mars Polar Lander*, launched on January 3, 1999, lost contact as it began its descent to the planet's surface. The tragic loss of both Martian spacecraft cast significant doubt on NASA's most recent design philosophy for planetary exploration. The envisioned decade-long scientific invasion of Mars has developed significant gaps in the planned study of the planet's climate and in the important search for microbial life (past or present).

Yet despite the cutbacks and disappointments of the 1990s, NASA developed an ambitious program of space exploration for the first decades of the twenty-first century. The United States and other nations planned missions to the Sun, the Moon, comets, the asteroid belt, and deep space. NASA's successful launch of the 2001 *Mars Odyssey* spacecraft on April 7, 2001, focused attention again on robot spacecraft exploration of Mars in preparation for a human mission to the planet as early as 2020.

At the dawn of a new millennium, our global civilization has started developing its first permanent outpost in the cosmos. The *International Space Station* (*ISS*) can serve as both an important space-technology focal point and an inspirational symbol of humankind's future in space. Although greatly tempered in the crucible of twentieth-century geopolitics, the early space-technology visions of Tsiolkovsky, Goddard, and Oberth have finally "launched" us, as a planetary species, on the pathway to the stars. The closing words from the 1935 film version of the classic 1933 H.G. Wells science-fiction story *The Shape of Things to Come* are especially relevant in summarizing the overall significance of space technology: "(For us now) the choice is the universe—or nothing. Which shall it be?"

Chapter 2

Chronology of Space Technology and Exploration

This chronology presents some of the key events, scientific concepts, and political developments that led to and then formed the age of space. Several of the entries correspond to events now obscured in antiquity, while others are associated with the emergence of modern science during the late Renaissance period in Western Europe. More recent entries highlight how stories of spaceflight in the science-fiction literature of the nineteenth century stimulated technical visionaries at the dawn of the twentieth century. Through the efforts of many others, the dreams of these visionaries then slowly became the framework for modern space technology.

The founders of aeronautics dreamed of spaceflight for peaceful purposes, but many of the most important developments in space technology emerged from war and international political conflict. The liquid-propellant rocket, which was key to future space travel, emerged from World War II as a novel, but deadly, weapon. It was then mated to a powerful new payload, the nuclear weapon warhead, and the combination became the major instrument of superpower confrontation during the Cold War. Throughout this period of intense political tension between the United States and the former Soviet Union, space-exploration achievements were directly linked to international prestige. Nowhere was this more vividly portrayed than during the race to land human explorers on the Moon. Fortunately, such politically motivated investments in space technology also produced an exponential growth in knowledge and technical capability. The period from approximately 1960 to 1989 is often called the first golden age of space exploration because it corresponds to that special moment in history when all the major bodies of the solar system except for tiny Pluto were initially visited and explored by spacecraft.

This chronology closes with the post–Cold War period—a time when competition in space has given way to peaceful international cooperation. Together as a planetary society and united by common space-exploration goals, we are now embarked on an even greater quest—to understand the beginnings of the universe and to answer one of humanity's most fundamental questions: Does life exist elsewhere in the solar system or in the universe?

Fourth Century B.C.E.	The ancient Greeks develop a model of the universe in which Earth is at the center and the planets and stars revolve about it embedded on a series of transparent, concentric spheres.
c. 150 C.E.	Ptolemy writes a compendium of contemporary astronomical knowledge drawn heavily from early Greek thinkers that presents a geocentric model of the universe. This work remains essentially unchallenged until the scientific revolution of the sixteenth and seventeenth centuries.
c. 850	The Chinese use gunpowder to make fireworks.
1232	The Chinese use rockets in warfare against the Mongols.
1379	Rockets are used in Western Europe for the first time during the siege of Chioggia (near Venice), Italy.
1420	In *Bellicorum Instrumentorum Liber* (Book of war machines), Italian engineer Joanes de Fontana suggests the use of a rocket-propelled battering ram and a torpedo delivered to its target by a rocket.
1543	*De Revolutionibus Orbium Coelestium* (On the revolutions of celestial orbs) by Polish astronomer Nicolaus Copernicus presents a heliocentric theory of the solar system.
1609	The German astronomer Johannes Kepler publishes *Astronomia Nova* (New astronomy), in which he modifies the Copernican model of the solar system by announcing that the planets have elliptical (not circular) orbits.
	Italian scientist Galileo Galilei begins telescopic observations of the Moon and planets that will confirm the Copernican theory.
1687	English scientist Sir Isaac Newton publishes *Philosophiae Naturalis Principia Mathematica*, providing the mathematical foundations for understanding motion.
1780s–1790s	The Indian rulers of Mysore employ rockets against the British army.

1804–1805 Learning from the British military's experiences in India, Sir William Congreve begins development of a series of British military rockets.

1865 French science-fiction writer Jules Verne publishes his classic novel *De la terre à la lune* (From the Earth to the Moon), a tale of a human flight to the Moon.

1869 American writer Edward Everett Hale publishes "The Brick Moon," the first fictional account of a human-crewed space station.

1897–1898 English novelist H.G. Wells publishes his classic science-fiction story *The War of the Worlds*, which chronicles the invasion of Earth by beings from Mars.

1903 Russian scientist Konstantin Tsiolkovsky publishes a paper entitled "Exploration of Space with Reactive Devices," which recognizes the potential of the rocket for space travel. Historians mark the publication of this paper as the birth date of astronautics.

1919 American physicist Robert H. Goddard publishes *A Method of Reaching Extreme Altitudes*, describing the important fundamental principles of modern rocketry and suggesting how the rocket might be used to get a modest payload to the Moon.

1923 German scientist Hermann Oberth publishes *The Rocket into Planetary Space*, providing a thorough discussion of the major aspects of space travel.

1926 Goddard successfully launches the world's first liquid-fueled rocket in a snow-covered farm field in Auburn, Massachusetts.

1929 Oberth publishes *Roads to Space Travel*, which helps popularize the concept of space travel.

German rocket scientists, including young physicist Wernher von Braun, found the Verein für Raumschiffahrt (German Society for Space Travel), which carries out innovative liquid-propellant rocket experiments that lead to the development of the V-2 rocket during World War II.

1933 Philip E. Cleator founds the British Interplanetary Society, which performs a detailed investigation of the technical aspects of landing human beings on the Moon during the 1930s.

1942 The modern military ballistic missile is born when German scientists successfully launch the A-4 rocket (later renamed the V-2).

1944 The Germans begin launching V-2 rockets against London and southern England.

1945 Wernher von Braun and other key German rocket scientists surrender to the American forces. Soviet military forces capture the German rocket facilities and haul away all remaining equipment and any lingering German rocket personnel.

The United States successfully detonates the world's first nuclear explosion.

British engineer and writer Arthur C. Clarke publishes the technical paper "Extra-Terrestrial Relays," suggesting the concept of the geostationary-orbit satellite for global communications.

1946 The U.S. Army launches the first American-adapted V-2 rocket from the White Sands Proving Ground in New Mexico.

1947 Under the direction of Sergei Korolev, Russian rocket engineers successfully launch a modified German V-2 rocket, beginning a program that will ultimately inaugurate the space age.

1950 The United States successfully launches a modified German V-2 rocket from the air force's newly established Long Range Proving Ground at Cape Canaveral, Florida. This modest event inaugu-

The Bumper 8, a modified, captured German V-2 rocket, was the first rocket launched from Cape Canaveral, Florida (July 24, 1950). Image courtesy of NASA.

rates the incredible sequence of military missile and space-vehicle launches soon to take place from Cape Canaveral.

1952 *Collier's* magazine helps stimulate a surge of American interest in space travel with a series of articles for which scientists such as Wernher von Braun and Willy Ley serve as technical consultants.

Wernher von Braun publishes *The Mars Project*, the first serious technical study of a human expedition to Mars.

1953 The Soviet Union detonates its thermonuclear device (hydrogen bomb). Because the Soviets also appear clearly ahead of the United States in heavy-payload ballistic missile technology, the fear of a "strategic nuclear missile gap" prompts President Dwight D. Eisenhower to give strategic ballistic missile development the highest national priority.

1955 Entertainment-industry visionary Walt Disney begins promoting space travel in a series of highly popular television programs.

1957 October 4—The Soviet Union successfully launches *Sputnik 1*, the first artificial Earth satellite.

December 6—The U.S. attempt to launch a satellite ends in disaster when the Vanguard rocket blows up after rising only a few centimeters from the pad at Cape Canaveral.

1958 January 31—The United States successfully launches *Explorer 1*, the first American satellite.

October 1—The National Aeronautics and Space Administration (NASA) becomes the civilian space agency for the U.S. government. A week later, it announces the start of the Mercury Project, America's pioneering program to put human beings into orbit.

December 18—The United States launches Project SCORE (Signal Communications Orbit Relay Experiment), which broadcasts a prerecorded Christmas-season message from President Eisenhower. This is the first time the human voice is broadcast from space.

1959 January 2—The Soviet Union launches a massive campaign to the Moon with the liftoff of *Luna 1*. Although the spacecraft misses the Moon by between 5,000 and 7,000 kilometers, it becomes the first human-made object to escape Earth's gravitation and orbit the Sun.

A malfunction in the first stage of the Vanguard launch vehicle caused the vehicle to lose thrust after just two seconds. The catastrophic destruction of this rocket vehicle and its small scientific satellite on December 6, 1957 temporarily shattered American hopes of effectively responding to the successful launches of two different Sputnik satellites by the former Soviet Union—a technical feat that ushered in the space age in late 1957. Image courtesy of U.S. Navy.

September 14—*Luna 2* successfully impacts the Moon, becoming the first space probe to (crash)-land on another world.

October 4—The Soviets launch *Luna 3*, which circumnavigates the Moon and takes the first images of the lunar farside.

1960

March 11—The United States launches *Pioneer 5*, NASA's first successful mission to place a spacecraft into orbit around the Sun.

April 1—The United States successfully launches the world's first weather satellite, *TIROS* 1 (Television and Infrared Observation Satellite).

April 13—The U.S. Navy successfully places the first experimental navigation satellite, *Transit 1B*, into Earth orbit.

August 10—The United States successfully orbits its first photoreconnaissance satellite, *Corona 13*. Publicly, the air force calls it *Discoverer 13* and identifies its mission as space research.

August 12—NASA successfully launches *Echo 1*, the world's first passive communications satellite.

August 19—The Soviet Union launches *Sputnik 5*, the test vehicle for the new Vostok spacecraft that will carry cosmonauts into space. It carries two dogs that become the first living creatures to successfully return from orbital flight.

1961 April 12—Soviet cosmonaut Yuri Gagarin becomes the first human to orbit the Earth inside the *Vostok 1* spacecraft.

May 5—NASA sends astronaut Alan B. Shepard, Jr., in a Mercury capsule on a suborbital flight.

May 25—President John F. Kennedy commits the United States to landing a man on the Moon by the end of the decade.

June 29—The United States successfully launches the *Transit 4A* navigation satellite, which uses a radioisotope thermoelectric generator to supplement electric power. The mission marks the first successful use of nuclear power in space.

1962 February 20—John H. Glenn, Jr., becomes the first American to orbit Earth.

July 10—A revolution in broadcasting and communications begins with NASA's launch of *Telstar 1*, the world's first commercially funded and constructed (American Telephone and Telegraph [AT&T]) communications satellite.

August 27—NASA successfully launches the *Mariner 2* spacecraft to the planet Venus. This is the world's first successful interplanetary probe.

1963 July 26—NASA successfully launches *Syncom 2*, the first communications satellite to operate in a synchronous orbit.

October 17—The U.S. Air Force successfully places the first pair of Vela nuclear-detonation-detection spacecraft into a high Earth orbit. These spacecraft monitor Earth and outer space for violations of the Limited Nuclear Test Ban Treaty signed by the Soviet Union, the United Kingdom, and the United States in August.

1964	July 28—NASA sends the *Ranger 7* spacecraft to the Moon. About 68 hours later, this robot probe successfully transmits more than 4,000 high-resolution television images of the lunar surface before crashing into the Sea of Clouds. The *Ranger 7*, 8, and 9 spacecraft greatly advance scientific knowledge about the lunar surface and help prepare the way for the Apollo human landing missions.
	July 14—*Mariner 4* encounters Mars and becomes the first spacecraft to fly by the Red Planet.
	August 20—The International Telecommunications Satellite Organization (INTELSAT) is formed to develop a global satellite communications system.
1965	March 18—The Soviet Union launches the *Voskhod 2* spacecraft, carrying two cosmonauts, Pavel Belyayev and Aleksei Leonov. During the second orbit, cosmonaut Leonov becomes the first human to leave an orbiting spacecraft.
	March 23—NASA launches the first two-person Gemini space capsule. The Gemini Project refines and expands the technology acquired from the Mercury Project and paves the way for the Apollo Moon-landing efforts.
	April 6—NASA successfully places *Early Bird 1* in orbit from Cape Canaveral. Also called *INTELSAT 1*, this spacecraft is the first commercial communications satellite placed in geosynchronous orbit.
1966	February 3—The Soviet *Luna* 9 spacecraft transmits the first panoramic television pictures ever received from the lunar surface.
	March 16—NASA launches the *Gemini* 8 mission, which accomplishes the first successful rendezvous and docking operation between a crewed spacecraft and an uncrewed target vehicle.
	March 31—The Soviet Union launches *Luna 10*, which becomes the first human-made object to achieve orbit around the Moon.
	May 30—NASA sends the *Surveyor 1* spacecraft to the Moon. The versatile robot spacecraft successfully lands on the lunar surface on June 1, becoming the first American spacecraft to achieve a soft landing on another celestial body.
	August 10—NASA sends the *Lunar Orbiter 1* spacecraft to the Moon to map the lunar surface in preparation for landings by the Apollo astronauts.

September 12—NASA launches the highly productive *Gemini 11* mission. It quickly rendezvous and docks with the Agena target vehicle. The crew then uses the Agena's restartable rocket motor to propel the crafts (in the docked configuration) to a record-setting altitude of 1,370 kilometers, the highest ever flown by an Earth-orbiting, human-crewed spacecraft. The astronauts also perform a successful tethered-spacecraft experiment. To further demonstrate emerging space technology, the reentry of the *Gemini 11* spacecraft is computer controlled.

1967 January 27—Astronauts Virgil Grissom, Edward White, and Roger Chaffee die when a flash fire sweeps through the capsule. NASA delays the Moon-landing program for 18 months, while the Apollo spacecraft receives major design and safety changes.

October 30—Two Soviet spacecraft, *Cosmos 186* and *188*, perform the first automated rendezvous and docking operation in space. The Soviets will use such operations to assist in the assembly and resupply of future space stations.

1968 December 21—NASA launches the *Apollo 8* spacecraft. The three crew members are the first people to leave Earth's gravitational influence. The astronauts go into orbit around the Moon and capture images of an incredibly beautiful Earth "rising" above the barren lunar horizon.

1969 May 18—In a full "dress rehearsal" for the first Moon landing, NASA's *Apollo 10* mission departs the Kennedy Space Center. The astronaut crew successfully demonstrates the complete Apollo mission profile.

July 16—The entire world watches as NASA's *Apollo 11* mission leaves for the Moon.

July 20—Astronaut Neil Armstrong cautiously descends the steps of the lunar module's ladder and contacts the lunar surface. He declares, "That's one small step for a man, one giant leap for mankind."

1970 August 17—The Soviet Union launches its *Venera 7* mission to Venus. When the spacecraft arrives at Venus on December 15, it ejects a capsule that transmits data back to Earth. The accomplishment represents the first successful transmission of data from the surface of another planet.

September 12—Having lost the undeclared race with the United States to send a human to the Moon, the Soviet Union begins a series of innovative robotic missions to explore Earth's natural

On July 16, 1969, American astronauts Neil Armstrong, Edwin (Buzz) Aldrin, and Michael Collins lifted off from Cape Canaveral, Florida, in the mammoth Saturn V rocket on their way to the Moon during the *Apollo 11* lunar landing mission. Image courtesy of NASA.

satellite with the launch of the *Luna 16* spacecraft. Once on the lunar surface, a teleoperated (i.e., remotely controlled from Earth) drill collects lunar dust and automatically places the material in a sample canister for the return to Earth. *Luna 16* is the first robotic spacecraft to successfully return a sample of material from another world.

November 10—The Soviets launch *Luna 17* to the Moon, where it achieves the first successful use of a mobile, remotely controlled (teleoperated) robot vehicle in the exploration of another planetary body.

1971 April 19—The Soviet Union launches the world's first space station, *Salyut 1*.

May 30—NASA launches the *Mariner 9* spacecraft to Mars, where it gathers data on the composition, pressure, temperature, and density of the Martian atmosphere.

July 26—NASA launches the *Apollo 15* mission to the Moon. It is the first Apollo "J"-series mission that deploys the Lunar Roving Vehicle. Astronauts David R. Scott and Alfred M. Worden become the first humans to drive a motor vehicle on another world.

1972 January 5—President Richard M. Nixon approves the space shuttle program.

March 2—An Atlas-Centaur vehicle successfully launches NASA's *Pioneer 10* spacecraft. This far-traveling robotic explorer is the first to transit the main asteroid belt and the first to encounter the gaseous giant planet Jupiter.

July 23—NASA successfully launches the *Earth Resources Technology Satellite–1* (*ERTS-1*) into a Sun-synchronous polar orbit. Later renamed *Landsat-1*, this spacecraft is the first civilian spacecraft to provide relatively high-resolution multispectral images of Earth's surface. These images help start a revolution in the way scientists study our home planet.

December 7—NASA launches *Apollo 17*, the last human expedition to the Moon in the twentieth century. The crew's safe return to Earth on December 19 brings to a close one of the epic periods of human exploration.

1973 April 5—NASA's *Pioneer 11* spacecraft departs on a trajectory that will send it through the asteroid belt to Jupiter and Saturn.

May 14—NASA launches *Skylab*, the first American space station.

November 3—NASA successfully launches *Mariner 10* on a trajectory to Venus and Mercury. It is the first spacecraft to use a planetary gravity-assist maneuver to reach another planet.

1974 May 30—NASA launches the *Applications Technology Satellite–6* (*ATS-6*). Its purpose is to demonstrate the feasibility of using a large antenna structure on a geostationary communications satellite to transmit good-quality television signals to small, inexpensive ground receivers. *ATS-6* helps pioneer the beaming of educational television programs to remote locations and villages and prepares the way for distance learning programs in the information-technology revolution.

1975 June 8—The Soviet Union launches *Venera* 9 to continue its scientific investigation of Venus. It goes into orbit around the planet on October 22 and releases a capsule that transmits the first television images of Venus's infernolike landscape.

July 15–24—The United States and the Soviet Union take the first steps toward cooperation in space with the Apollo-Soyuz Test Project (ASTP), which performs the first international spacecraft-docking exercise.

August 20—NASA begins a major scientific assault on Mars with the launch of *Viking 1*. Its identical twin, *Viking 2*, is launched on September 9. *Viking 1* reaches the Red Planet in June 1976 and, on July 20, 1976, becomes the first American spacecraft to soft-land on another planet. The spacecrafts' primary objective is to determine whether microbial life existed on Mars. The evidence they return is inconclusive.

October 16—NASA launches the *Geostationary Operational Environmental Satellite 1* (*GOES-1*) for the U.S. National Oceanic and Atmospheric Administration. This spacecraft is the first in a long series of operational meteorological satellites that operate in geostationary orbit and monitor weather conditions on a hemispheric scale.

1976 May 4—NASA successfully launches the first Laser Geodynamics Satellite (LAGEOS), demonstrating the feasibility of using ground-to-satellite laser ranging systems to study tiny movements (centimeters per year or less) of Earth's surface.

1977 August 20—NASA launches the *Voyager 2* spacecraft on an epic "grand tour" in which it successfully encounters all four gaseous giant outer planets and then leaves the solar system on an interstellar trajectory. A second craft, *Voyager 1*, is launched on September 5 on the same trajectory. (NASA named the second probe *Voyager 1* because it would eventually overtake *Voyager 2*.)

September 29—The Soviet Union launches the *Salyut* 6 space station, which contains several important design improvements, including the addition of a second docking port and the use of automated Progress resupply spacecraft. With this second-generation space station, the Soviet human spaceflight program evolves from short-duration to long-duration stays in space.

1978 May 20—NASA successfully launches *Pioneer 12* to Venus. Also called the *Pioneer Venus Orbiter*, it arrives at Venus on December

4 and starts using its radar mapping system to image the surface of the cloud-enshrouded planet. It is the first American spacecraft to orbit Venus, and its numerous accomplishments pave the way for NASA's more sophisticated *Magellan* mission.

June 27—NASA launches *Seasat-1*, the first NASA satellite devoted exclusively to the scientific study of the world's oceans. *Seasat-1* successfully demonstrates the use of passive and active microwave instruments to perform oceanography from space.

August 8—NASA launches the second spacecraft in the Pioneer Venus Program, *Pioneer 13* (also called the *Pioneer Venus Multiprobe*), to a successful encounter with the planet. On December 9, four probes released by the spacecraft enter into the Venusian atmosphere and collect scientific data as they plunge toward the surface.

1979 September 1—*Pioneer 11*, now called *Pioneer Saturn*, becomes the first spacecraft to view the magnificently ringed world at close range. Its successful encounter with the planet also demonstrates a safe flight path through the rings for the more sophisticated Voyager spacecraft that will follow.

December 24—In close cooperation with France, the European Space Agency (ESA) successfully launches the *Ariane 1* rocket. ESA vigorously pursues and captures more than 50 percent of the world's commercial launch-vehicle market by the close of the twentieth century.

1980 July 18—India's Space Research Organization successfully launches a modest 35-kilogram test satellite called *Rohini* into low Earth orbit. The launch vehicle is a four-stage, all-solid-propellant rocket manufactured in India.

1981 April 12—NASA inaugurates the era of "resuable" space transportation with the launch of the space shuttle *Columbia*. At the end of the successful two-day STS-1 (Space Transportation System–1) test mission, *Columbia* becomes the first spacecraft to return to Earth by gliding through the atmosphere and landing like an airplane. This mission also achieves two additional "firsts" within the American space program: the first use of solid-propellant rockets in a crewed mission and first crewed spacecraft to return to Earth on land.

1982 November 11—NASA launches the space shuttle *Columbia* on the first operational flight of the U.S. Space Transportation System (STS). A crew of four astronauts is another space-technology first. The STS-5 mission involves the successful launching of two commercial communications satellites (*SBS 3* and *Anik C-3*) from the shuttle's large cargo bay.

1983 January 25—NASA launches the *Infrared Astronomy Satellite* (*IRAS*). Unhindered by the absorbing effects of Earth's atmosphere, *IRAS* completes the first "all-sky" scientific survey of the universe in the infrared portion of the electromagnetic spectrum.

April 4—NASA launches the space shuttle *Challenger*. During the STS-6 mission, astronauts perform the first extravehicular activity (EVA) from the space shuttle.

June 13—*Pioneer 10* crosses the orbit of Neptune (which at the time is the farthest of the planets from the Sun) and becomes the first human-made object ever to leave the planetary boundaries of the solar system.

June 18—NASA launches the space shuttle *Challenger* on its second voyage into space. The STS-7 mission is also the first with a five-person crew and the first to use the shuttle's robot arm (remote manipulator system) to deploy and then retrieve a small co-orbiting satellite (called *SPAS-01*).

November 28—NASA launches the space shuttle *Columbia* on the STS-9 mission, the first orbital flight of the European Space Agency's *Spacelab*.

1984 January 25—President Ronald Reagan calls for a permanent American space station so "we can follow our dreams to distant stars." His call meets with little enthusiasm from the public.

April 6—NASA launches the *Challenger* (STS-41C) to deploy the *Long Duration Exposure Facility* (*LDEF*), a spacecraft the size of a bus that studies the harsh environment of outer space. Then the crew accomplishes an important space-technology milestone by retrieving, repairing, and redeploying the *Solar Maximum Satellite*. This is the first time a malfunctioning satellite is repaired in orbit.

1985 Spring—The European Space Authority, Japan, and the United States sign agreements for participation in the development of an *International Space Station*.

1986 January 28—The space shuttle *Challenger* explodes during ascent, killing the crew of the STS-51L mission. One of the seven-member crew is Christa McAuliffe, a schoolteacher flying the shuttle as part of NASA's Teacher-in-Space program.

February 20—The Soviets introduce their third-generation space station, a sophisticated modular facility called *Mir* (peace). This new station contains more extensive automation, more spacious crew accommodations, and an important new design feature—a multiport docking adapter.

March 14—An international armada of spacecraft encounter Comet Halley. The European Space Agency's *Giotto* spacecraft makes and survives the hazardous flyby of Comet Halley as it streaks within 610 kilometers of the comet's nucleus. Two Japanese spacecraft and two Soviet spacecraft also study this famous comet, but at greater encounter distances than *Giotto*.

1988 September 19—Israel launches its first satellite from a site south of Tel Aviv. The launch vehicle is a Shavit (comet) three-stage rocket, produced in Israel. To avoid the political problems of flying a rocket over the populated regions of the Arab countries to the east, Israeli space authorities launch the vehicle in a westerly (overwater) direction. The *Ofeq-1* (*Horizon-1*) satellite becomes the first object to orbit Earth from east to west—a disadvantageous orbital trajectory that is actually opposite to the direction of Earth's rotation.

September 29—NASA successfully launches the space shuttle *Discovery* (STS-26). The four-day mission marks a return to human spaceflight by the United States after a 32-month suspension following the fatal *Challenger* accident.

1989 May 4—The space shuttle *Atlantis* (STS-30) deploys NASA's *Magellan* spacecraft on a flight to Venus. *Magellan* is the first spacecraft deployed from the space shuttle on an interplanetary trajectory.

July 20—President George Bush helps celebrate the 20th anniversary of the *Apollo 11* Moon landing by declaring that the United States should commit to a permanent presence in space.

August 25—NASA's far-traveling *Voyager 2* spacecraft encounters Neptune. Space scientists consider this encounter the end of the first golden age of space exploration.

October 18—NASA launches the space shuttle *Atlantis* (STS-34) to deploy the *Galileo* spacecraft on a mission to study Jupiter.

November 18—NASA launches the *Cosmic Background Explorer* (*COBE*) into polar orbit. The scientific spacecraft carefully measures the cosmic-wave background spectrum and helps scientists answer some of their most important questions about the primeval "big-bang" explosion that started the expanding universe.

1990 January 1—NASA officially begins the Voyager Interstellar Mission (VIM). In this extended mission, both nuclear-powered Voyager spacecraft search for the heliopause—the location in deep space that forms the boundary between the outermost extent of the solar wind and the beginning of interstellar space.

April 24—NASA launches the space shuttle *Discovery* (STS-31 mission) to deploy the Hubble Space Telescope (HST), NASA's powerful space-based optical observatory.

October 6—Space shuttle *Discovery* lifts off from the Kennedy Space Center to deploy the *Ulysses* spacecraft on a journey that will make the nuclear-powered spacecraft the first to investigate the third dimension of space over the Sun's poles.

1991 April 7—The space shuttle *Atlantis* deploys the Compton Gamma Ray Observatory (CGRO), a major space-based observatory that explores the universe in the very high-energy (gamma-ray) portion of the electromagnetic spectrum. Once operational, the Compton Gamma Ray Observatory detects hundreds of mysterious gamma-ray bursts from all over the celestial sky—an important discovery that impacts current models of the universe.

1992 February 11—The National Space Development Agency of Japan successfully launches the country's first Earth resources satellite, the *JERS-1* (*Japanese Environmental Resource Satellite–1*), into an operational Sun-synchronous polar orbit around Earth.

September 25—NASA successfully launches the *Mars Observer* spacecraft. This sophisticated 1,000-kilogram-mass spacecraft is the first in NASA's new Observer series of planetary missions. Unfortunately, for unknown reasons, contact with the *Mars Observer* is lost on August 22, 1993, just before it is to go into orbit around Mars.

1994 January 25—The joint Department of Defense and NASA *Clementine* spacecraft lifts off for the Moon. While its primary mission is to test spacecraft components under extended exposure to the space environment, *Clementine* also provides more than 1.6 million high resolution images of the lunar surface.

February 3—NASA launches the space shuttle *Discovery* (STS-60). The six-person crew includes cosmonaut Sergei Krikalev, the first Russian to leave Earth on an American space vehicle. Krikalev's presence on the space shuttle signifies the beginning of a new era of cooperation in space between the United States and Russia.

The space shuttle *Atlantis* departs from the Russian space station *Mir* on July 4, 1995. This image was taken during shuttle mission STS-71 by cosmonauts aboard their *Soyuz TM* transport space vehicle. The Earth's limb serves as a backdrop. Image courtesy of NASA.

1995 February 3—NASA launches the space shuttle *Discovery* (STS-63), which approaches but does not dock with the *Mir* space station as a prelude to developing the *International Space Station*.

June 27—NASA launches the space shuttle *Atlantis* on the STS-71 mission. This mission is the 100th U.S. human spaceflight. During the mission, *Atlantis* docks with the Russian *Mir* space station for the first time. In another historic first, *Atlantis* delivers the *Mir 19* crew to the Russian space station and then returns the *Mir 18* crew back to Earth. The shuttle-*Mir* docking program is the first phase of the *International Space Station*.

1996 August 7—A NASA research team from the Johnson Space Center announces that it has found evidence in a Martian meteorite called ALH84001 that "strongly suggests primitive life may have existed on Mars more than 3.6 billion years ago." The Martian microfossil hypothesis touches off a great deal of technical debate and rekindles both scientific and public interest in a renewed search for (microbial) life on Mars. NASA responds by placing the Red Planet at the center of its space-exploration activities for the next decade.

November 7—NASA launches the *Mars Global Surveyor* (MGS) mission. This is the first mission in NASA's ambitious Mars Surveyor Program, intended to fully explore Mars with an armada of robotic orbiter and lander/rover spacecraft. The MGS enters orbit around Mars in September 1997 and begins full operational mapping in March 1999.

December 4—NASA successfully launches the *Mars Pathfinder* mission to the Red Planet. The lander spacecraft touches down on the surface of Mars on July 4, 1997. Guided by human controllers at NASA's Jet Propulsion Laboratory, a robotic minirover deploys from the lander and explores the surface. The minirover is the first mobile robot teleoperated on another planet. *Mars Pathfinder* also demonstrates the first use of a new airbag technique to deliver a spacecraft safely to a planetary surface.

1997 October 15—NASA uses a powerful Titan IV/Centaur rocket to send the *Cassini* mission to Saturn.

1998 June 2—NASA launches the space shuttle *Discovery* on the ninth and final docking mission with the Russian *Mir* space station. The STS-91 mission successfully concludes Phase 1 of the *International Space Station* program.

December 4—NASA launches the space shuttle *Endeavour* on the first assembly mission of the *International Space Station*.

December 11—NASA launches the *Mars Climate Orbiter* (MCO). Unfortunately, as the MCO spacecraft approaches Mars on September 23, 1999, and begins its aerobraking maneuvers, all contact is lost. The suspected reason is a human-induced navigational error that causes the spacecraft to enter the Martian atmosphere too steeply and either burn up or crash. The loss of the spacecraft casts some doubt on NASA's new approach to planetary exploration (with more frequent but less expensive missions) and leaves significant gaps in the detailed study of the planet's climate.

1999 January 3—NASA launches the *Mars Polar Lander* mission. However, all contact is lost with this robotic spacecraft just before it is scheduled to touch down in the southern polar-cap region of Mars on December 3, 1999, to begin its search for subsurface water ice. This is the second Mars mission lost by NASA in 1999 and forces a major reevaluation of the agency's overall Mars-exploration program.

May 27—NASA launches the space shuttle *Discovery* (STS-96) to accomplish the first docking mission with the international space station.

July 23—The space shuttle *Columbia* (STS-93) carries NASA's Chandra X-Ray Observatory (CXRO) into space. This powerful new observatory is designed to observe X-rays from high-energy regions of the universe, such as remnants of exploded stars.

2000 June 4—NASA mission controllers safely deorbit the massive Compton Gamma Ray Observatory (CGRO) at the end of its useful scientific mission, and any pieces that survive splash harmlessly into a remote area of the Pacific Ocean.

2001 March 23—Russian mission controllers safely deorbit the *Mir* space station. Any pieces of the huge decommissioned space derelict that survive atmospheric reentry fall without incident into a remote region of the South Pacific Ocean.

April 7—NASA successfully launches the 2001 *Mars Odyssey* spacecraft on its mission to Mars.

July 24—NASA launches the space shuttle *Atlantis* (STS-104) to install the joint airlock module on the *International Space Station*, providing the station's permanent crew independent access to space in the absence of a docked space shuttle vehicle.

2002 March 1—Space shuttle *Columbia* takes off from Kennedy Space Center with a crew of seven astronauts on Servicing Mission 3B to the Hubble Space Telescope. The STS-109 shuttle mission is the fourth servicing mission to the orbiting optical telescope and results in a more powerful instrument for further astronomical discovery.

May—Instruments on NASA's 2001 *Mars Odyssey* spacecraft detect large quantities of water ice buried just below the surface in a large region of the planet's south pole.

May 4—NASA's *Aqua* satellite is successfully launched into polar orbit by a Delta II rocket from Vandenberg Air Force Base, California. This sophisticated Earth-observing spacecraft joins its sibling, NASA's *Terra* spacecraft, in performing integrated studies of Earth as a dynamic, complex and interconnected system.

October 1—The United States Department of Defense forms the U.S. Strategic Command as the control center for all American strategic (nuclear) forces. USSTRATCOM also conducts military space operations, strategic warning and intelligence assessments, and global strategic planning.

2003

February 1—Gliding back to Earth after a very successful 16-day scientific research mission in low Earth orbit, NASA's space shuttle *Columbia* experienced a catastrophic accident at an altitude of about 63 kilometers over the state of Texas. Traveling at approximately 18 times the speed of sound, the orbiter vehicle disintegrated, taking the lives of all seven crew members: six American astronauts (Rick Husband, William McCool, Michael Anderson, Kalpana Chawla, Laurel Clark, and David Brown) and the first Israeli astronaut (Ilan Ramon). Disaster struck the STS-107 mission when *Columbia* was just 15 minutes from its landing site at the Kennedy Space Center in Florida.

Chapter 3

Profiles of Space Technology Pioneers, Visionaries, and Advocates

In this chapter, we meet some of the most important space-technology pioneers, visionaries, and advocates. We also discuss several key organizations that helped create modern space technology.

Interest in the phenomena of space is not recent. From antiquity, curiosity has driven human beings to study, chart, and debate the mysteries of the celestial spheres. Out of this ancient interest in space, the scientific revolution eventually emerged—a revolution in thought and cosmic perspective that continues to unfold today. The first three visionaries (Nicolaus Copernicus, Johannes Kepler, and Galileo Galilei) boldly challenged conventional beliefs and championed a new, Sun-centered (heliocentric) model of the universe. Their pioneering work in the sixteenth and seventeenth centuries started the great scientific revolution. It also prepared the way for Isaac Newton to invent the mathematical and physical framework with which to predict the motion of most objects in the nonrelativistic universe.

In the middle of the nineteenth century, Jules Verne started writing technically believable accounts of space travel. Through such popular stories as *From the Earth to the Moon*, Verne moved the concept of travel to other worlds out of the realm of pure fantasy and into the realm of technical possibility. He and other nineteenth-century writers worked hard to integrate contemporary science into their stories. At the time, however, no one knew enough about the basic physics of rocket propulsion to fully comprehend what it would take to launch a spacecraft.

Excitement about possible "canals" on Mars inspired another gifted technical prophet, H.G. Wells, to create equally influential science-fiction stories. In particular, his classic space invasion tale *The War of the Worlds* was a story that raised alarming new questions about the possible behavior of alien (extraterrestrial) life.

Early in the twentieth century, the marvelous science-fiction works of Jules Verne, H.G. Wells, and others continued to inspire some of the brightest young minds. Konstantin Tsiolkovsky, Robert Goddard, and Hermann Oberth provided the key technical insights that would transform the dream of space travel into reality. Through their pioneering efforts, large liquid-propellant rockets became possible. With these rockets, other space advocates could get on with the business of creating the age of space in the second half of the twentieth century.

While Tsiolkovsky, Goddard, and Oberth firmly connected the rocket to space travel, Albert Einstein and Edwin Hubble reshaped our understanding of the universe. Following the devastation of World War II, Wernher von Braun, Sergei Korolev, Nikita S. Khrushchev, Walt Disney, John F. Kennedy, Arthur C. Clarke, and Krafft Ehricke each made unique contributions to the development of modern space technology. Cosmonauts and astronauts like Yuri A. Gagarin and John H. Glenn, Jr., pioneered the new "space frontier" by going where no human beings had ever gone before and blazing a trail so other equally brave men and women could follow. Certain organizations also played a very important role in rapidly developing and applying the technology necessary for spaceflight. These organizations included the National Aeronautics and Space Administration (NASA), the National Reconnaissance Office (NRO), and the European Space Agency (ESA). In its own special way, each helped shape our modern planetary society through the innovative use of space technology.

Space technology continues to exert significant influence on us in the twenty-first century. Through advances in space technology, we will establish permanent outposts in space and carefully explore the most distant corners of our solar system. We will also monitor and understand our home planet in ways never before possible. As the human race ventures into the solar system on its way to the stars, are you prepared to become the next space-technology pioneer?

NICOLAUS COPERNICUS (1473–1543)

Nicolaus Copernicus, who proposed that all the planets in the solar system actually revolved around the Sun in circular orbits (a heliocentric hypothesis), helped create a new era in astronomy. Science historians also regard his pioneering work as the beginning of the scientific revolution.

Nicolaus Copernicus was born on February 19, 1473, in Torun, Poland. Raised by his uncle, a powerful prince-bishop, he studied mathematics at the University of Cracow before traveling to Italy in 1496 to continue his education in medicine and canon law. In Italy, he developed an interest in astronomy and became fascinated with the little-known theory of Aristarchus of Samos (c. 320–250 B.C.E.) that the Earth revolved around the Sun (heliocentric hypothesis). This ran contrary to the accepted Ptolemaic system that the Earth was the center of the universe (geocentric hypothesis). In 1505, Copernicus returned to Poland and became a canon at his uncle's cathedral in Frombork. While performing his ecclesiastical duties, Copernicus enthusiastically worked to prove the heliocentric theory through careful observation of planetary motion and mathematical calculation. He found that if he assumed that the Earth and the other planets actually revolved around the Sun, he could predict planetary motions, an exercise not possible using the geocentric model.

To avoid open conflict with church authorities who considered the Sun-centered model heresy, Copernicus cautiously circulated his handwritten notes to a few close friends. In 1540, one of his students, the Austrian mathematician Rheticus (Georg Joachim von Lauchen), published a summary of these notes but was very careful not to specifically mention Copernicus by name. This trial exposure of the heliocentric model actually occurred without angering church authorities. On the contrary, scientific excitement about the Copernican hypothesis spread rapidly.

As a result, Copernicus finally agreed to have Rheticus supervise the publication of his complete book, *De Revolutionibus Orbium Coelestium* (On the revolutions of celestial spheres). To avoid any potential doctrinal problems, Copernicus dedicated the book to Pope Paul III. Unfortunately, Rheticus left the final publication steps to a Lutheran minister named Andreas Osiander. The minister, mindful that Martin Luther firmly opposed the new Copernican theory, added an unauthorized preface to weaken the impact of its contents. In effect, Osiander's unauthorized preface stated that Copernican theory was not being advocated as a description of the physical universe but only as a convenient way to calculate the tables of planetary motions. The book (so modified) finally appeared in 1543. Historic legend suggests that Copernicus received the first copy as he lay on his deathbed.

After his death on May 24, 1543, in Frombork, Poland, church authorities aggressively attacked the Copernican model and banned *On the Revolutions of Celestial Spheres* as heretical. This book remained on the church's official list of forbidden books until 1835.

JOHANNES KEPLER (1571–1630)

German astronomer and mathematician Johannes Kepler discovered three major laws of planetary motion that now bear his name. Kepler's laws are used extensively in the age of space to describe the motion of natural and artificial satellites as well as that of unpowered spacecraft in orbit around planets.

Kepler was born on December 27, 1571, in Weil der Stadt in the duchy of Württemberg (in present-day Germany). His father was a mercenary soldier who abandoned the family, and his mother was once tried for witchcraft and acquitted. Kepler attended the University of Tübingen in hopes of becoming a Lutheran minister. He graduated in 1588 and received a master's degree in 1591. At Tübingen he became interested in mathematics and astronomy, particularly the new Copernican (heliocentric) model, which he embraced. By 1594, he abandoned his plans for the ministry and became a mathematics instructor at the University of Graz in Austria.

In 1596, Kepler published *The Cosmographic Mystery*, an intriguing work in which he tried to analytically relate the five basic geometric solids from Greek mathematics to the distances of the six known planets from the Sun. The work attracted the attention of Europe's greatest observational astronomer, Tycho Brahe, who was carrying out detailed observations of the motion of the planets and stars. In 1600, the Danish astronomer invited Kepler to become his assistant. When Brahe died the following year, Kepler succeeded him as the imperial mathematician to Holy Roman Emperor Rudolf II. In 1604, Kepler published *The New Star*, in which he described a supernova in the constellation Ophiuchus that he first observed on October 9, 1604. Today this supernova is called Kepler's star.

From 1604 until 1609, Kepler's main interest involved a detailed study of the orbit of Mars, a task Brahe had assigned him. Kepler began his study with the widely held belief that orbits had to be circular but found from Brahe's observations and his own calculations that Mars did not conform to the theory. He eventually concluded that the movement of Mars and other planets could not be explained unless he assumed that the orbit was an ellipse. He published this discovery in 1609 in *New Astronomy*. In this book, Kepler presented his first two laws of planetary motion: the orbits of the planets are ellipses, with the Sun as a common focus; and as a planet orbits the Sun, the radial line joining the planet to the Sun sweeps out equal areas within the ellipse in equal times. *New Astronomy* permanently shattered two thousand years of geocentric Greek astronomy and produced a major advance in understanding the solar system.

When he published *Concerning the Harmonies of the World* in 1619, Kepler continued his great work involving the orbital dynamics of the planets. Although this book extensively reflected Kepler's fascination with mysticism, it also provided a very significant insight that connected the mean distances of the planets from the Sun with their orbital periods. This discovery became known as Kepler's third law of planetary motion. Between 1618 and 1621, Kepler summarized his planetary studies in the publication *Epitome of Copernican Astronomy*, which became the most influential text on the Copernican model of the day.

In 1627, Kepler published *Rudolphine Tables* (named after his benefactor, Emperor Rudolf), which provided astronomers detailed planetary position data. The tables remained in use until the eighteenth century. Kepler, himself a skilled mathematician, used the logarithm (newly invented by the Scottish mathematician John Napier) to help perform the extensive calculations. This was the first important application of the logarithm.

Kepler also worked in the field of optics. He communicated with the great Italian scientist Galileo throughout his career and was one of the first scientists Galileo informed about his creative application and improvement of the telescope (invented in 1608 by the Dutch optician Hans Lippershey) for use in astronomy. According to one historic anecdote, Kepler refused to believe that Jupiter had four moons that behaved like a miniature solar system unless he personally observed them. A Galilean telescope somehow arrived at his doorstep. Kepler promptly used the device and immediately described the four major Jovian moons as satellites, a term he derived from the Latin word *satelles*, meaning a person who escorts or loiters around a powerful person. In 1611, Kepler improved the design of Galileo's original astronomical telescope by introducing two convex lenses in place of the arrangement of one convex lens and one concave lens used by the Italian astronomer.

Before his death in 1630, Kepler wrote *The Dream*, a novel about an Icelandic astronomer who travels to the Moon. While the tale contained demons and witches (who help get the hero to the Moon's surface in a dream state), Kepler's description of the lunar surface was quite accurate. Consequently, many historians treat this story, published posthumously in 1634, as the first genuine piece of science fiction.

Kepler constantly battled financial difficulties. He died of fever on November 15, 1630, in Regensburg, Bavaria, while searching for new funds from government officials there.

GALILEO GALILEI (1564–1642)

Galileo made fundamental contributions to the study of motion and the development of the scientific method, changing science from a philosophical discipline to one based on observation and mathematics. His work in astronomy prepared the way for the acceptance of the Copernican heliocentric system.

Galileo Galilei was born in Pisa on February 15, 1564, into a poor patrician family. He entered the University of Pisa in 1581 to study medicine, but his inquisitive mind soon became more interested in physics and mathematics. Lack of money forced him to leave the university in 1585 without receiving a degree, and he became a lecturer at the academy of Florence. There he focused his activities on the physics of solid bodies, particularly the motion of falling objects and projectiles. In 1589, Galileo became a mathematics professor at the University of Pisa. A brilliant lecturer, he attracted students from all over Europe but alienated many faculty members with the sharp wit and biting sarcasm he used to win philosophical arguments.

At that time, professors taught physics (then called natural philosophy) as an extension of Aristotelian philosophy. Galileo initiated the scientific revolution in Italy by making physics and astronomy observational, experimental sciences. His activities constantly challenged the 2,000-year tradition of ancient Greek learning. For example, Aristotle had stated that heavy objects would fall faster than lighter objects. Galileo held the opposite view that, except for air resistance, the two objects would fall at the same speed regardless of their masses. It is not certain whether he personally performed the legendary experiment in which a musket ball and a cannonball were dropped from the Leaning Tower in Pisa to prove this point. However, he did conduct a sufficient number of experiments with objects on inclined planes to upset Aristotelian physics and to create the science of mechanics.

By 1592, Galileo's anti-Aristotelian research and abrasive personality had sufficiently offended his colleagues at the University of Pisa that they "invited" him to teach elsewhere. Later that year, he moved to the University of Padua, which had a more lenient policy of academic freedom, encouraged in part by the progressive Venetian government.

In 1597, the German astronomer Johannes Kepler provided Galileo a copy of Copernicus's banned book *On the Revolutions of Celestial Spheres*, which postulated a heliocentric theory of the solar system. Although Galileo had not previously been interested in astronomy, he immediately embraced the Copernican model. Between 1604 and 1605, Galileo performed his first (pretelescope, naked-eye) astronomical observations. He witnessed the supernova of 1604 in the constellation Ophiuchus and used it to refute the cherished Aristotelian belief that the heavens were immutable (unchangeable).

In 1609, Galileo learned that a new optical instrument, the telescope, had just been invented in Holland. Within six months, he had devised his own version of the instrument, a telescope three times more powerful than earlier models. In 1610, he turned this improved telescope to the heavens and started the age of telescopic astronomy. With his crude instrument, he made a series of astounding discoveries, including mountains on the Moon, many new stars, and the four major moons orbiting Jupiter. Galileo published these important discoveries in *Starry Messenger* (1610). In this work, he used the Jovian moons to disprove the accepted Ptolemaic theory that all heavenly bodies revolve around Earth and provided observational evidence for the heliocentric Copernican model.

Unwilling to continue teaching "old doctrine" at the university, Galileo left Padua in 1610 and went to Florence, where he became chief mathematician and philosopher to the grand duke of Tuscany, Cosimo II. He resided in Florence for the remainder of his life.

In 1613, Galileo published his *Letters on Sunspots*, using the existence and motion of sunspots to demonstrate that the Sun itself changes, again attacking Aristotle's doctrine of the immutability of the heavens. In so doing, he also openly endorsed the heliocentric model. Late in 1615, Galileo went to Rome and publicly argued in support of Copernicus. Galileo's public advocacy angered Pope Paul V, who formed a special commission to review the theory of Earth's motion. Dutifully, the (unscientific) commission concluded that the Copernican theory was contrary to biblical teachings and possibly a form of heresy. In late February 1616, the Church officially admonished Galileo never to teach or to write again about the Copernican model.

Galileo remained silent for a few years. In 1623, he published *The Assayer*, in which he discussed the principles for scientific research but carefully avoided support for Copernican theory. Nine years later, Galileo received papal permission to publish a work comparing the old and new astronomy provided that he present a balanced description of each. In the masterful but satirical *Dialogue on the Two Chief World Systems*, Galileo had two people present scientific arguments to an intelligent third person concerning the Ptolemaic and Copernican worldviews. The Copernican cleverly won these debates. Galileo represented the Ptolemaic system with an ineffective character called Simplicio. For a variety of reasons, Pope Urban VIII regarded Simplicio as an insulting, personal caricature. Within months after the book's publication, the Inquisition summoned Galileo to Rome. Under threat of execution, the aging Italian scientist publicly retracted his support for the Copernican model on June 22, 1633. The Inquisition then sentenced him to life in prison, a term that he actually served under house arrest at his villa in Arcetri (near Florence). Church authorities also banned *Dialogue*, but Galileo's supporters smuggled copies out of Italy, and the Copernican message again spread across Europe. Blindness struck the brilliant scientist in 1638. He died while imprisoned at home on January 8, 1642. Three and a half centuries later, on October 31, 1992, Pope John

Paul II formally retracted the sentence of heresy passed on Galileo by the Inquisition.

SIR ISAAC NEWTON (1642–1727)

Sir Isaac Newton's brilliant work in physical science and mathematics fulfilled the scientific revolution and dominated science for two centuries. His universal law of gravitation and his three laws of motion allow scientists to explain in precise mathematical terms the motion of almost every object observed in the universe, from an apple falling to the ground to planets orbiting the Sun.

Newton was born in Woolsthorpe, England, on December 25, 1642. His father had died before his birth, and his mother placed her three-year-old son in the care of his grandmother when she remarried. Separated from his mother, Newton had a very unhappy childhood. When his hated stepfather died in 1653, Newton returned home to help his mother on the family farm. He failed miserably as a farmer and was sent to grammar school to prepare for the university. In 1661, Newton entered Cambridge, which was still a bastion of Aristotelian science. Nevertheless, he developed a deep interest in the emerging scientific revolution and privately studied the works of such men as René Descartes. Following his graduation in 1665, he returned to his mother's farm to avoid the plague in Cambridge. For the next two years, he laid the foundation for his theory of gravitation and motion and his work in optics. He also invented calculus (independently invented by the German mathematician Gottfried Leibnitz).

By 1667, the plague epidemic had subsided, and Newton returned to Cambridge as a minor fellow at Trinity College. Three years later, he became professor of mathematics. In 1671, Newton was elected a member of the prestigious Royal Society, and the following year he produced his first public paper on the nature of color. His research was bitterly attacked by physicist Robert Hooke, an influential member of the society. The criticism deeply affected Newton, and he withdrew into virtual seclusion. His experience made him reluctant to publish throughout his career.

In August 1684, Newton's friend, astronomer Edmund Halley, convinced the introverted genius to address one of the major scientific questions of the day: "What type of curve does a planet describe in its orbit around the Sun, assuming an inverse square law of attraction?" To Halley's delight, Newton immediately responded, "An ellipse." Halley pressed on and asked Newton how he knew the answer to this important question. Newton nonchalantly informed Halley that he had done the calculations years ago. The absent-minded Newton could not find the old calculations but promised to send Halley another set as soon as he could. In partial fulfillment of that promise, Newton sent Halley his *De Motu Corporum* (1684). In this document, Newton demonstrated that the force of gravity between two bodies is directly proportional to the product of their masses and inversely proportional to the square of the distance between them. This later became known as Newton's universal law of gravitation. Three years later, with Halley's patient encouragement and financial support, Newton carefully documented all of his work on gravitation and orbital mechanics in *Philosophiae Naturalis Principia Mathematica* (Mathematical principles of natural philosophy). In the *Principia*, Newton gave the world his universal law of gravitation and his three laws of motion. This monumental work transformed physical science and completed the scientific revolution started by Copernicus, Kepler, and Galileo.

For all his brilliance, Newton was also emotionally fragile. After the *Principia*, he drifted into public affairs and eventually suffered a serious nervous disorder in 1693. Upon recovery, he left Cambridge in 1696 to assume a government post in London as warden (then master) of the Royal Mint. During his years in London, Newton enjoyed power and worldly success and in 1704 became president of the Royal Society. Queen Anne knighted him in 1705. Newton continued to rule the scientific landscape until his death in London on March 20, 1727.

When asked where all his scientific insights came from, Newton wrote this rather humble comment in a response letter to his rival Hooke in 1676: "If I have seen further than other men, it is because I stood on the shoulders of giants."

JULES VERNE (1828–1905)

Jules Verne lit the flame of imagination for those who would develop the technology to free humankind from the bonds of Earth. Because of Verne's vision, space travel became the technical dream of the twentieth century.

The son of a lawyer, Verne was born in Nantes, France, on February 8, 1828. He entered the University of Paris in 1847 to study law but found the subject uninteresting and turned his attentions to the theater. In 1850, under the patronage of Alexander Dumas, he produced a successful play, *The Broken Straws*. Outraged that his son had abandoned the law, his father cut off all financial support, forcing Verne to earn a living as a writer.

To prepare himself for this new career, Verne spent many hours in the libraries of Paris, studying astronomy, engineering, geology, and other technical subjects. These diligent efforts helped him become the first modern science-fiction writer, making technology and scientific developments a significant part of the plot or story background. In 1863, he published his first novel, *Five Weeks in a Balloon*, based on an essay he had written describing the exploration of Africa in a balloon. This story made Verne quite popular and gave him the winning formula for his numerous romantic fantasies about science and technology. This success was followed in 1864 by his subterranean fantasy, *Journey to the Center of the Earth*. For more than 40 years he produced (on average) about one novel per year, covering a wide range of exciting topics.

Verne triggered the modern dream of space travel with his classic 1865 novel *From the Earth to the Moon*. In it, his travelers are blasted on a journey around the Moon in a special hollowed-out capsule fired from a large cannon. Verne correctly located the cannon at a low-latitude site on the west coast of Florida called Tampa Town. (By coincidence, this fictitious site is about 120 kilometers west of Launch Complex 39 at the NASA Kennedy Space Center from which Apollo astronauts actually left for journeys to the Moon between 1968 and 1972.) We recognize today that the acceleration of Verne's capsule down the barrel of this huge cannon would have crushed the intrepid explorers. In addition, the capsule would have burned up in Earth's atmosphere. But Verne's famous story correctly prophesied the use of small reaction rockets to control the attitude of this capsule during spaceflight. Despite the obvious limitations, this story made spaceflight appear technically possible for the first time.

Each of the three great pioneers of space technology, Konstantin Tsiolkovsky, Robert Goddard, and Hermann Oberth, identified the works of Jules Verne as a key stimulus in their lifelong interest in space travel. The great French novelist died in Amiens, France, on March 24, 1905.

HERBERT GEORGE (H.G.) WELLS (1866–1946)

H.G. Wells was a novelist and science-fiction writer whose works popularized the idea of space travel and inspired the pioneers of astronautics. The son of a shopkeeper, Wells was born on September 21, 1866, in Bromley, England. In 1884, he entered the Normal School of Science in South Kensington under scholarship, but uninspired by routine academics, he left that institution in 1887 without a degree. Most of his education came from omnivorous reading, a habit he developed as a child while convalescing from a broken leg. He taught in private schools for four years and eventually received a degree from the Normal School of Science in London. He settled in London, where he worked as a teacher and wrote extensively on educational matters.

Wells's career as a science-fiction writer began in 1895 with the publication of his immensely successful book *The Time Machine*, the story of a man who travels 800,000 years into the future. The work embodied Wells's fascination with technological innovation and social change, which were themes of many of his later works. At the turn of the century, he focused his attention on the consequences of contact with aliens from other planets and the possibility and problems of space travel. *The War of the Worlds*, which appeared as a magazine serial and then a book between 1897 and 1898, was the classic tale of an invasion of Earth from space. In his story, hostile Martians land in nineteenth-century England and prove to be unstoppable, conquering villains until they themselves are destroyed by tiny terrestrial microorganisms.

Wells was most likely influenced by the popular (but incorrect) theory of Martian "canals" that was very fashionable in late-nineteenth-century astronomy. In 1877, the Italian astronomer Giovanni Schiaparelli reported the linear features he observed on the Martian surface as "canali"—a word that means "channels" in Italian, but that became misinterpreted when translated into English as "canals." As a result, other notable astronomers, like the American Percival Lowell, soon searched for and discovered these "canals"—features they immediately misinterpreted as signs of an intelligent Martian civilization.

Wells cleverly solved (or perhaps ignored) the technical aspects of space travel in his 1901 novel *The First Men in the Moon* by creating "cavorite," a fictitious antigravity substance. This story inspired many young readers to think about space travel at the beginning of the century. However, space-age missions to the Moon have now completely vanquished the delightful (though incorrect) products of this writer's fertile imagination, including giant moon caves that contained a variety of lunar vegetation and even bipedal Selenites.

Wells's works frequently anticipated technological advances, winning for him the status of scientific prophet. In *The War in the Air* (1908), he foresaw the military use of aircraft, and in *The World Set Free* (1914), he foretold the splitting of the atom. Following his period of successful fantasy and science-fiction writing, Wells focused on social issues and the problems associated with emerging technologies. His 1933 novel *The Shape of Things to Come* warned about the problems facing Western civilization. In 1935, Alexander Korda produced a dramatic movie version of this futuristic tale. The movie closes with a memorable philosophical discussion on technological pathways for the human race. Sweeping an arm, as if to embrace the entire universe, one principal character in the movie asks his colleague: "Can it really be our destiny to conquer all this?" As the scene fades out, his companion replies:

"The choice is simple. It is the whole Universe or nothing. Which shall it be?" (H.G. Wells, *The Shape of Things to Come*). The famous novelist and visionary died in London on August 13, 1946.

KONSTANTIN EDUARDOVICH TSIOLKOVSKY (1857–1935)

> Mankind will not remain tied to Earth forever.
>
> Konstantin E. Tsiolkovsky
> (engraved on his tombstone in Kaluga, Russia)

Konstantin Tsiolkovsky, one of the founders and pioneers of the science of astronautics and rocket dynamics, also wrote visionary books on space travel. Tsiolkovsky, the son of a Polish nobleman who had been deported to eastern Russia, was born on September 17, 1857, in Izhevskoye, Russia. Scarlet fever left him almost totally deaf at the age of nine and forced him to end his formal education. Undaunted, he educated himself at home, becoming interested in mathematics, physics, and the possibility of space travel. Recognizing his brilliance, his family sent him to Moscow to study mathematics, chemistry, and mechanics, but fearing for his son's health, his father called him home before he had completed his education. Tsiolkovsky earned a teaching certificate and spent his life as a provincial schoolteacher in remote areas of Russia. Despite his isolation from major scientific research centers, Tsiolkovsky made significant contributions to the fields of chemistry, physics, and astronautics. In 1881, he independently worked out the kinetic theory of gases. He then proudly submitted a manuscript concerning this work to the Russian Physico-Chemical Society, only to be informed that the theory had been developed a decade earlier. Nevertheless, the originality and quality of Tsiolkovsky's work impressed the reviewers, who invited him to become a member of the society and encouraged him to continue his research.

Tsiolkovsky's early interest in aeronautics stimulated his more visionary work involving the theory of space travel. As early as 1883, he accurately described the weightlessness conditions of space. In his 1895 science-fiction work *Dreams of Earth and Sky*, he described his vision of humans living in space colonies constructed using materials from the solar system. By 1898, he correctly linked the rocket to space travel and concluded that the rocket would have to be propelled by liquid fuel to achieve the necessary escape velocity.

Tsiolkovsky presented many of the fundamental principles of astronautics in his seminal work, "Exploration of Space with Reactive Devices"

(1903). The paper contained what came to be known as the "Tsiolkovsky formula," establishing the relationships between a rocket's thrust (T), the velocity (V_e) of the exhaust gases, and the mass flow rate ($\dot{m}$) of the expelled propellant (also called the flow rate of reaction mass). This formula, written in its most basic form as thrust (T) = mass flow rate ($\dot{m}$) $\times$ exhaust velocity (V_e), is basic to rocketry. The paper also contained a design for a liquid-propellant rocket that used liquid hydrogen and liquid oxygen. The publication of this paper is considered the birth date of astronautics.

Because Tsiolkovsky was a village teacher in rural tsarist Russia, his important work went essentially unnoticed by the world. Few in Russia cared about space travel in those days, and he never received funding to pursue any type of practical demonstration of his innovative concepts. These suggestions included the space suit, space stations, multistage rockets, large habitats in space, the use of solar energy, and closed life-support systems.

Following the Russian Revolution of 1917, the new Soviet government grew interested in rocketry and rediscovered Tsiolkovsky's work. He was honored for his previous achievements and encouraged to continue his research. During the 1920s, he recognized that a single-stage rocket would not be powerful enough to escape Earth's gravity, so he developed the concept of a staged rocket, which he called a rocket train in his 1924 book *Cosmic Rocket Trains*.

Tsiolkovsky's contribution to astronautics remained theoretical. Because of limited resources and political instability in the former Soviet Union, he personally never built or tested any of his proposed rockets. Nevertheless, his work and his visions of space travel inspired many future Soviet aerospace engineers, including Sergei Korolev, the driving force behind the Soviet Union's space program. Tsiolkovsky continued to make significant contributions to astronautics until his death on September 19, 1935, in Kaluga, USSR.

ROBERT H. GODDARD (1882–1945)

> It is difficult to say what is impossible, for the dream of yesterday is the hope of today and the reality of tomorrow.
>
> Robert H. Goddard

Often called the "father of modern rocketry," Robert H. Goddard successfully launched the world's first liquid-fueled rocket and, like Konstantin Tsiolkovsky and Hermann Oberth, recognized the liquid-propellant rocket as the enabling technology for space travel. The son of a bookkeeper, Goddard was born in Worcester, Massachusetts, on October 5, 1882. He

Robert H. Goddard with a steel combustion chamber and rocket nozzle (1915). He is considered the "father of American rocketry" and one of the pioneers in the use of rockets for the exploration of outer space. Photograph courtesy of NASA.

became interested in space travel during his childhood after reading the works of Jules Verne and H.G. Wells. He graduated from Worcester Polytechnic Institute in 1908 and received a doctorate in physics from Clark University in 1911. He joined the faculty at Clark in 1914. Goddard developed the mathematical theory of rocket propulsion in 1912 and received patents for liquid-fuel and multistage solid-fuel rockets in 1914. The following year he demonstrated that rocket engines could develop thrust in a vacuum, a discovery that proved that spaceflight was possible, since the rocket as a self-contained reaction engine could operate in a vacuum and did not need to "push off" the ground or anything else, as was commonly believed at the time.

Goddard was one of the few scientists of the early twentieth century to tie rocketry to space travel. In January 1918, he boldly explored the far-reaching consequences of space technology in the article "The Ultimate Migration." In this work, Goddard postulated that an atomic-propelled space ark, possibly constructed using a small moon or asteroid, might someday carry human civilization away from a dying Sun to the safety of a new star system. However, fearing ridicule, he sealed the manuscript in an envelope, where it remained undiscovered and unpublished until 1972.

In 1919, Goddard summarized his work in a monograph entitled *A Method of Reaching Extreme Altitudes*, which outlined the fundamental principles of modern rocketry. He also included a final chapter on how the rocket might be used to get a modest payload to the Moon. Unfortunately, the press missed the true significance of his work and, instead, sensationalized his suggestion about reaching the Moon with a rocket, giving him such unflattering nicknames as "Moony" and the "Moon Man." Offended by this negative publicity, Goddard worked in seclusion for the rest of his life, avoiding further controversy by publishing as little as possible. As a consequence, much of his work went unrecognized during his lifetime.

On March 16, 1926, Goddard made space-technology history by successfully launching the world's first liquid-propellant rocket. This primitive ancestor to all modern liquid-propellant rockets was just 1.2 meters (4 feet) tall and 15.2 centimeters (6 inches) in diameter. Gasoline and liquid oxygen served as its propellants. The historic rocket rose to a height of only about 12 meters, and its engine burned for only two and one-half seconds. Nevertheless, humanity now had a technical pathway to space. Yet despite its great technical significance, the world would not learn about Goddard's invention for some time. The publicity-shy Goddard invited only his wife and two colleagues to the launch.

Goddard continued his rocket work during the 1920s and 1930s, anticipating much of the technology used by German rocketeers in World War II. In July 1929, he successfully launched the first rocket to carry a payload. The launch created a major disturbance, and local authorities ordered him to cease his rocket-flight experiments in Massachusetts. With the help of Charles Lindbergh, he moved his work to Roswell, New Mexico. There, undisturbed and well out of public view, Goddard conducted experiments that led to the development of steering devices for rockets, self-cooled rocket motors, power-driven fuel pumps, and other devices. During his lifetime, he registered 214 patents on various rockets and their components. Goddard offered to develop rockets for the military during World War II, but the U.S. government had little interest in the technology. Instead, the navy assigned Goddard to developing jet thrusters for seaplanes. He died on August 10, 1945, in Baltimore, Maryland, and did not see the dawn of the space age.

HERMANN J. OBERTH (1894–1989)

> To make available for life every place where life is possible. To make inhabitable all worlds as yet uninhabitable, and all life purposeful.
>
> Hermann J. Oberth,
> in response to the question "Why space travel?"

With Konstantin Tsiolkovsky and Robert H. Goddard, Hermann Oberth founded the science of astronautics. The son of a physician, Oberth was born on June 25, 1894, in Nagyszeben, a German enclave in the Transylvanian region of Romania, then part of the Austro-Hungarian Empire. As a child, he was deeply influenced by the works of Jules Verne, especially *From the Earth to the Moon*, and became intrigued by the thought of space travel. He taught himself the mathematics he thought he would need for spaceflight, and by the age of 14 he had envisioned a multistage rocket, fueled by liquid propellants, capable of space travel.

In 1912, Oberth entered the University of Munich to study medicine, but after being wounded in World War I, he turned his attention to astronautics. Near the end of the war, he tried to interest the German War Ministry in the development of a long-range military rocket, but the government rejected his proposal as fantasy. Undaunted, Oberth returned to the university and investigated the theoretical problems of rocketry. In 1922, he presented his doctoral dissertation on the theory of rocketry to the University of Heidelberg. Unfortunately, the university committee rejected his dissertation as too utopian. Still inspired by space travel, he revised this work and published it in 1923 as *The Rocket into Planetary Space*. This modest-sized book provided a thorough discussion of the major aspects of space travel, and its contents inspired many young German scientists and engineers, including Wernher von Braun, to explore rocketry. Oberth's work in the 1920s became the foundation for the practical application of rocketry developed in Germany. He served as a leading member of Verein für Raumschiffahrt (VfR), the German Society for Space Travel, whose members conducted critical experiments in rocketry in the late 1920s and early 1930s.

In 1929, Oberth expanded his ideas concerning the rocket for space travel and human spaceflight in the prize-winning book titled *Roads to*

Space Travel. This book helped popularize the concept of space travel for both technical and nontechnical audiences. In this visionary book, Oberth also anticipated the development of electric rockets and ion propulsion systems. Oberth used some of the prize money to fund rocket-engine research within the VfR. Young engineers like Braun had a chance to experiment with liquid-propellant engines, including one of Oberth's own concepts, the Kegeldüse (conic) engine design. He received a Romanian patent for a liquid-propellant rocket in 1931 and launched his first rocket near Berlin the same year.

During World War II, Oberth worked briefly with Braun's military rocket team, which developed the V-2 rocket. Following the war, he became a writer and lecturer in Switzerland and then moved to Italy, where he developed solid-propellant rockets for the Italian military. In 1955, he joined Braun's team of German rocketeers at the U.S. Army's Redstone Arsenal near Huntsville, Alabama. He worked there for several years before retiring to Germany in 1958 to devote himself to writing philosophy. Of the three founding fathers of astronautics, only Hermann Oberth lived to see some of their pioneering visions come true, including the dawn of the space age (1957), human spaceflight (1961), the first human landing on the Moon (1969), and the first flight of the space shuttle (1981). He died in Nuremberg, Germany, on December 29, 1989.

ALBERT EINSTEIN (1879–1955)

Albert Einstein was the preeminent scientist of the twentieth century. He proposed new ways of thinking about time, space, and gravitation that revolutionized our understanding of the universe.

Einstein was born into a middle-class Jewish family in Ulm, Germany, on March 14, 1879. He graduated with a degree in physics and mathematics from the prestigious Federal Polytechnic Academy in Zürich, Switzerland, in 1900. Shortly thereafter, he became a clerk in the Swiss Patent Office in Berne. This rather unchallenging job provided him with the time he needed to continue his theoretical work in physics.

In 1905, while he was still a clerk in the Swiss Patent Office, Einstein presented a paper titled "Zur Elektrodynamik bewegter Körper" (On the electrodynamics of moving bodies). In this paper, he offered the special theory of relativity, which deals with the laws of physics as seen by observers moving relative to one another at constant velocity, that is, by observers in nonaccelerating or inertial reference frames. In formulating special relativity, Einstein proposed two fundamental postulates:

First postulate of special relativity: The speed of light (c) has the same value for all (inertial-reference-frame) observers, independent and regardless of the motion of the light source or the observers.

Second postulate of special relativity: All physical laws are the same for all observers moving at constant velocity with respect to each other.

From the theory of special relativity, Einstein concluded that only a "zero-rest-mass" particle, like a photon, could travel at the speed of light. Another major consequence of special relativity is the equivalence of mass and energy, which is expressed in Einstein's famous formula $E = mc^2$, where E is the energy equivalent of an amount of matter (m) that is annihilated or converted completely into pure energy and c is the speed of light. Among many other important physical insights, this equation was the key astronomers needed to understand energy generation in stars. The theory he developed completely transformed twentieth-century physics and became one of the modern foundations of the discipline.

Einstein published several other important papers during 1905 and received his doctoral degree in physics from the University of Zürich in that year as well. Nevertheless, he could not obtain a university teaching job until the University of Zürich finally offered him a low-paying position in 1909. He received a special professorship at the Kaiser Wilhelm Physics Institute in Berlin in 1913.

Two years later, Einstein introduced his general theory of relativity. He used this development to describe the space-time relationships of special relativity for cases where there was a strong gravitational influence such as white dwarf stars, neutron stars, and black holes. One of Einstein's conclusions was that gravitation is not really a force between two masses (as Newtonian mechanics suggests) but rather arose as a consequence of the curvature of space and time. In a four-dimensional universe (described by three spatial dimensions [x, y, and z] and time), space-time became curved in the presence of matter, especially large concentrations of matter. The fundamental postulate of general relativity states that the physical behavior inside a system in free fall is indistinguishable from the physical behavior inside a system far removed from any gravitating matter (that is, the complete absence of a gravitational field). This very important postulate is also called Einstein's principle of equivalence.

With the announcement of his general theory of relativity, Einstein's scientific reputation grew. In 1921, he received the Nobel Prize in physics for his "general contributions to physics and his discovery of the law of the photoelectric effect." At the time, his work on relativity was too sensational and "cutting edge" for the conservative Nobel Prize Committee to officially recognize. By 1930, his best physics work was behind him, but he continued to influence the world as a scientist-diplomat. When Hitler rose to power in Germany in 1933, Einstein sought refuge in the United States, where he was influential in convincing President Franklin D. Roosevelt to develop an atomic bomb before Nazi Germany did.

In 1940, Einstein accepted a position at the Princeton Institute for Advanced Study. He remained there until his death on April 18, 1955.

EDWIN POWELL HUBBLE (1889–1953)

Edwin Hubble revolutionized our understanding of the physical universe. He determined that galaxies existed beyond the Milky Way and calculated that the universe appears to be expanding at a constant rate.

Born in Marshfield, Missouri, on November 20, 1889, Hubble earned a degree in mathematics and astronomy from the University of Chicago in 1910 but abandoned these disciplines to study law at Oxford as a Rhodes scholar. In 1913, he set up practice in Kentucky. Quickly bored with the law, he returned to the study of astronomy, accepting a research position at the Yerkes Observatory of the University of Chicago in 1914. There, on the shores of Lake Geneva at Williams Bay, Wisconsin, Hubble started studying nebulae. By 1917, the year he received his doctorate from Chicago, he had concluded that the spiral-shaped ones (now called galaxies) were quite different from the diffuse nebulae (which are actually giant clouds of dust and gas).

Following military service in World War I, Hubble joined the staff at the Carnegie Institute's Mount Wilson Observatory, located in the San Gabriel Mountains northeast of Los Angeles. He remained affiliated with this observatory for the remainder of his life. Once at Mount Wilson, he resumed his careful investigation of nebulae. In 1923, he discovered a Cepheid variable star in the Andromeda "nebula," now known as the Andromeda galaxy, or M31. A Cepheid variable is one of a group of important very bright, supergiant stars that pulsate periodically in brightness. By studying this Cepheid variable in M31, Hubble concluded that it belonged to a separate collection of stars far beyond the Milky Way galaxy. This important discovery provided observational evidence that galaxies existed beyond the Milky Way. The size of the known universe expanded by incredible proportions.

Hubble continued to study other galaxies and in 1925 introduced the classification system of spiral galaxies, barred spiral galaxies, elliptical galaxies, and irregular galaxies. In 1929, Hubble studied the recession velocities of galaxies (i.e., the rate at which galaxies are moving apart) and their distances away. He discovered that the more distant galaxies were receding (going away) faster than the galaxies closer to us. This very important discovery revealed that the universe continues to expand. Hubble's work provided the initial observational evidence that supported big-bang cosmology. Today, this discovery is known as Hubble's law in his honor.

Hubble's law describes the expansion of the universe. As Hubble initially observed and subsequent astronomical studies have confirmed, the apparent recession velocity (v) of galaxies is proportional to their distance (r) from an observer. The proportionality constant is H_0, the Hubble constant. Currently proposed values for H_0 fall between 50 and 90 kilometers per second per megaparsec (km s^{-1} Mpc^{-1}). The inverse of the Hubble constant ($1/H_0$) is called the Hubble time. Hubble time is a measure of the age of the universe. If H_0 has a value of 50 km s^{-1} Mpc^{-1}, then the universe is about 20 billion years old, but an H_0 value of 80 km s^{-1} Mpc^{-1} suggests a much younger universe, ranging in age between 8 and 12 billion years. Despite space-age investigations of receding galaxies, there is still much debate today within the astrophysical community as to the proper value of the Hubble constant.

Hubble served as a ballistics expert during World War II and remained an active researcher until his death on September 28, 1953, in San Marino, California. More than 40 years later, NASA named its powerful orbiting astronomical observatory after him. One of the major goals of the Hubble Space Telescope is refining the value of the Hubble constant.

WERNHER VON BRAUN (1912–1977)

Wernher von Braun championed space exploration and served as one of the world's most important rocket developers from the 1930s to the late 1970s. His aerospace-engineering skills, leadership abilities, and unwavering technical vision formed a unique bridge between the dreams of the founding fathers of astronautics and the first golden age of space exploration.

Braun was born into a prosperous, aristocratic family in Wirsitz, Germany, on March 23, 1912. The science-fiction novels of Jules Verne and H.G. Wells and his mother's gift of a telescope kindled his lifelong interest in space exploration, while Hermann Oberth's book *The Rocket into Planetary Space* set him on a career in rocketry. In 1929, he became a founding member of Verein für Raumschiffahrt (VfR), the German Society for Space Travel. Within this privately funded organization, Braun worked closely with Oberth and other German rocket enthusiasts, carrying out innovative liquid-propellant rocket experiments near Berlin in the early 1930s.

In 1932, the German army hired a reluctant Braun to head a team developing ballistic missiles. While initially engaged in military rocket development, Braun received his doctorate in physics in 1934 from the University of Berlin. By 1941, his team produced the prototype missile that eventually became the world's first operational ballistic missile, the V-2, first flown experimentally in October 1942. The German army began firing operational V-2 rockets against London and Allied forces on the Continent in September 1944. After World War II, this innovative, liquid-propellant rocket became the direct technical ancestor to the rockets used for space exploration by both the United States and the Soviet Union.

Braun and many of his colleagues surrendered to the Americans at the end of the war and were sent to the American Southwest to continue their work. In 1950, the Army moved Braun and his team to the Redstone Arsenal, near Huntsville, Alabama, where he supervised development of early army ballistic missiles, such as the Redstone and the Jupiter, which were later used in the U.S. space program. As the Cold War missile race heated up, the U.S. Army made Braun chief of its ballistic weapon program.

Although Braun was working on military missiles, he never lost his childhood dream of space exploration and, during the 1950s, became a well-known advocate of space travel. Starting in the fall of 1952, he served as a technical advisor to the editor of *Collier's* magazine and supported the production of a beautifully illustrated series of visionary articles dealing with space travel and humans in space. In the mid-1950s, the charismatic scientist promoted space travel through his frequent appearances on television. With Walt Disney, he cohosted an inspiring three-part television series on human spaceflight and space exploration that appeared between 1955 and 1957 on Disney's very popular family television show. Through these shows, Braun helped introduce millions of Americans to the excitement and promise of space exploration.

Braun became an American citizen in 1955. Within months, his new country demanded almost the impossible. Following the successful Soviet *Sputnik 1* and *Sputnik 2* satellite launches in late 1957 and the disastrous failure of the first American Vanguard satellite mission, Braun's army rocket team was given less than 90 days to develop and launch the first U.S. satellite. On January 31, 1958, under Braun's direction, a hastily converted military rocket successfully propelled *Explorer 1* into Earth orbit.

In 1960, the U.S. government transferred Braun's rocket-development center to the National Aeronautics and Space Administration (NASA), where, as director of NASA's Marshall Space Flight Center, Braun developed the successful Saturn family of large and complex expendable launch vehicles that sent man to the Moon.

Wernher von Braun in front of the Saturn IB launch vehicle on the pad at Cape Canaveral, Florida (circa January 1968). Photograph courtesy of NASA (Marshall Space Flight Center).

Just after the first human landings on the Moon in 1969, NASA appointed Braun deputy associate administrator for planning. However, in less than two years, Braun left NASA. The rapidly declining U.S. government interest in human space exploration following the first Moon landing clearly disappointed him. He then worked for Fairchild Industries, a major aerospace company, until his death in Alexandria, Virginia, on June 16, 1977.

SERGEI PAVLOVICH KOROLEV (1906–1966)

Sergei Korolev was the driving force behind the Soviet Union's ballistic missile program and space program. Under his visionary supervision, the Soviet Union ushered in the space age.

Korolev was born on December 30, 1906, in Zhitomir, Russia, where his father was a teacher. He graduated from the prestigious Moscow Higher Technical School (now the Bauman Technological Institute) in Moscow in 1930. Inspired by the ideas of rocket pioneer Konstantin Tsiolkovsky, he became an aeronautical engineer and maintained the lifelong goal of spaceflight. He early recognized rocketry as the technical means of achieving this goal and helped found the Moscow rocketry organization, Group for Investigation of Reactive Motion (GIRD), which tested liquid-fueled rockets during the early 1930s. The military quickly saw the potential of rocketry and established the Rocket Science Research Institute (RNII). There Korolev supervised a series of rocket-engine tests and winged-rocket flights culminating in the RP-318, the Soviet Union's first rocket-propelled aircraft, in 1933.

Despite his technical brilliance, Korolev, along with most of the staff at RNII, became politically suspect and was imprisoned in Siberian labor camps during the Stalin purges of the late 1930s. He initially worked in one of the most dreaded parts of the gulag, the Kolyma gold mines, but as Russia prepared for war with Germany, he was sent to a scientific labor camp to develop jet-assisted takeoff systems for aircraft. He was freed after the war.

A brilliant engineer and a charismatic manager, Korolev resumed his work on rockets and ultimately organized the Soviet ballistic missile and space programs. His first rockets were improved versions of the captured German V-2 rocket, but in 1954 he began work on the first Soviet intercontinental ballistic missile (ICBM). Because the Soviet Union had only primitive nuclear devices, Korolev had to develop very powerful rockets to carry the heavy weapons. The first ICBM, the R-7, could carry a 5,000-kilogram payload across continental distances (i.e., more than 5,000 kilometers). Korolev also realized that his powerful new rocket could easily place massive payloads into low Earth orbit.

As the space race developed between the United States and the Soviet Union in the late 1950s, the Soviets used Korolev's missile to achieve important space firsts. On October 4, 1957, the Soviets placed the world's first artificial satellite, *Sputnik 1*, into orbit around Earth. On April 12, 1961, using one of Korolev's mighty military rockets and his *Vostok 1* spacecraft, they sent the first human into orbit. His work led to a robot mission to Mars in 1962, to the development of the Voskhod spacecraft capable of carrying multiple cosmonauts in 1964, and to the technology that permitted humans to walk in space in 1965. Korolev's rockets also propelled large Soviet spacecraft to the Moon and Venus. Appropriately, one of the largest features on the lunar farside now bears Korolev's name.

Because of political pressures, Korolev frequently was not allowed to pursue these developments in a logical, safe fashion. For example, in 1964 he was forced to improvise *Voskhod 1* when Premier Nikita Khrushchev wanted to fly a three-person crew in space before the flight of NASA's two-person Gemini capsule. Recognizing the danger and over strong opposition from his own design engineers, Korolev removed the ejection seat of a Vostok spacecraft and replaced it with three couches. Without an ejection seat, the cosmonaut crew could not eject from the capsule during the final stages of reentry descent. On October 12, 1964, the Soviets launched the improvised craft carrying three cosmonauts without spacesuits under very cramped conditions. Western intelligence experts knew little of the improvisation, and the mission created Khrushchev's intended impression that the Soviet Union was once again far ahead of the United States in space technology. From 1962 to 1964, Khrushchev's continued political use of space diverted Korolev from working on more important projects such as new boosters, the Soyuz spacecraft, and the Soviet space station.

Because of Soviet paranoia, Korolev received no public recognition for his work. Throughout his life, the Soviet government chose to hide his identity and publicly referred to him only as the "chief designer of carrier rockets and spacecraft." On January 14, 1966, the man who ushered in the space age died suddenly during routine surgery.

NIKITA S. KHRUSHCHEV (1894–1971)

Nikita S. Khrushchev was the provocative and often boisterous premier of the former Soviet Union who used early Russian space-technology achievements as a major political weapon in Cold War competition with the United States. His permission and enthusiastic encouragement allowed Soviet scientists to open the space age with the launching of *Sputnik 1* on October 4, 1957.

Khrushchev was born on April 17, 1894, in the village of Kalinovka in the Ukraine, which was then part of the empire of the Russian czar. He was a workingman of humble beginnings whose father was a miner and grandfather a serf. Throughout his political career, Khrushchev always kept in touch with these working-class roots by enjoying visits to farms and factories at home and when traveling abroad.

Following the carnage of World War II, he and other rising members of the Soviet leadership walked a political tightrope trying to please Joseph Stalin, the brutal Russian dictator who was famous for his bloody purges and palace intrigues. After Stalin's death on March 5, 1953, Khrushchev used his own peasant toughness and political skills to emerge from the ensuing political power struggle as both the new Soviet premier and the First Secretary of the Communist Party of the (former) USSR. As the leader of an emerging superpower, he shocked the old-guard Soviet political structure by attacking Stalin's programs and encouraging an era of technology development that he felt would draw the Soviet Union equal to and then ahead of the United States. Unlike Stalin, he did not believe that war with the United States was inevitable, and he even introduced a policy of peaceful coexistence. Khrushchev believed that a period of peaceful coexistence would provide the Soviet Union the time necessary to develop and demonstrate superior levels of technology.

He also quickly recognized the international political power to be harvested from successful space-technology accomplishments. However, he also maintained a policy of secrecy by ensuring that any potentially embarrassing failures were kept neatly tucked away as well-hidden state secrets. In 1957, he gave Sergei Korolev permission to use the powerful new Soviet intercontinental ballistic missile, called the R-7, to place the world's first artificial satellite into orbit. His decision and that event on October 4, 1957, changed history. As *Sputnik 1* circled Earth, its beeping signal heralded the dawn of the space age—an age brought suddenly into the center of the world stage by the crafty actions of a plebeian Ukrainian politician in charge of a supposedly technically inferior nation. The technical shock of *Sputnik 1* rippled throughout the U.S. government and the American population for months, only to be reinforced by additional Soviet space achievements and initial American failures. These conditions started the great space race of the Cold War, a race enthusiastically embraced by Khrushchev, whose powerful boosters gave the Russian space program a clear early technical advantage.

Recognizing the power of space technology as an instrument of politics, Khrushchev pushed Korolev and other Russian rocket scientists to keep coming up with impressive space-technology achievements, and they responded to his wishes. On April 12, 1961, Khrushchev's aggressive style of space politics allowed the Soviet Union to send the first human being (cosmonaut Yuri Gagarin) into orbit around Earth. Khrushchev openly bragged about Gagarin's flight while speaking derisively of the modest suborbital flight taken by American astronaut Alan Shepard on May 5, 1961. But Khrushchev's blustering also goaded a newly elected American president, John F. Kennedy, to make a daring decision in late May 1961. America would respond to Khrushchev's space-technology challenge by doing the impossible and sending human beings to the surface of the Moon and returning them safely to Earth within a decade. Thus Khrushchev's use of space politics ultimately stimulated the greatest technical accomplishment in human history, but not quite the way he had intended, since it was an American space accomplishment that the world would celebrate in July 1969.

Khrushchev also pursued a very dangerous game of military brinkmanship with the United States by placing nuclear-armed intermediate-range ballistic missiles in Communist Cuba in 1962. His actions precipitated the Cuban missile crisis of October 1962, which brought the world to the very edge of nuclear warfare. Khrushchev badly underestimated the response of the United States and was forced to withdraw his missiles in a resounding political defeat.

Despite continued space achievements (often dangerous, prematurely forced technical activities intended to "beat the United States"), Khrushchev's political star fell, and he was stripped of power on October 15, 1964. The "retired" premier was the victim of failing economic programs at home and the lingering political consequences of his Cuban missile crisis defeat. He lived quietly on pension in Moscow until his death on September 11, 1971. The passing of the politician who ushered in the age of space was not even acknowledged by a state funeral.

WALTER ELIAS "WALT" DISNEY (1901–1966)

Legendary American motion-picture animator and producer Walt Disney introduced millions of people to the excitement of space travel. Disney was born on December 5, 1901, in Chicago, Illinois, and was raised on his family's farm near Marceline, Missouri. There he began his cartooning career by sketching farm animals. In 1917, the family returned to Chicago, where Disney attended high school, but evening art classes were his real interest. Without graduating from high school, he volunteered to serve as a Red Cross ambulance driver during World War I.

After the war, Disney began producing advertising films in Kansas City, Missouri, and eventually turned to animation. Enjoying only limited success, he moved to Hollywood, California, in 1923. Five years later, he produced the first animated cartoon to use synchronized sound. This cartoon, called *Steamboat Willie*, introduced the world to a charming new cartoon character, Mickey Mouse. Soon Minnie Mouse, Pluto, Goofy, and Donald Duck joined Mickey, to the delight of millions of fans around the world. Under his inspirational leadership, Disney's cartoon and motion-picture studios continued to produce a wide range of award-winning cartoon shorts and innovative feature films. After World War II, he also produced award-winning "true-life adventure" films.

Disney was fascinated with the possibility of space travel and used his growing entertainment empire to popularize space for the American public. In the early 1950s, he started planning an entirely new form of entertainment, a family-oriented amusement complex, which he called a "theme park." As part of his "Disneyland," he constructed Tomorrowland, emphasizing future technology, particularly space travel, which scientists such as Wernher von Braun had been discussing in popular journals. Disneyland opened in the summer of 1955. The Tomorrowland section of the park featured a Space Station X-1 exhibit and a simulated rocket ride to the Moon. A large (25-meter-tall), needle-nosed rocket ship designed by Willy Ley and Wernher von Braun greeted visitors to the Moon-mission attraction.

Disney also developed a series of three television shows that introduced millions of Americans to space travel. Each show combined careful research and factual presentation with incredibly beautiful visual displays and a splash of Disney humor for good measure. The first show, "Man in Space," aired on March 9, 1955. It presented a history of rocketry by historian Willy Ley, a discussion of the hazards of human spaceflight featuring aerospace-medicine expert Heinz Haber, and a detailed presentation of a large, four-stage rocket that could carry six humans into space and safely return them to Earth by Wernher von Braun. "Man in Space" proved so popular with audiences that Disney rebroadcast the show twice. President Dwight D. Eisenhower liked the show so much that he personally called Walt Disney and borrowed a tape of the show to use as a space-education primer for the Pentagon.

Disney's second space-themed television show, "Man and the Moon," aired on December 28, 1955. In this show, Braun enthusiastically described his wheel-shaped space-station concept and explained how it could serve as the assembly platform for a human voyage around the Moon. Braun emerged from this show as the premier space advocate in the United States.

At the dawn of the space age, Disney aired the third space-themed show, "Mars and Beyond," on December 4, 1957. Braun appeared only briefly in this show because he was very busy trying to launch the first successful American satellite (*Explorer 1*). Through inputs from Braun and his colleague, Ernst Stuhlinger, the highly animated show featured an armada of nuclear-powered interplanetary ships heading to Mars. It also contained amusing (though highly speculative), cartoon-assisted discussions about the possibility of life in the solar system. The series popularized the dream of space travel for millions of youngsters, including many of those who would become America's astronauts and space scientists in the years to come.

Disney lived to see the United States take its first steps into space and set its course for the Moon. He died on December 15, 1966, in Burbank, California.

JOHN FITZGERALD KENNEDY (1917–1963)

President John F. Kennedy responded to the Soviet civilian space-technology challenge of the early 1960s by boldly committing the United States to land humans on the Moon within a decade. Through his bold and decisive leadership, human beings traveled through interplanetary space and walked on another world for the first time in history.

Born on May 29, 1917, in Brookline, Massachusetts, John F. Kennedy graduated from Harvard University in 1940 and then served in the U.S. Navy during World War II. He was a member of the U.S. House of Representatives from 1947 to 1953 and won a seat in the U.S. Senate in the 1952 election. Kennedy narrowly defeated his Republican opponent, Richard M. Nixon, in the 1960 election to become the 35th president of the United States.

Kennedy became president at a time when the Soviet Union's achievements in space had led many to question U.S. technical superiority. During Kennedy's presidency, Soviet premier Nikita Khrushchev constantly flaunted his nation's space-technology accomplishments as an illustration of Communist superiority over Western capitalism. Driven by political circumstances early in his presidency, Kennedy took steps to respond to this challenge. Kennedy boldly selected a Moon-landing project to symbolize American strength. During a special "Urgent National Needs" message to the U.S. Congress on May 25, 1961, he announced the Moon-landing mission with these immortal words: "I believe that this nation should commit itself to achieving the goal, before this decade is out, of landing a man on the Moon and returning him safely to Earth. No single space project in this period will be more impressive to mankind, or more important for the long-range exploration of space; and none will be so difficult or expensive to accomplish."

When Kennedy made his decision, the United States had not yet successfully placed a human being in orbit around Earth. Kennedy's mandate galvanized the American space program and marshaled incredible levels of technical and fiscal resources. Science historians often compare NASA's Project Apollo to the Manhattan Project (the World War II atomic bomb program) or the construction of the Panama Canal in extent, complexity, and national expense. On July 20, 1969, two Apollo astronauts, Neil Armstrong and Buzz Aldrin, stepped on the lunar surface and successfully fulfilled Kennedy's bold initiative.

Often forgotten in the glare of the Moon-landing announcement were several other important space-technology initiatives Kennedy called for in his historic "Urgent National Needs" speech. Kennedy requested additional funding to accelerate the use of communications satellites to expand worldwide communications and the use of satellites for worldwide weather observation. Both of these initiatives quickly evolved into major areas of space technology that now serve the global community. He also accelerated development of the ROVER nuclear-rocket program as a means of preparing for more ambitious space-exploration missions beyond the Moon. Due to dramatic change in space-program priorities, the Nixon administration canceled this program in 1973.

Sadly, the young president who launched the most daring space-exploration project of the Cold War did not personally witness its triumphant conclusion. An assassin's bullet took his life in Dallas, Texas, on November 22, 1963.

ARTHUR C. CLARKE (1917–)

Arthur C. Clarke is one of the most celebrated science-fiction/space-fact authors of all time. Many of his more than 60 books have predicted the development and consequences of space technology and have suggested exciting extraterrestrial pathways for human development.

Born in Minehead, England, on December 16, 1917, Clarke was interested in science from childhood but could not afford to pursue his studies at the university level. In 1934, he moved to London, where he worked as a government auditor and joined the British Interplanetary Society (BIS), established the previous year to promote space exploration and astronautics. Clarke served as a radar instructor in the Royal Air Force (RAF) during World War II. Immediately after the war, he published the pioneering technical paper, "Extra-Terrestrial Relays" in the October 1945 issue of *Wireless World*. In this paper, he described the principles of satellite communications and recommended the use of geostationary orbits for a global communications system.

Clarke received a bachelor's degree in physics and mathematics from King's College, London, in 1948. In 1951, he began a career writing about space travel. He produced technical books and articles about rocketry and space, as well as prize-winning science-fiction novels that explored the impact of space technology on the human race. In 1951, he published *The Exploration of Space*, in which he suggested the inevitability of spaceflight. He also mentioned that the exploration of the Moon would serve as a stepping-stone for human voyages to Mars and Venus. Clarke closed this book with an even bolder, far-reaching vision, namely, that humanity would eventually contact (or be contacted by) intelligent life beyond the solar system.

In 1953, Clarke published the immensely popular science-fiction novel *Childhood's End.* This pioneering novel addressed the consequences of Earth's initial contact with a superior, alien civilization that decided to help humanity "grow up." In 1968, he collaborated with film producer Stanley Kubrick to develop *2001: A Space Odyssey.* This award-winning film, adapted from Clarke's 1951 short story "The Sentinel," tells of the discovery of an ancient monument on the Moon and the realization that human beings are not alone in the universe. Clarke turned the script into a book that same year. He later published several sequels, titled *2010: Odyssey Two* (1982), *2061: Odyssey Three* (1988), and finally, *3001: The Final Odyssey* (1997).

During the 1950s, Clarke continued to push for the use of satellite technology for weather forecasting. His dream was realized in 1960 when the United States launched the world's first weather satellite, TIROS 1.

In 1962, Clarke wrote a very perceptive and delightful book, *Profiles of the Future.* In this nonfiction work, he explored the impact of technology on society and described how a technical visionary often succeeded or failed within a particular organization or society. To support his overall theme, Clarke introduced his three laws of technical prophecy, the third of which is useful in looking forward to twenty-first-century space technology: "Any sufficiently advanced technology is indistinguishable from magic."

Clark received many awards for his work on communications satellites, including the prestigious Marconi International Fellowship in 1982. In his honor, the International Astronomical Union (IAU) also named the geostationary orbit at altitude 35,900 kilometers above Earth's surface the Clarke orbit. He was knighted in 1998. Clarke has lived in Colombo, Sri Lanka, since 1956.

KRAFFT A. EHRICKE (1917–1984)

Krafft Ehricke was a dedicated space visionary who not only designed advanced rocket systems that greatly supported the first golden age of space exploration, but also addressed the important but often ignored social and cultural impacts of space technology. Born in Berlin, Germany, on March 24, 1917, Ehricke grew up in the political and economic turbulence of post–World War I Germany. By chance, at the age of 12, Ehricke saw Fritz Lang's 1929 motion picture *The Woman in the Moon.* This film introduced him to the concept of rockets and space travel, and Ehricke knew immediately what career he wanted to pursue. In the early 1930s, he was still too young to participate in the German Society for Space Travel (VfR), so he experimented in a self-constructed laboratory at home. During World War II, he worked on the German V-2 rocket program and, in 1942, obtained a degree in aeronautical engineering from the Technical University of Berlin. Near the end of World War II, Ehricke joined the majority of the German rocket scientists who surrendered to the United States and immigrated to America to join the army's growing rocket program.

In the early 1950s, Ehricke joined the newly formed Astronautics Division of General Dynamics. There he worked as a rocket concept and design specialist. He participated in the development of the first American intercontinental ballistic missile, the Atlas, and, as director of the Centaur missile program, helped develop the liquid-hydrogen liquid-oxygen–propellant upper stage of the rocket. In 1965, he joined the advanced-studies group at North American Aviation in Anaheim, California. From 1968 to 1973, Ehricke worked as a senior scientist in the North American Rockwell Space Systems Division in Downey, California. These positions allowed him to explore a wide range of military, scientific, and industrial applications of space technology. After departing Rockwell International, he continued his visionary space advocacy efforts through his own consulting company, Space Global, located in La Jolla, California. Throughout his career, Ehricke championed the peaceful exploitation of space. Pointing out that humanity had always progressed through exploration, he introduced the concept of the "extraterrestrial imperative," the idea that humanity had no rational alternative but to explore space if it were to continue to develop. Civilization would progress through the judicious management of Earth's resources and the development of extraterrestrial resources. He championed the use of the Moon as the first step in what was to be an unbounded "open-world civilization." Until his death on December 11, 1984, in La Jolla, California, he spoke and wrote tirelessly and eloquently about how space technology provided humanity with the ability to create such a civilization.

When asked, "What is the value of the Moon?" during a National Lumar Base Symposium in Washington, D.C., in October 1984, Krafft Ehricke eloquently responded: "The Creator of our Universe wanted human beings to become space travelers. We were given a Moon that was just far enough away to require the development of sophisticated space technologies, yet close enough to allow us to be successful on our first concentrated attempt."

YURI A. GAGARIN (1934–1968)

Yuri A. Gagarin was the Russian cosmonaut who was the first human being to fly in outer space and orbit Earth, a feat accomplished in April

1961. Gagarin was born on a collective farm near Gzhatsk, Russian Federation, on March 9, 1934. He was the son of a carpenter and developed an interest in flying at a young age. In 1957, he graduated as a lieutenant from the Soviet Air Force Cadet Training School at Orenburg, Russia. Before being selected to join a special group of cosmonaut trainees, he served as a test pilot. On April 12, 1961, he was chosen to make the first human flight into outer space. He rode a powerful military rocket into orbit aboard a *Vostok 1* spacecraft, circled Earth once, and then returned safely, assisted by a parachute on the very last portion of the flight.

His relatively brief (89-minute) orbital journey made him an instant global celebrity. Premier Nikita Khrushchev proclaimed him a "Hero of the Soviet Union" and sent him as a shining example of Soviet space accomplishments to many countries throughout the world. Although he never flew in space again, Gagarin took an active role in training other cosmonauts. Unfortunately, he met an untimely death in an aircraft accident near Moscow during a routine aircraft flight on March 27, 1968. He was given a state funeral, and his ashes were enshrined in the Kremlin wall beside other Soviet heroes. As another tribute to his memory, his home town of Gzhatsk was renamed "Gagarin."

JOHN HERSCHEL GLENN, JR. (1921–)

John Glenn, Jr., was the world-famous NASA astronaut and U.S. Marine Corps officer who had the "right stuff" to be selected by NASA as one of the original seven Mercury astronauts in April 1959 and then become the first American to orbit Earth in his Mercury Project *Friendship 7* space capsule in February 1962. Undaunted by age, he also set the record as the oldest person (77 years old) ever to fly in space by serving as a crew member on the space shuttle *Discovery* (1998).

Born in Cambridge, Ohio, on July 18, 1921, John H. Glenn, Jr., attended Muskingum College in Ohio and graduated with a bachelor of science degree in engineering. He was commissioned as an officer in the U.S. Marine Corps in 1943 and served his country as a marine combat pilot in both World War II and the Korean War. Returning from Korea, Glenn attended Test Pilot School at the Naval Air Station, Patuxent River, Maryland. In April 1959, he was selected as one of NASA's seven original Mercury Project astronauts. On February 20, 1962, he became the first American to orbit Earth as a modified Atlas-Mercury 6 rocket lifted his *Friendship 7* space capsule into orbit from Cape Canaveral, Florida. Glenn's flight—an American first—lasted 4 hours and 55 minutes, during which he circled Earth three times, observing everything from a dust storm in Africa to Australian cities from an altitude of about 260 kilometers. Glenn was the first American to see a sunrise and sunset from space, and he was the first photographer in orbit. Despite some anxious moments about a potentially faulty heat shield, Glenn safely returned to Earth and became a national hero, having achieved the Mercury Project's primary goal of demonstrating the feasibility of human spaceflight.

Following his historic flight, Glenn continued to support NASA's spaceflight program, specializing in space-capsule layout and control systems. He resigned from the NASA Manned Spacecraft Center in January 1964 and retired as a colonel from the U.S. Marine Corps in January 1965. In November 1974, he was elected to the U.S. Senate from Ohio and served his nation in that capacity until he retired in January 1999.

Prior to retiring from the U.S. Senate, John Glenn demonstrated that he still had the "right stuff" when he flew again in outer space on board the space shuttle *Discovery*. During the STS-95 shuttle mission (October 29–November 7, 1998), the 77-year-old Glenn made 134 orbits of Earth and set the record as the oldest human ever to fly in outer space in the twentieth century. His unusual record may hold well into the twenty-first century.

NATIONAL AERONAUTICS AND SPACE ADMINISTRATION (NASA)

The National Aeronautics and Space Administration is responsible for planning, directing, and conducting civilian (including scientific) aeronautical and space activities for the United States. NASA was formed in response to the Soviet Union's launch of the world's first artificial satellite, *Sputnik 1*, in October 1957 and the highly publicized failure of America's first space effort, Vanguard, three months later. In the aftermath and the growing worldwide belief in Soviet space-technology superiority, American leaders recognized the need for a special organization to direct the space program if the United States were to win the emerging "space race." On October 1, 1958, the National Aeronautics and Space Act created the National Aeronautics and Space Administration, which absorbed the National Advisory Committee for Aeronautics and most of the fragmented (civilian/scientific) American space program then being pursued by the American military.

Over the years, NASA's dedicated men and women made the United States the world's greatest spacefaring nation. Through Project Apollo, the United States landed astronauts on the Moon six times between 1969 and 1972. During what is known as the first golden age of space exploration

(1960–1989), NASA was responsible for American space probes that traveled to the edges of the solar system and beyond, revolutionizing our understanding of the planets, the Sun, and the interplanetary medium. NASA's experimental communications satellites transformed the global telecommunications industry, while an armada of pioneering Earth-observing and weather-monitoring spacecraft stimulated the growth of the important multidisciplinary field of Earth system science.

To meet the demands for reusable, lower-cost space transportation systems, NASA developed the space shuttle, launched in 1981. At the end of the twentieth century, the agency took responsibility for developing the U.S. components of the *International Space Station*.

NASA's focus and fortunes changed from the heyday of the Apollo program in the 1960s to the beginning of the twenty-first century. During the race to be first on the Moon, the agency had tremendous public support and government backing. At the height of the Apollo program, NASA accounted for 4 percent of the national budget. But following the Moon landings, public interest in the agency declined, and funding dried up as American priorities changed in light of the Vietnam War and the growing crisis in U.S. cities. The agency struggled to answer the question, "Where can NASA go after it has been to the Moon?" and developed no goals around which to unite the nation. NASA continued to explore space, exciting the American public with extraordinary images of distant planets, and it built the first reusable spacecraft, the space shuttle. However, it was not able to generate support for visionary projects such as a human expedition to Mars or more easily achievable programs such as a permanent American space station, which President Ronald Reagan proposed in 1984. For many, the Mars project was too costly and the space station a step backward.

Toward the end of the 1990s, the agency redirected its space-probe program. It abandoned the complex, costly probes that had surveyed the solar system, such as Voyager and Pioneer, and redirected its efforts to producing smaller, cheaper probes designed to perform very specific missions, such as the *Mars Pathfinder*. It entered the twenty-first century with an ambitious program of missions to explore planets, particularly Mars, and expand scientists' understanding of the beginning of the universe.

Six fundamental questions of science and research currently guide NASA's goals: How did the universe, galaxies, stars, and planets form and evolve? Does life in any form, however simple or complex, carbon based or other, exist elsewhere beyond planet Earth? How can we (NASA) use the knowledge of the Sun, Earth, and other planetary bodies to develop effective Earth system models that support sustainable development and improve the quality of life? What is the fundamental role of gravity and cosmic radiation in vital biological, physical, and chemical systems in space, on other planetary bodies, and on Earth? How can we develop revolutionary technological advances in air and space travel? How can we most effectively transfer the knowledge gained from space research and discoveries to commercial ventures on Earth and in the aerospace environment?

NATIONAL RECONNAISSANCE OFFICE (NRO)

The National Reconnaissance Office (NRO), an agency of the Department of Defense, is responsible for U.S. spaceborne reconnaissance. The technology and expertise the NRO developed made a significant contribution to the early U.S. space program.

In the 1950s, rapid Soviet nuclear weapons developments along with that nation's long-range bomber and missile-delivery systems created the possibility of a devastating surprise attack—a "nuclear Pearl Harbor." To monitor Soviet activity, the United States used U-2 reconnaissance aircraft to collect data, but the sheer size of the Soviet land mass and rapid advancements in Soviet air defense systems began to limit the effectiveness of these spy planes. To overcome this growing limitation, the U.S. Air Force began developing a family of spy satellites in 1956. This low-priority effort became a high-priority program following the Soviet launch of *Sputnik 1* on October 4, 1957. Under the program, the Central Intelligence Agency and the U.S. Air Force jointly developed satellites to photograph denied areas from space. An interagency struggle for control of the program led the Eisenhower administration to create the NRO in August 1960.

From August 1960 to May 1972, the NRO oversaw the development and deployment of almost 150 Corona reconnaissance satellites designed to be America's eyes in space. These spacecraft were placed in polar orbits about 180 kilometers above Earth to image targets primarily in the Soviet Union and the People's Republic of China. To maintain the program's secrecy, the air force, which launched the spacecraft, labeled the satellites "Discoverer" and identified space research and technology as their primary mission.

The Corona program revolutionized intelligence activities and served as the prototype for future American photoreconnaissance-satellite programs. From a technical perspective, Corona achieved many space firsts that eventually led to advancements in other areas of the national space program. It was the first space program to recover an object from orbit, to de-

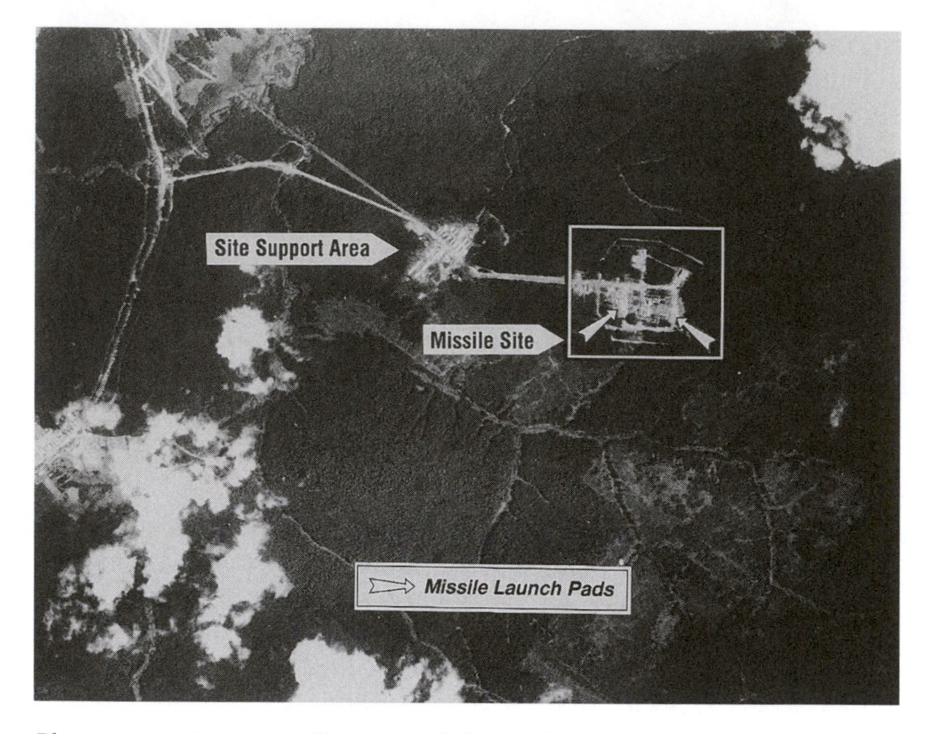

Photoreconnaissance satellites provided critical information about the status of the Soviet missile program and helped maintain global stability during the most dangerous portions of the Cold War nuclear arms race. Shown here is a declassified image of the Yurya ICBM complex in the former Soviet Union with the construction of an SS-7 missile launch site clearly noted by American intelligence analysts. A *Corona* photoreconnaissance satellite captured this image in June 1962. Photograph courtesy of National Reconnaissance Office (NRO).

liver photoreconnaissance information from a satellite, to use multiple reentry vehicles, to pass the 100-mission mark, and to produce stereoscopic space imagery. Its most remarkable technological advance, however, was the improvement in its ground resolution from an initial capability of 7.6 to 12.2 meters to an eventual resolution of 1.8 meters. The smaller the spatial resolution of an image, the more detailed the information it contains for a trained analyst and, therefore, the more useful it becomes in national defense applications.

Many of the technical advances developed under the program helped change the modern world. Corona provided a fast and relatively inexpensive way to map Earth from space, leading to the development of geographic information systems. The massive data-handling requirement of military reconnaissance satellites also promoted advances in space-based communications technology, which, in turn, helped create the modern information revolution.

But the most important contribution of the Corona system to national security came from the intelligence it provided. Satellite imagery gave American leaders the confidence to enter into negotiations and to sign important arms-control agreements with the Soviet Union. Successor programs continued to monitor ICBM sites and verify strategic-arms agreements and the Nuclear Nonproliferation Treaty. Cold War historians draw an interesting parallel. The flight of one satellite, *Sputnik 1*, created the famous "missile and space-technology gap" crisis that swept through the United States, while the secret flight of another, *Corona 14*, proved to a limited number of American officials that these fears were quite unfounded.

The Department of Defense declassified the existence of the NRO on September 18, 1992. Until then, the U.S. government did not even publicly acknowledge this very important organization. A presidential executive order dated February 24, 1995, declassified more than 800,000 images collected by these early photoreconnaissance systems during the period from 1960 to 1972. This historic, declassified imagery (some with a resolution of 1.8 meters) now helps environmental scientists improve their understanding of global environmental processes and serves as a well-documented baseline from the 1960s for assessing global change.

Because the agency operates in deep security, little is known of its current missions. In recent years, the NRO has moved beyond its military function. Today its imaging satellites support other urgent issues at home and abroad, including disaster-relief operations and the study of critical environmental problems.

EUROPEAN SPACE AGENCY (ESA)

The European Space Agency (ESA) is an international organization that promotes the peaceful applications of space technology and research for its European member states. The European Space Agency, formed in 1975, emerged out of, and took over the obligations and rights of, the two earlier European space organizations: the European Space Research Organization (ESRO) and the European Launcher Development Organization (ELDO), both founded in 1964. ESRO was a cooperative scientific research program; ELDO was responsible for the development of an independent space launch-vehicle capability. The driving force behind the decision of the European governments to coordinate and pool their efforts in joint space endeavors was primarily an economic one. In comparison to the ambitious space programs of the United States and the Soviet Union, no individual Western European country could afford to independently sponsor a complete range of space projects and all the necessary technology infrastructure. The nations of Western Europe also collectively recognized that they could not be left out of the "space race." European government officials reasoned that an independent space program would serve as a powerful economic and technical stimulus to European industry.

Notable space-exploration contributions by ESA include the *Giotto* flyby mission to Comet Halley (1986) and the *Ulysses* spacecraft mission to the Sun's polar regions (encounters in 1994–1995). ESA cooperated closely with NASA on many important space-exploration projects, including the development of *Spacelab* (launched 1983), the *Cassini* mission to Saturn (launched 1997), the *Galileo* mission to Jupiter (launched 1989), and the Hubble Space Telescope (launched 1990). ESA has its own corps of astronauts who have participated on U.S. space shuttle missions involving *Spacelab*. They have flown onboard the Russian *Mir* space station (EuroMir program) and are participating in the *International Space Station*.

Current ESA member states are Austria, Belgium, Denmark, Finland, France, Germany, Ireland, Italy, the Netherlands, Norway, Portugal, Spain, Sweden, Switzerland and the United Kingdom. Canada is a cooperating state. Today, ESA's involvement spans the fields of space science, Earth observation, telecommunications, space segment technologies (including orbital stations and platforms), ground infrastructures, and space transportation systems, as well as basic research in microgravity.

While individual European nations still avoid sponsoring expensive independent space projects, some countries (like France and Germany) also maintain thriving national space programs in addition to extensive participation in ESA. But it is primarily through ESA that the countries of Western Europe now fully participate in space exploration, technology application, and commerce.

Chapter 4

How Space Technology Works

This chapter presents the basic design features, functions, and principles of operation of the key elements of space technology. It will introduce you to the most basic scientific laws on which spaceflight is based. You then will learn how a rocket works and about the different types of rockets that have been used in, or at least proposed for, space travel. You will also understand how aerospace workers launch modern rockets and the major differences between expendable and reusable launch vehicles.

The next portion of the chapter describes spacecraft, their basic components, and general functions. You will discover that there are actually many different types of spacecraft, each carefully designed to perform a particular mission, such as monitoring Earth's environment, watching for enemy missile attacks, providing navigational assistance, exploring other worlds, or providing global communications. Some spacecraft can sustain astronauts and cosmonauts while they travel in orbit around Earth. In the past, specially designed interplanetary spacecraft carried human explorers to the Moon; in the future, new interplanetary spacecraft will carry human explorers to Mars and interesting destinations in the solar system.

You will also learn about the basic physical laws that describe how small objects move in orbit around much larger objects and how human-made objects fly on specific trajectories through interplanetary space. This is the field of orbital mechanics. While the mathematics can be quite complex, the basic physical concepts are fairly easy to grasp. One particularly interesting aspect of space travel is the microgravity environment experienced on board an Earth-orbiting spacecraft.

The chapter concludes with a discussion about how people live and work in space. A description of astronaut activities onboard NASA's space shuttle and the *International Space Station (ISS)* provides some insight into conditions during relatively short journeys through space. The discussion of anticipated crew activities during a proposed three-year-duration human expedition to Mars provides a glimpse into the realm of long-duration, deep-space travel by human explorers in the twenty-first century.

FUNDAMENTAL PRINCIPLES OF ASTRONAUTICS: THE PHYSICS OF ROCKETRY

While people have used rockets and rocket-powered devices for the last thousand years or so, scientists have understood the physical basis for their operation only in the last three hundred years. Rocketry, as a science, actually began when Sir Isaac Newton published his famous book *Philosophiae Naturalis Principia Mathematica* in 1687. In the *Principia*, Newton stated three important scientific principles that describe the motion of almost any object, including rockets. Knowing these basic principles, called Newton's laws of motion, aerospace engineers can predictably design the powerful rockets that deliver spacecraft and human crews into space.

Newton's first law of motion introduces the concept of inertia. It states that objects at rest will stay at rest and objects moving in a straight line will keep moving in a straight line unless acted upon by an unbalanced external force. Physicists also call this statement the principle of conservation of linear momentum. In physics, the linear momentum of an object is equal to the product of the object's mass and its velocity.

Newton's second law states that the rate of change of momentum of a body is proportional to the force acting upon the body and is in the direction of the applied external force. For an object of constant mass, this results in the familiar statement that force is equal to mass times acceleration. During their propellant-burning operations, chemical rockets continuously undergo a loss of mass, so a slightly more complex mathematical form of Newton's second law becomes appropriate.

Newton's third law of motion is the action-reaction principle, the physical basis of all rockets. It states that for every force acting upon a body, there is a corresponding force of the same magnitude that the body exerts in the opposite direction.

These deceptively simple statements form the basis of Newtonian mechanics, an extremely powerful and useful tool in science and engineering. Mass and energy are treated as separate, conservative mechanical properties in Newtonian mechanics. Early in the twentieth century, another brilliant scientist, Albert Einstein, extended Newtonian mechanics into the realm of relativistic physics—a realm in which Einstein treated mass and energy as equivalent, as expressed in the famous equation $E = mc^2$. However, until we start designing starships that travel at 90 percent or more of the speed of light, Newtonian mechanics works just fine helping us design and operate powerful (but nonrelativistic) rocket vehicles. Because of their importance in helping us understand the performance of rocket vehicles, we will examine the significance of each law in a little more detail.

Newton's First Law of Motion

Newton's first law of motion is just a statement of easily observable physical fact. It states that if a body in motion is not acted upon by an external force, its momentum remains constant:

linear momentum (p) = mass (m) × velocity (v), or $p = mv$.

But to really know what this implies, you must understand the scientific meaning of the terms *rest*, *motion*, and *external unbalanced force*. Think of rest and motion as the opposite of each other. Rest is the state of an object when it is not changing position *in relation to its surroundings*. For example, if you are sitting still in a chair, you are at rest with respect to the chair.

Let us now assume that the chair in which you sit motionlessly is actually one of many seats on a high-flying commercial airplane. Relative to Earth's surface, you and the chair are certainly moving. However, you are still at rest with respect to the chair, since you are not moving in relation to your immediate surroundings. If we defined rest as the total absence of motion, it would be a meaningless definition and one that could not exist in nature. Even if you were sitting comfortably in your chair at home (and not on the airplane), you would still be moving, because your chair is located on the surface of a spinning planet that is orbiting a star. The star is moving through a rotating galaxy that is itself moving through the universe. (Remember, according to "big-bang" cosmology, the entire universe is in motion and expanding.) While you are sitting "still" at home, you are actually traveling through space at a speed of hundreds of kilometers per second.

Of course, you do not feel this motion, and for practical problem solving here on Earth, engineers often "neglect" our planet's motion through space. In aerospace engineering, as well as other technical fields, it is important for you always to understand the assumptions and constraints a particular scientific model of the physical world contains. Later in this chapter, we will see how planetary motions do become an important part of some problems, as, for example, when aerospace engineers plan to send a spacecraft from Earth to Mars.

Motion is also a relative term. All matter in the universe is moving all the time. However, in the context of Newton's first law, the term *motion* specifically means that an object is changing its position in relation to its immediate surroundings. A ball is at rest if it is motionless on the ground, but when it is rolling on the ground, scientists say that the ball is in motion. The rolling ball keeps changing its position in relation to its surroundings. Similarly, a rocket vehicle blasting off its launch pad changes from a state of rest to a state of motion.

The third term necessary to understand Newton's first law is that of an external, unbalanced force. *External* means that the force comes from outside the object. If you hold a ball in your hand and keep your hand still, the ball is at rest with respect to both your hand and the immediate surroundings. While you hold the ball motionless in the air, the force of Earth's gravity keeps trying to pull the ball downward. The ball does not move because your hand provides a lift force that pushes just the right amount against the ball to hold it up and motionless. In this case, the external forces acting on the ball are balanced. The downward force of gravity and the lift force provided by your hand keep the ball suspended in the air. But if you quickly tilt your hand and let the ball go, the external forces on the ball become unbalanced. Now, the unopposed force of gravity makes the ball drop to the ground. In this case, the ball changes from a state of rest to a state of motion.

During the flight of a rocket, forces change between balanced and unbalanced all the time. During the countdown, for example, a rocket rests motionless on its launch pad, and all the external forces on the rocket are balanced. The force of gravity pulls it downward, while the surface of the launch pad pushes it upward. When the countdown reaches zero, the launch director sends a special command signal that ignites the rocket's engines. Now, the thrust from the engines creates an unbalanced force, and the rocket travels upward. Later, when its propellant supply runs out, the rocket slows down, stops at the highest point in its flight, and then starts falling back to Earth.

Objects in space also react to forces. A spacecraft moving through the solar system is in constant motion. The spacecraft will travel in a straight line if all the forces acting on it are in balance. However, this happens only when the spacecraft is very far away from any massive object (source of gravity) such as Earth or the other planets and their moons. If the spacecraft comes near a massive planetary body, the gravitational force of that body will unbalance the forces acting on the spacecraft and curve its path. This also happens when a rocket sends a satellite into space on a path (trajectory) that is tangent to the planned orbit around the planet. The unbalanced gravitational force causes the satellite's path to follow an arc. The arc is a combination of the satellite's inward fall toward the planet and its forward (outward) motion. When these two motions are just right, the satellite travels in a closed path (called an orbit), "continually falling" around the larger body. This closed orbital path can either be circular or elliptical, depending on certain physical factors.

For now, we shall consider that this satellite is in a circular orbit around the massive planetary object. Since the planet's gravitational force changes with altitude (height) above its surface (as predicted by another famous Newtonian law, the law of gravitation), each orbital altitude has an associated value of satellite velocity that creates a circular orbit. For example, to achieve a circular orbit at an altitude of approximately 200 kilometers (low Earth orbit), a spacecraft needs an orbital velocity of about 7.7 kilometers per second. Aerospace engineers recognize that maintaining this orbital velocity is extremely important if the spacecraft is to operate in a circular orbit during the mission. Due to a combination of several almost imperceptible phenomena that occur in the space environment, an orbiting spacecraft's velocity does change (usually declining) a bit over time. Therefore, human controllers will occasionally fire tiny thruster rockets on the spacecraft to make the necessary fine adjustments in its orbital velocity. This process is called stationkeeping.

We can now restate Newton's first law. If an object (such as a rocket) is at rest, it requires an unbalanced force to move it. If the object is already moving, then it requires an unbalanced force to stop it, change its direction from a straight-line path, or alter its speed.

Newton's Second Law of Motion

For our purposes, Newton's second law of motion is essentially a statement of the equation that the force (F) equals the mass (m) times the acceleration (a), that is,

$$F = ma. \qquad (4\text{-}1)$$

To explore the significance of this physical principle, we will use the old-fashioned cannon shown in Figure 4.1. When we fire this cannon, the explosive charge releases gases that propel the cannonball out the open end of the barrel. Depending on the mass of the cannonball and the energy

Figure 4.1 An old-fashioned cannon helps explain Newton's second law of motion. Image courtesy of NASA.

released in the explosion, the cannonball might travel a kilometer or so to its target. At the same time, the cannon itself recoils (jumps backward) a meter or two. (The recoil action happens because of Newton's third law, which we will discuss next.) The explosive force acting on the cannon and the cannonball is the same. Newton's second law describes what happens to each object after the explosion. First, we will write the second law for each object:

$$F_{\text{explosion}} = m_{(\text{cannon})} a_{(\text{cannon})}, \tag{4-2a}$$

$$F_{\text{explosion}} = m_{(\text{ball})} a_{(\text{ball})}. \tag{4-2b}$$

As indicated in the equations, the first equation refers to the cannon and the second to the cannonball. In the first equation (equation 4-2a), the mass is that of the cannon, and the acceleration involves the recoil movement of the cannon after the explosion and departure of the cannonball. In the second equation (equation 4-2b), the mass is that of the cannonball, and the acceleration relates to its high-speed flight. Because we have assumed that the explosive force is the same for both equations, we can combine and rewrite these equations (4-2a and 4-2b) as follows:

$$m_{(\text{cannon})} a_{(\text{cannon})} = m_{(\text{ball})} a_{(\text{ball})}. \tag{4-3}$$

To keep both sides of this new equation balanced, the accelerations experienced by the cannon and the cannonball must vary with their respective masses. As a result, the cannon with its very large mass experiences a small, modest acceleration, while the cannonball with its small mass experiences a very large acceleration during the explosive event.

To help us understand how the second law helps describe the performance of a rocket engine, we will perform a very simple "thought experiment" by placing this same cannon somewhere in interplanetary space. The cannon now represents the mass structure of a very unusual "rocket ship," and the cannonball represents the total mass of the combustion gases ejected to space through the nozzle of our unusual "rocket ship." The explosive charge represents the high pressure created by the combustion of the chemical propellants that takes place inside a chemical rocket's engine. In this simple model of rocket performance, the "exhaust gases" (cannonball) leave the rocket at very high acceleration, and the "rocket ship" (cannon) responds by accelerating in the opposite direction. A few important features of rocket design should become apparent from this example. The more cannonballs we can fire, the more total acceleration we give to the cannon ("rocket ship"). The lower the total mass of the cannon ("rocket ship"), the greater its final acceleration. The greater the acceleration of each cannonball fired, the greater the total recoil acceleration of the cannon ("rocket ship"). These are precisely the reasons why rocket engineers try to make the mass of the rocket vehicle (including the payload) as low as possible, burn as much propellant as possible, and expand the combusted gases through the nozzle at the highest possible exit velocity.

Of course, the previous "thought experiment" represents only an instantaneous picture (snapshot) of how a rocket engine works. Other interesting things happen with real rockets that do not occur with the cannon and ball in this example. Perhaps most important is the fact that when we fire the cannon and the ball flies out, the entire event lasts only a brief moment and produces an instantaneous, impulsive (reaction) thrust. For chemical rockets, the generation of thrust is a reasonably continuous process that lasts as long as the engines have propellant to burn (typically many seconds to several minutes). A second important difference is that the mass of the rocket keeps changing during the powered (thrusting) portion of its flight. The rocket vehicle's mass is the sum of all its parts, including engines, propellant tanks, payload, control system, and propellants. By far the largest part of the rocket's mass is its propellants (typically, 85 to 92 percent of the total takeoff mass of a modern rocket is propellant). But that amount constantly changes (decreases) as the engines fire. That is why a rocket vehicle starts off climbing slowly and goes faster and faster as it ascends into space. On the basis of this very simple, but useful analogy, we can restate Newton's second law as follows: the greater the mass of rocket fuel burned, and the faster the combustion gases can escape through the nozzle, the greater the thrust of the rocket.

Newton's Third Law of Motion

Newton's third law of motion states that for every action, there is an equal and opposite reaction. The action-reaction principle is the basis of the operation of all rockets. The rocket engine expels mass at high velocity, and the reaction thrust drives the rocket vehicle in the opposite direction.

The *thrust equation* is the fundamental equation for rocket-engine performance. For reaction engines (i.e., rockets) that generate thrust by expelling a stream of internally carried mass, scientists express this equation as

$$T = \dot{m} V_e + (p_e - p_a)A_e, \qquad (4\text{-}4)$$

where T is the thrust force (newtons), $\dot{m}$; is the mass flow rate of ejected materials (kilograms per second), V_e is the exhaust velocity of the ejected mass (meters per second), p_e is the exhaust pressure at the nozzle exit (newtons per meter squared), p_a is the ambient pressure (newtons per meter squared), and A_e is the nozzle's exit area (square meters).

Many people make the common mistake of believing that a rocket is propelled through the air by its exhaust gases pushing against the outside air. Nothing could be further from the truth. In fact, as shown in equation 4-4, rockets work much better in outer space, where the ambient pressure (p_a) is zero.

Understanding Newton's laws of motion is basic to understanding how a rocket works. An unbalanced force must be applied to a rocket if it is to rise from the launch pad and climb into space. The same is true if a space vehicle changes its speed or direction while traveling through space (first law). The rate at which a rocket consumes the propellant mass and the exhaust speed of the ejected materials determine the amount of thrust (force) produced by a rocket engine (second law). The reaction (or forward) motion of a rocket vehicle is equal to and opposite of the action (or thrust) from the engine (third law).

ROCKET FUNDAMENTALS

A rocket is a completely self-contained projectile, pyrotechnic device, or flying vehicle propelled by a reaction engine. In its simplest form, it is just a chamber enclosing a gas under pressure. A small opening at one end of the chamber, called a nozzle, allows this pressurized gas to escape. As the gas rushes out through the nozzle, it produces a reaction thrust that propels the rocket in the opposite direction.

An inflated balloon can help us understand this concept. Within its elastic limits, a balloon's stretched rubber wall confines the high-pressure air that keeps the balloon inflated. The inflated balloon's wall pushes in and the confined air pushes back so that these inward (stretched balloon wall) and outward (internal air pressure) forces balance. Physicists call this balanced condition a state of mechanical equilibrium. If we release the pressurized air through the balloon's narrow opening (a nozzle), the air rushes away from the high-pressure region inside and escapes to the lower-air-pressure environment outside. As the air flows through the nozzle, a reaction force occurs that propels the balloon in the opposite direction. Have you ever tried to blow up a balloon for a party, only to have it slip away at the last minute and scoot around the room? Your erratic flying balloon obeyed the same basic law (Newton's third law of motion) as the powerful rockets that send spacecraft into orbit.

Of course, there is a major difference in the way rockets and balloons acquire the thrust-producing pressurized gas. In chemical rockets, aerospace engineers generate this high-pressure gas by burning propellants inside the rocket's combustion chamber. In the case of a balloon, we provide the pressurized gas by pumping (or blowing) air into an elastic (often rubber) enclosure that expands as the air pressure increases up to some safe limit.

Engineers avoid designing rockets that have a lot of unnecessary mass. In fact, they do everything they can to slim down a rocket vehicle to the bare essentials. This often involves using specially created, low-mass materials, clever structural designs (that do not compromise the vehicle's integrity under operational loads), and staging, discarding useless mass during the flight as a clever way of improving a rocket's mass fraction (MF). Rocketeers define the mass fraction (MF) as the mass of the propellants a rocket carries divided by its total mass (including propellant load):

$$\text{mass fraction (MF)} = \frac{\text{mass of propellants}}{\text{total mass of rocket (including propellants)}} \qquad (4\text{-}5)$$

From this equation, we would think that the perfect rocket has a mass fraction of unity, but this would mean that the entire rocket is nothing more than a big lump of propellant. Whether we are building a solid-fueled or a liquid-fueled rocket, there are some physical limits on the minimum mass of structural components and control hardware that we must use to contain and then burn a given mass of propellant. Aerospace engineers like to define an "ideal rocket" as one for which the total mass of the vehicle is distributed roughly as follows: chemical propellants, 91 percent of the total initial mass; structure and control hardware (including engines, tanks, casings, fins, pumps, and so on), 3 percent of the total initial mass; and payload, 6 percent of the total initial mass. The larger the mass fraction (MF) value, the less payload the rocket vehicle carries. The smaller the MF value, the less a rocket's range becomes (because of propellant supply limitations). Remember that some rockets are used in military missions to carry warheads (explosive payloads) against enemy targets, near and far away. An intercontinental ballistic missile (ICBM) has a range of 5,500 kilometers or more. Aerospace engineers, therefore, consider an MF number of 0.91 to represent a good balance between payload-carrying capability and range. As a point of interest, NASA's space shuttle has an MF of 0.82, although this number varies from mission to mission and orbiter vehicle to orbiter vehicle.

Large rockets, the kind needed to carry spacecraft into orbit, have serious MF problems. To achieve orbital or escape velocities, such large rockets must consume a great deal of propellant. The hardware to carry and burn all this propellant becomes excessively massive as the rocket vehicle becomes larger. Why carry all the extra structural mass when a great deal of the propellant is gone? The answer is simple: Discard mission-useless mass as the rocket climbs in altitude. Aerospace engineers call the process *staging*, and modern "step rockets" have opened outer space to both robot and human exploration. The first-stage rocket is the largest and propels itself and the companion upper stages to some altitude and velocity. Upon depleting its propellant supply, the first stage then separates and falls away from the rest of the flight vehicle. Next, the second-stage engine fires. The process continues with excess structural mass being discarded at every step in the sequence. Eventually, the payload reaches orbital (or even escape) velocity in an efficient manner (with a favorable mass fraction).

Stability

Building an efficient rocket engine is only part of the problem in producing a successful rocket vehicle. The rocket must also be stable in flight. A stable rocket is one that flies in a smooth, uniform direction. An unstable rocket takes an erratic path, sometimes tumbling, spinning, or changing direction without warning. Unstable rockets are dangerous because launch personnel cannot predict where they will go. They can even flip over and suddenly head back for the launch pad.

Aerospace engineers use a control system to make a rocket stable. Control techniques can either be passive or active. Passive controls are fixed (nonmoving) devices, such as fins, that keep a rocket stabilized by their simple presence on the vehicle's exterior. Active controls, as the name implies, move while the rocket is in flight to help stabilize and steer the vehicle.

Before describing the basic control techniques that aerospace engineers use, we need to understand several fundamental physical ideas that involve the behavior of moving objects. All matter, despite size, mass, or shape, has a special point within called the *center of mass* (CM). The center of mass is the precise location inside an object at which engineers and scientists assume all of the mass of the object lies. This *point-mass* or *lumped-mass* model of a chunk of matter makes it easier to do force-balance and motion calculations with Newton's laws. Sometimes the location of the center of mass corresponds to the geometric center of an object, but other times (when the object is not uniform in density or shape) the center of mass lies elsewhere.

The center of mass is important in rocket flight because it is around this point that an unstable rocket tumbles. In flight, spinning or tumbling takes place around one or more of the three intersecting axes that pass through the center of mass. Aerospace engineers call these axes *pitch*, *roll*, and *yaw* (see Figure 4.2). For control of a rocket's flight through the atmosphere,

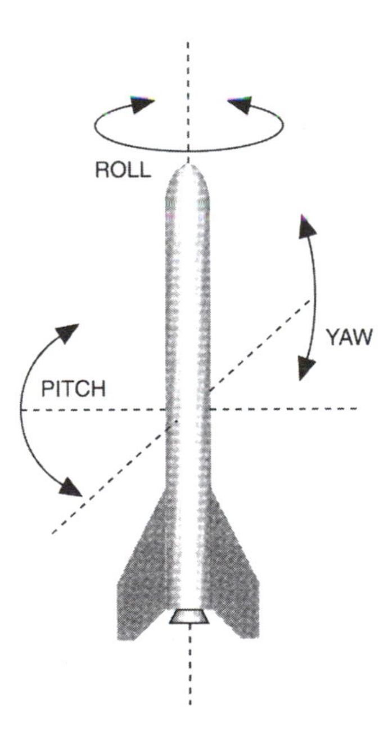

Figure 4.2 The pitch, roll, and yaw axes of a rocket. Image courtesy of NASA.

engineers consider the pitch and yaw axes the most important. Any unwanted movement in either of these two directions can cause the rocket to go off its planned course. The roll axis is less important, because movement along this axis does not affect the flight path directly. In fact, a rolling or spinning motion often helps stabilize a rocket. That is why military engineers design cannons to put a spin on artillery shells and successful quarterbacks put a spin on footballs when they toss long, spiraling touchdown passes. However (as we will discover shortly), aerospace engineers initially consider all three axes of equal importance when they design the control system for a spacecraft that must travel through interplanetary space. Depending on the particular mission, spinning or rolling around a particular spacecraft axis may or may not be desirable.

At this point, we will perform another "thought experiment" that should help make these important concepts a little more understandable. Consider a long, thin, rectangularly shaped ruler with a length of 100 centimeters. We assume that the ruler (consisting of wood, plastic, or metal) has uniform composition and density throughout. If this is true, the ruler will balance nicely at the 50-centimeter mark when you carefully place it on your horizontally extended pointing finger. We say that the ruler's 50-centimeter mark corresponds to the lengthwise location of its center of mass. If the ruler is 2 centimeters wide and 1 centimeter thick, we can quickly calculate the exact internal position of the center of mass. Engineers often find it very helpful to locate the origin of a Cartesian coordinate system (that is, the location of zero in *x*-, *y*-, and *z*- space) at the center of mass of an object. (A Cartesian coordinate system, named after French philosopher and mathematician René Descartes [1596–1650], is one in which the locations of all points in space are expressed by reference to three mutually perpendicular planes that intersect in three straight lines called the *x*-, *y*-, and *z*-coordinate axes.) This choice makes their control and stability calculations easier. We shall do likewise and place the origin of our spatial reference system at the center of mass of the ruler. Now, the *x*-direction (lengthwise) edges are 50 centimeters away, the *y*-direction (widthwise) edges are 1 centimeter away, and the *z*-direction (thicknesswise) edges are $\frac{1}{2}$ centimeter away from the center of mass.

Now we will make this "thought experiment" a little more interesting. Attach a large metal clip to one end of the ruler (your choice) and see what happens. The ruler no longer balances on your finger at the 50-centimeter mark. Instead, the balance point shifts down the ruler closer to the metal clip that you attached. Why? The answer is quite simple. By attaching a large metal clip to the ruler, we changed its mass and now have a system with a nonuniform mass distribution. Of course, we can still locate the center of mass. (Trial-and-error balancing should work just fine.) But with the metal clip attached at one end, the CM of the ruler no longer corresponds to its geometric center. Rocket engineers also have to deal with the "shifting" location of the center of mass of a flying vehicle. However, instead of dealing with CM shifts due to added mass, they must handle continuous CM shifts due to the loss of internal mass from a system that keeps its outside shape the same. During powered flight, a rocket burns propellant and expels combustion gases (mass) through its nozzle.

Before we leave this "thought experiment," let us push our imagination just a little further. Now imagine that our ruler has become a long, slender, "arrowlike" rocket (complete with pointed nose and finned tail) and that it is flying through Earth's atmosphere on the way to space. We recognize that the fin-surface area at the tail is larger than the head-surface area. The outside shape of our "ruler-rocket" remains the same as it picks up speed and expels propellant mass. The atmosphere also exerts aerodynamic forces on its surfaces as it flies faster and faster. Now we need some way of examining how any unbalanced combination of these aerodynamic forces might cause our rocket to tumble or spin. Engineers use another concept, called the *center of pressure* (CP), to study how air flowing past a moving object affects its motion. As air rubs and pushes against the outer surface of the rocket, it can start revolving around one or more of its three axes. The center of pressure is the imaginary point inside the rocket where the atmospheric forces acting on the surface area of the vehicle balance. We can also say that the surface area of the rocket is the same on one side of the center of pressure as on the other. In this example, the center of pressure is not in the same place as the center of mass because the tail fins have much more surface area than the tip of the rocket. This is actually as it should be in good design practice. Aerospace engineers know that it is extremely important to locate the center of pressure in a rocket toward the tail and the center of mass toward the nose. If these two centers occur in the same place or very near each other, the rocket will become very unstable in flight. In response to aerodynamic forces, the rocket will try to rotate about the center of mass in the pitch and yaw axes, producing a very unstable and dangerous flight situation. However, with the center of pressure located in the proper place (between the CM and the tail), the rocket will remain stable as it flies through the air.

The first attempts at designing a passive control system were made by ancient Chinese rocketeers who attached long sticks to the end of their fire-arrow rockets. Although the sticks helped keep the center of pressure behind the rocket's center of mass during flight, the fire-arrow rockets still proved terribly inaccurate. The use of heavy, long sticks also reduced the range of these early rockets. During the Renaissance, European rocketeers helped guide artillery rockets to their targets by launching them from tubes or from troughs aimed in the proper direction. Mid-nineteenth-century engineers came up with a dramatic improvement in control by mounting clusters of low-mass, nonmoving fins around a rocket's lower end near the nozzle. This gave the solid-propellant rocket a dartlike appearance, and the large surface area of the fins easily kept the center of pressure behind the center of mass. Some creative rocketeers even bent the lower tips of the fins in a pinwheel fashion to promote rapid spinning in flight. These "spin fins" made nineteenth-century rockets much more stable in flight. However, as in many engineering design efforts, there was also a negative side to the "spin-fin" flight-control approach. This design produced much more aerodynamic drag and, therefore, reduced the range of a particular rocket.

In the twentieth century, engineers needed to improve the stability of modern rockets and reduce the mass required for the control system. Active control systems provided the answer. Active control techniques include movable fins, canards, vanes, gimbaled (swiveled) nozzles, vernier rockets, fuel injection (for liquid-propellant engines), and attitude-control rockets. Movable (tilting) fins and canards are quite similar to each other. The major difference between them is their location on the rocket. Engineers place canards at the front end of a modern rocket, while tilting fins are at the rear. In flight, the movable fins and canards tilt like rudders to deflect the air flow and cause the rocket to change its course. A rocket's onboard guidance system can detect unplanned directional changes and issue commands that produce the required tilting of the fins and canards to correct the course. Military rockets, like an air-to-air missile system, carry sensors that detect and track a target aircraft. Once its sensors lock onto the enemy aircraft, the air-to-air missile uses the movable fins and canards to steer itself to the hostile target (even if it is trying to evade the missile) and destroy it. Movable fins and canards are smaller and lighter than large fins and produce much less aerodynamic drag.

Engineers can also use other active control systems to eliminate fins and canards entirely. By tilting the angle at which the exhaust gas leaves the rocket engine, the rocket vehicle can change its course during flight. Several exhaust-gas "tilting" techniques are possible.

First, the engineer can use vanes. These are small, finlike devices that sit inside the exhaust of the rocket engine just beyond the nozzle exit. Tilting such vanes deflects the exhaust gas and points the rocket in the opposite direction by virtue of the action-reaction principle of Newton's third law.

Another interesting engineering technique to change the direction of a rocket's exhaust gases is to gimbal (swivel) the entire nozzle. If the rocket's

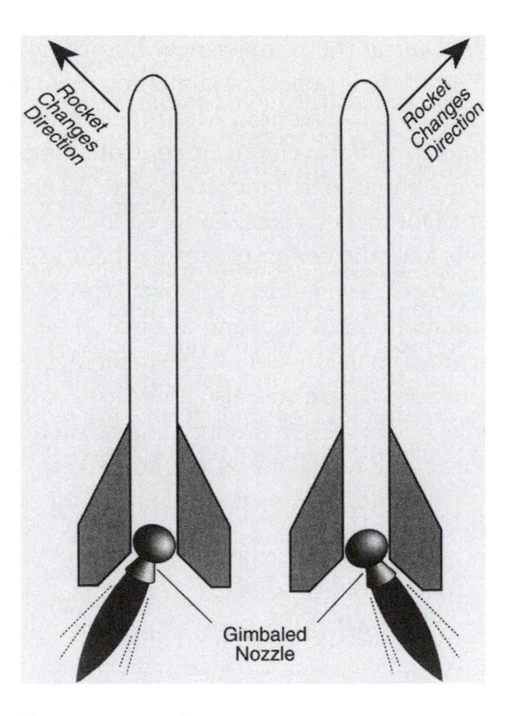

Figure 4.3 Aerospace engineers sometimes use the controlled swiveling of a gimbaled nozzle to help steer a rocket vehicle. Image courtesy of NASA.

nozzle is tilted in the proper direction, the rocket responds and changes its course to the desired direction. Because of the gimbaled nozzle's complexity, engineers generally use this control technique with large solid- or liquid-propellant rockets (see Figure 4.3).

Engineers also employ tiny vernier rockets to change a large rocket's direction. They mount these small engines at strategic positions on the outside of the large rocket. When needed, the vernier rockets fire and produce the desired course change.

Of course, once a rocket is beyond Earth's atmosphere or in outer space, fins and canards are useless for control-system purposes, since there is no air. You can look with enjoyment on both old and new science-fiction movies that show rocket ships with fins zipping through space. Such movies may be exciting with regard to their story, but they are definitely lacking in rocket science.

Thrust

If the rocket is to successfully leave the launch pad, its engine (or engines) has to generate enough initial thrust to overcome the weight of the fully loaded vehicle, including payload and crew (if any), propellant, structure, and supporting flight equipment. Weight is the downward force Earth's gravity exerts on an object. Physicists define this *weight force* as mass (m) times the acceleration of gravity ($a_{gravity}$). Often, engineers and physicists use the symbol g to represent the value of the acceleration of gravity on Earth's surface. Numerically, one g is about 9.8 meters per second squared (m/s^2) at sea level on Earth. If a small rocket has a fully loaded, total mass of 10,000 kilograms, its engine would have to generate at least 98,000 newtons of thrust just to barely lift it off the launch pad.

How did we get that value? For any motion upward to occur, the thrust force (F_{thrust}) of the engine must be equal and opposite to the force of gravity on the rocket's total initial mass. We can calculate the initial *weight* of this rocket from Newton's second law (see equation 4-1), namely,

$$\text{weight}_{rocket} = m_{rocket} \times \text{acceleration}_{gravity} = m_{rocket} \times g,$$

$$\text{weight}_{rocket} = (10{,}000 \text{ kg})\ (9.8 \text{ meters/second}^2),$$

$$\text{weight}_{rocket} = 98{,}000 \text{ newtons.}$$

(The unit of force in the international system is the *newton*, which corresponds to one kilogram-meter per second squared.) Of course, if our rocket vehicle is to reach orbital velocity, its engine must provide a lot more thrust for a sustained period of time.

How much thrust would this same 10,000-kilogram-mass rocket need to barely lift itself off the surface of Mars? Because the acceleration of gravity on Mars is much less than on the surface of Earth, only 3.7 m/s^2, the answer is at least 37,000 newtons. Consequently, although the rocket's mass is the same on both planets, its weight actually changes in value. Aerospace engineers have to keep this in mind when they design a rocket system to lift astronauts from the surface of the Moon or Mars and send them safely on their way back home to Earth.

The *specific impulse* is a very important parameter that describes rocket-engine performance as a function of propellant combination. Engineers define the specific impulse (I_{sp}) as the thrust force of an engine divided by the mass flow rate of propellant. We write the specific-impulse equation as specific impulse (I_{sp}) = (thrust force)/(mass flow rate of propellant). (4-6)

In the international (SI) system, the specific impulse has the unit newtons/(kilogram/second), which simplifies to meters/second. The very best (theoretical) specific impulse value we can obtain with chemical propellants is about 4,300 meters per second (m/s). This value represents a chemical rocket engine that perfectly burns liquid hydrogen (as fuel) and liquid oxygen (as oxidizer). Later in this chapter, we will see that a nuclear-thermal rocket, using just liquid hydrogen as the propellant, has a theoretical specific-impulse value between 9,600 and 66,000 meters per second (m/s), depending upon the design of the nuclear-reactor rocket.

Controlling the thrust of a rocket engine is very important when we attempt to launch payloads into orbit or place spacecraft on interplanetary trajectories. If the rocket engine thrusts too long, the space vehicle and/or its payload will end in the wrong orbit or depart from Earth on the wrong escape trajectory. If the rocket engine shuts off too soon or operates below the needed level of thrust, the space vehicle might not get into orbit and would simply fall back to Earth. Thrusting in the wrong direction or at the wrong time will also result in a similar undesirable situation.

Nozzles

As with other rockets, the nozzle for a solid-propellant rocket engine is the specially designed passageway (opening) at the back of the vehicle. Its function is to expand and accelerate the hot combustion gases and allow them to escape to the environment. One very efficient nozzle is the converging-diverging (C-D) design (see Figure 4.4). During operation, the exhaust-gas velocity in the converging portion of the nozzle remains subsonic. The gas velocity increases to sonic speed at the throat and then expands to supersonic speeds as it flows through and exits the diverging portion of the nozzle. The throat is the narrowest part of the nozzle. Escaping gases flow through this constricted region with sonic velocity. Excessive heat transfer to the nozzle wall at the throat is often a problem. Engineers provide thermal protection for the throat by using a liner that either can withstand high temperatures (for a brief period of operation) or else ablates (intentionally erodes away). Long-burning, large solid-rocket motors (like the space shuttle's solid-rocket booster) generally rely on ablative materials to protect the nozzle's throat. Smaller solid-rocket motors that burn only briefly frequently use high-temperature materials for the nozzle.

The purpose of the nozzle is to accelerate the combustion gases to very high velocity as they escape. This maximizes the thrust produced by the

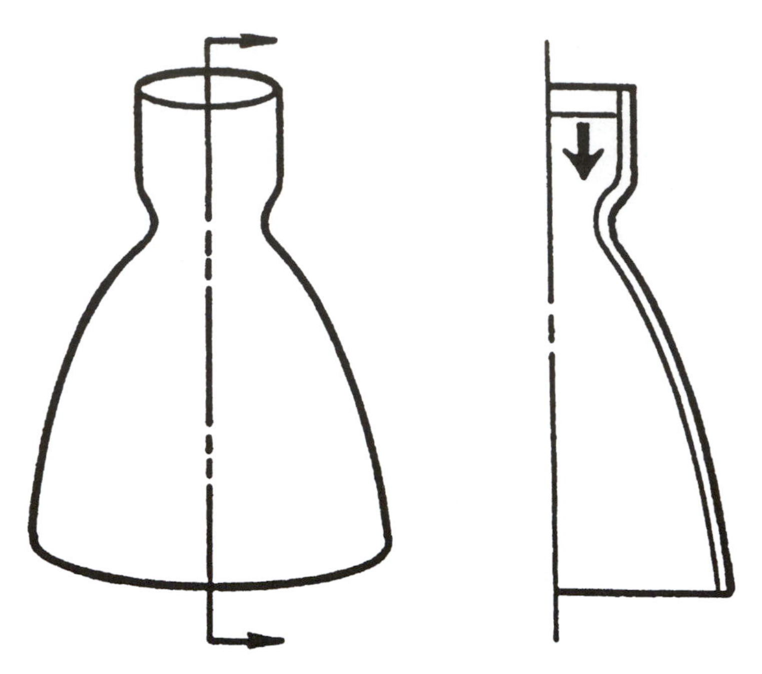

Figure 4.4 A full and cutaway view of the bell-shaped, converging-diverging nozzle. Note that the expanding (exit) portion of the nozzle, which lies downstream of the narrow throat region, is very large to accommodate the expansion of the exiting gases at very high velocity. NASA image modified by author.

rocket. From classical thermodynamics, we treat the rocket nozzle as a simple flow device that extracts no work and takes little (if any) thermal energy from the flowing fluid. The hot combustion gas enters the nozzle at high pressure and low velocity. As the gas escapes from the rocket, it flows through the converging (decreasing-area) section, the throat (smallest area), and then the diverging (increasing-area) section of the nozzle. As it flows, the pressure of the gas decreases, while its velocity greatly increases.

If you look back at the thrust equation (equation 4-4), you will see why rocket engineers want the combustion gases to leave at as high a velocity as possible. The higher the exit velocity of the gas, the greater the thrust. If the nozzle completely expands the exiting gas to ambient pressure conditions (vacuum conditions in space), then it produces even more thrust. Large rockets (both solid and liquid fueled) that start at Earth's surface and rise to space operate in an environment of continuously decreasing ambient pressure. Engineers must design the converging-diverging nozzle of these rocket vehicles with some optimized exit-area value. They accept the fact that they cannot achieve complete exhaust-gas expansion throughout powered ascent up through the lower atmosphere. (Variable-exit-area C-D nozzles are too heavy and cumbersome.)

ROCKET TYPES

Aerospace engineers often classify rockets according to the energy source, or propellant, used by the reaction engine to accelerate the ejected matter that creates the vehicle's thrust. For example, there are chemical rockets, nuclear rockets, and electric rockets. Chemical rockets, in turn, come in two general subclasses: solid-propellant rockets and liquid-propellant rockets. Most modern rockets operate with either solid or liquid chemical propellants.

Contrary to what one might think, the term *propellant* does not simply mean fuel; it refers to both *fuel* and *oxidizer*. The fuel is the chemical propellant the rocket engine burns, but an oxidizer is also needed to supply the oxygen necessary for combustion.

Chemical Rockets

Solid-Propellant Rockets

Solid-rocket propellants are dry to the touch and contain the chemical fuel and oxidizer blended together in some appropriate mixture, which aerospace engineers commonly call the *grain*. Usually the solid fuel is a mixture of hydrogen compounds and carbon, and the oxidizer consists of oxygen compounds.

Aerospace engineers often divide solid propellants into three basic types: monopropellants, double-base propellants, and composites. Monopropellants are energetic compounds, such as nitroglycerin or nitrocellulose, that contain both fuel (carbon and hydrogen) and oxidizer (oxygen). Because of potential storage and safety problems, engineers seldom use monopropellants in modern rockets. Instead, they create double-base propellants that are actually special mixtures of monopropellants. Engineers usually combine double-base propellants with additives that improve the handling and burning characteristics of the grain. The mixture often resembles a puttylike material that can easily be loaded into the solid rocket's case and allowed to cure (harden).

Engineers also form composite solid propellants from mixtures of two or more unlike chemical compounds that by themselves do not make good

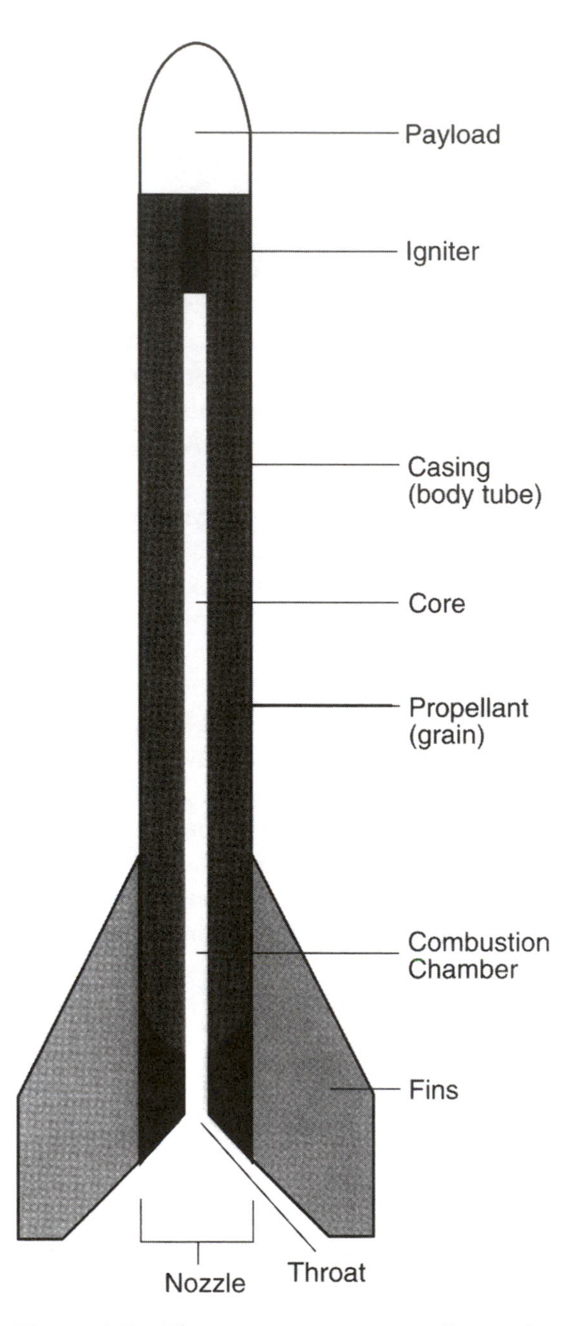

Figure 4.5 The major components of a simplified solid-propellant rocket. Image courtesy of NASA.

solid-rocket propellants. Usually one compound serves as the fuel and the other as the oxidizer. The propellants used in the solid-rocket boosters (SRBs) of NASA's space shuttle fall into this category. Rocket technicians call the space shuttle booster composite PBAN, aerospace-engineering shorthand for polybutadiene acrylic acid acrylonitrile terpolymer. In addition to the PBAN (which serves as a binder), this composite propellant consists of approximately 70 percent ammonium perchlorate (oxidizer), 16 percent powdered aluminum (fuel), and a trace of iron oxide (to control the burning rate). The cured propellant looks and feels like a hard rubber eraser.

A *solid-propellant rocket* is the simplest type of rocket. It consists of a nozzle, a case, insulation, propellant, a payload compartment, and an igniter (see Figure 4.5). The rocket engine's case is usually a relatively thin-walled, hollow metal cylinder that has its interior lined with thermal insulation to keep the reacting propellant from burning through and destroying the vehicle. The propellant itself is packed inside the insulation layer. The case is an inert part of the rocket, and its mass is an important factor in determining how much payload the rocket can carry and how far the vehicle can travel. To produce efficient, high-performance solid rockets, aerospace engineers make the casing out of the lightest materials possible. Alloys of steel and titanium are often used for solid-rocket casings.

In a solid-propellant rocket, the energetic nature of the propellant chemicals and the shape of their exposed burning surfaces determine the production of thrust as a function of time. A modern solid propellant will burn at any point exposed to the right amount of intense heat or hot gases at the proper (high) temperature. This "low-temperature flame resistance" is a major aerospace safety feature. Engineers intentionally created such relatively insensitive solid-propellant compounds so that a stray spark or inadvertent exposure to a low-temperature flame would not prematurely ignite a large solid-rocket booster.

Aerospace engineers design end-burning and internal-burning solid-propellant grains. Launch personnel ignite an end-burning solid-propellant rocket near the nozzle and allow combustion to go gradually up the length of the propellant load. Since only the surface of the propellant burns, the area of the burning surface is always at a minimum in an end-burning design. As a result, this is the slowest burning of any grain design. For the same amount of propellant mass, the thrust produced by an end-burning design is lower than for other grain designs, but thrust continues over longer periods.

To get higher thrust, rocket engineers use internal-burning grain, a solid-propellant load with a hollow core. This hollow core is often cylindrical in shape and runs the total length of the solid-propellant rocket. The major advantage of this internal-burning grain is that the exposed surface of the hollow core is much larger than the smaller propellant surface exposed in an end-burning grain. At launch, a special pyrotechnic device simultaneously ignites the entire internal-burning core, and combustion proceeds rapidly from the inside out over the total exposed surface area. To further increase the propellant surface available for burning at any given moment, rocket engineers sometimes use a cruciform or star design instead of a cylindrical hollow core.

By varying the geometry of the core design, rocket engineers customize the thrust produced by a large internal-burning grain as a function of time to accommodate specific mission needs. For example, the massive solid-rocket boosters (SRBs) used by NASA's space shuttle feature a core that has an 11-pointed-star design in the forward segment. At approximately 65 seconds into the launch, the star points burn away, and the thrust temporarily diminishes. This reduction in solid-booster thrust coincides with the passage of the space shuttle vehicle through the sound barrier. Buffeting occurs during this passage, and the reduced SRB thrust helps alleviate strain on the vehicle.

To ignite the propellants of a solid rocket, rocketeers use many kinds of devices, called igniters. In the early days of gunpowder rocketry, a "volunteer" would use a small torch to light a crude fuse that extended from the nozzle. Sometimes the rocket launched correctly after the individual retreated to a safe distance, but other times the unpredictable device ignited too quickly and burned the startled rocketeer as it flew away or exploded. Today, aerospace launch personnel use a much safer and more reliable approach to solid-rocket ignition—a technique that employs electricity. They remain a safe distance away and send an electric current through a special wire inside the solid-propellant rocket. As the current flows, this special wire heats up and raises the temperature of the solid-propellant surface it is in contact with. When the propellant surface gets hot enough, combustion starts.

For launch safety and reliability reasons, today's rocket ranges actually use more advanced versions of this "hot-wire" igniter technique. To achieve successful ignition of a modern solid-propellant rocket, engineers want to quickly saturate its grain surface with hot gases. One way of achieving this is to encase the electric (hot-wire) igniter in a clump of special, easily ignited chemicals. Upon receiving the proper electric signal, these chemicals quickly ignite and shower the surface of the solid-propellant grain with hot gases. Sometimes, for very large solid-rocket motors, launch personnel use a small rocket motor as the igniter. They firmly place this small rocket inside the upper end of the hollow core, at the lower end of the core, or even completely outside the big solid motor at the exit of its nozzle. As before, an electric circuit with a special hot wire starts the small rocket motor. Burning in place, it sends a stream of flames and hot gas down the hollow core, igniting the entire exposed solid-propellant surface in a fraction of a second.

Liquid-Propellant Rockets

The other major type of chemical rocket is one that uses liquid propellants. Compared to the solid-propellant rocket, the liquid-propellant rocket is a very complex device. Liquid-propellant rockets have three principal components in their propulsion system: propellant tanks, the rocket engine's combustion chamber and nozzle assembly, and turbopumps (see Figure 4.6). The propellant tanks are the load-bearing structures that contain the liquid propellants. There are separate tanks for the fuel and for the oxidizer. The combustion chamber is the region into which engineers pump or pressure-feed the liquid propellants. Once inside the combustion chamber, the liquid propellants vaporize and react (combust), creating the hot, high-pressure exhaust gases that then expand through the nozzle, generating thrust. The liquid-propellant rocket's turbopumps are specially designed pieces of fluid-flow machinery that deliver the propellants from the tanks to the combustion chamber at high pressure and sufficient flow rate. Modern liquid rockets generally use powerful, lightweight turbopumps to take care of this task. However, engineers sometimes choose to eliminate the turbopumps in a particular liquid-rocket design by using a gaseous "overpressure" in the propellant tanks to pressure-feed the propellants into the combustion chamber.

The function of the propellant tanks is simply the storage of one or two propellants until they are needed in the combustion chamber. Depending on the type of liquid propellant the rocket uses, a tank may be nothing more than a low-pressure envelope, or it may be a well-engineered containment vessel capable of containing propellants under high pressure. For cryogenic (very low-temperature) propellants, engineers must design the tanks as extremely well-insulated structures to prevent the very cold liquids from boiling away (evaporating). The most important cryogenic propellants are liquid oxygen (LO_2) (an oxidizer) and liquid hydrogen (LH_2) (a fuel). At normal (sea-level) pressure, oxygen remains in a liquid state when its temperature is kept at −183° Celsius (C) or below, while hydrogen remains liquid at −253° C or below. As a point of reference, absolute zero temperature (0 Kelvin [K]) corresponds to −273.16° C. If the propellant temperature rises beyond these critical values, the cryogenic liquid will boil off as a vapor. Rocket engineers do not want vapor pressures building up in the cryogenic-liquid tanks, nor do they want to store the pro-

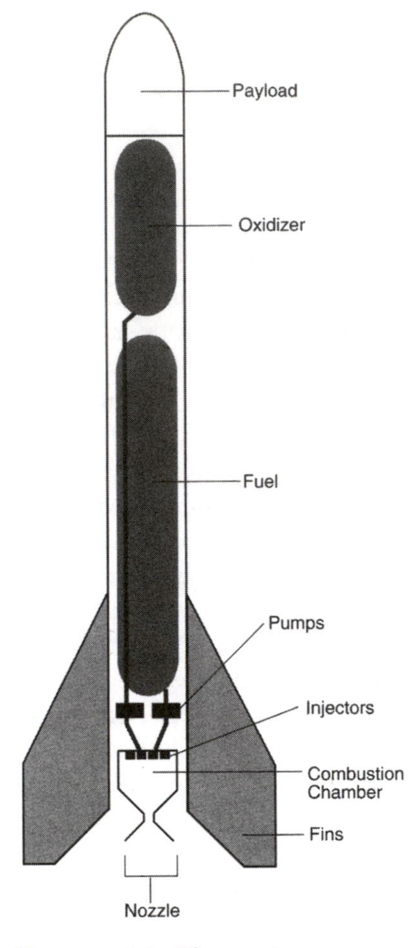

Figure 4.6 The major components of a simplified liquid-propellant rocket. Image courtesy of NASA.

pellants as high-pressure gases. In addition, venting gaseous hydrogen from a propellant tank at the launch site represents a major safety hazard. (The ill-fated German zeppelin *Hindenburg*, which exploded at its mooring on May 6, 1937, in Lakehurst, New Jersey, used gaseous hydrogen for buoyancy.) The rocket engineer's solution: keep it cool, very cool.

Minimizing the mass of the propellant tanks presents a major design challenge. Aerospace engineers fully recognize that the lighter they can make the propellant tanks, the more payload the rocket can carry or the farther it can travel. Often, engineers construct liquid-propellant tanks out of very thin sheets of metal or possibly thin metal layers wrapped with high-strength fibers and cements. The internal pressure of the propellants helps stabilize the loaded tanks in much the same way that the wall of an inflated balloon gains strength from the gas inside (up to a certain level of internal pressure). However, very large propellant tanks and tanks that contain cryogenic propellants require additional strengthening. Engineers use structural rings and ribs to strengthen tank walls, giving the tanks the appearance of an aircraft frame. Rocket designers also employ large quantities of special, low-mass insulating materials to keep cryogenic propellants in their liquefied form. Unfortunately, even with the best available insulation, cryogenic propellants are difficult to store for a long period of time and eventually boil off. The difficulties encountered in using cryogenic propellants is often balanced in the better performance of space launch vehicles that enjoy an increase in reaction thrust for the equivalent mass of chemical propellants. This is why military rocket designers do not select cryogenic propellants for long-range missiles that must stay launch ready for months at a time.

Turbopumps provide the required flow of propellants from the low-pressure propellant tanks to the high-pressure combustion chamber. Engineers often generate the power needed to operate the turbopumps by combusting a small fraction of the propellants in a preburner. Expanding gases from these burning propellants drive one or more turbines that, in turn, drive the main turbopumps.

The combustion chamber of a liquid-propellant rocket is a bottle-shaped container with openings at opposite ends. The openings at the top inject propellants into the chamber. Each opening consists of a small nozzle that squirts in either fuel or oxidizer. The main purpose of the injectors is to vaporize and mix the liquid propellants, thereby ensuring smooth and complete combustion and avoiding detonations. Combustion-chamber injectors come in many designs, and a particular liquid-propellant engine may require hundreds of properly engineered injectors. To get a general idea of how an injector looks and works, examine the showerhead fixture in a modern bathroom. Most designs have many tiny holes through which a mixture of hot and cold water streams out under pressure. The injected stream of water usually provides a pleasant bathing experience (unless the hot-water supply depletes too quickly).

There are other conditions that must be satisfied to achieve the efficient operation of a liquid-propellant rocket engine. Of primary importance is the requirement that the vaporized propellants ignite when they enter the combustion chamber. Hypergolic propellant combinations ignite spontaneously on contact, but other types of liquid propellants need a starter device, like a spark plug. Once combustion starts with such nonhypergolic propellants, the process remains self-sustaining.

The opening at the opposite (lower) end of the combustion chamber is the throat. It is the narrowest part of the rocket engine's nozzle. Combustion of the propellants raises gas pressure inside the chamber, and the pressurized gas exhausts through this nozzle. By the time the gas leaves the exit cone (the widest part of the nozzle),it travels at supersonic velocity and imparts forward thrust to the rocket vehicle.

Because of the high-temperature environment encountered during sustained propellant combustion, engineers must cool the chamber and nozzle of large liquid-propellant rockets. For example, the combustion chambers of the three liquid-propellant main engines on NASA's space shuttle reach 3,590 K (3,317° C) during firing. Since these engines operate together continuously for about eight minutes, engineers must protect all their exposed surfaces from the eroding effects of the high-temperature, high-pressure combustion gases.

Aerospace engineers choose either of two general approaches to cool the combustion chamber and nozzle of a long-burning liquid-propellant rocket. The first approach is identical to the cooling technique they use with many solid-propellant rocket nozzles. Rocket engineers simply cover the surface of the nozzle with an ablative material that intentionally erodes away when it experiences a stream of high-temperature gas. Since the ablated material carries away a large amount of thermal energy (heat) per unit mass, its intentional sacrifice protects the surface underneath and keeps the nozzle walls cool. However, the ablative-surface cooling approach adds extra mass to a liquid-propellant engine and so reduces the payload and/or range capability of the rocket vehicle. Engineers prefer to use the ablative-surface cooling technique only when the liquid-propellant engine is small or when a simplified engine design is more important than high-efficiency performance.

Rocketeers call the second general method of liquid-propellant engine cooling *regenerative cooling*. They construct combustion-chamber and nozzle walls with a complex plumbing arrangement of small tubes inside the walls. This design arrangement lets cold fuel circulate within the walls before the fuel goes through the preburner and into the combustion chamber. By absorbing heat (thermal energy) from the walls of the combustion chamber and the nozzle, the circulating fuel provides an important level of cooling and prevents the heated walls from melting. Engineers use the term *regenerative cooling* because the flowing fuel actually becomes a bit more energetic on its way to the combustion chamber as it absorbs heat from the chamber and nozzle walls. Although regenerative cooling is more complicated than ablative cooling, it significantly reduces the overall mass of large rocket engines and improves flight performance.

Propellants for liquid rockets generally fall into two categories: monopropellants and bipropellants. Monopropellants consist of a fuel and an oxidizing agent stored together in one container. The monopropellants can be two premixed chemicals, such as alcohol and hydrogen peroxide, or a homogeneous chemical such as nitromethane. Another chemical, hydrazine, becomes a monopropellant after it is brought into contact with a catalyst. The catalyst initiates a reaction that generates heat and gases from the chemical decomposition of the hydrazine. Because of their thrust-producing limitations, engineers usually restrict the use of most monopropellants to the small attitude-control or steering rocket engines on spacecraft and aerospace vehicles. With its need for only a single propellant-storage tank, the monopropellant greatly simplifies the design and operation of a low-thrust liquid-rocket engine that can reliably fire often.

By far the most common liquid-propellant rocket is the bipropellant system. Engineers store the bipropellants (a fuel and an oxidizer) separate from each other until they combine them in the combustion chamber. Commonly used bipropellant combinations include liquid oxygen (LO_2) and kerosene, liquid oxygen (LO_2) and liquid hydrogen (LH_2), and monomethylhydrazine (MMH) and nitrogen tetroxide (N_2O_4). The last bipropellant combination (MMH and N_2O_4) is hypergolic, meaning that these two propellants ignite spontaneously when brought into contact with each other. Hypergolic propellants are especially useful for attitude-control rockets that must fire frequently with a high degree of reliability.

Rocket engineers must consider many factors in selecting the appropriate bipropellant combination for a particular rocket system. For example, liquid hydrogen (LH_2) and nitrogen tetroxide (N_2O_4) would make a good combination based on propellant performance, but their widely divergent storage temperatures (cryogenic and room temperature, respectively) would require the use of large quantities of thermal insulation between the two tanks, adding considerable mass to the rocket vehicle. Another important factor is the toxicity of the chemicals used. Monomethylhydrazine (MMH) and nitrogen tetroxide (N_2O_4) are both highly toxic. Rocket vehicles that use this propellant combination require special propellant handling and prelaunch preparation.

Modern launch vehicles use three types of liquid propellants: petroleum-based (hydrocarbons), cryogenic, and hypergolic. One commonly used hydrocarbon rocket fuel is a highly refined form of kerosene, called RP-1 (Refined Petroleum). Because the combination is relatively inexpensive and readily available, rocket engineers like to burn RP-1 with liquid oxygen, although this combination delivers considerably less specific impulse (I_{sp}) than cryogenic bipropellant combinations.

While solid-propellant rockets do not stop thrusting until they exhaust their propellant supply, aerospace engineers (or more correctly an onboard computer) can stop the liquid-propellant engine any time by simply cutting off the flow of propellants into the combustion chamber. A microprocessor in the rocket's guidance system decides when the vehicle needs thrust and communicates with the computer that turns the rocket engine on or off accordingly. On more complicated flights, such as an interplanetary trajectory to the Moon or Mars, human mission controllers or the rocket's onboard computer will start and stop a spacecraft's liquid-propellant engine(s) several times. Engineers call this type of engine a *restartable engine*.

We can control the amount of thrust produced by a liquid-propellant engine during powered flight by varying the amount of propellant that enters the combustion chamber. Engineers refer to this rocket-engine process as *throttling*. Typically, they will vary engine thrust during ascent to help control the level of acceleration experienced by astronauts or to keep the aerodynamic forces acting on the launch vehicle within acceptable limits.

Comparison of Modern Solid- and Liquid-Propellant Rockets

Compared to liquid-propellant rocket systems, solid-propellant rockets offer the advantages of simplicity and reliability. Except for stability controls, solid-propellant rockets have no moving parts. When they are loaded with propellant and erected on a launch pad, solid rockets stand ready for firing at a moment's notice. As part of its mutual assured destruction (MAD) strategy, the U.S. military deploys solid-propellant strategic missiles in underground silos and in launch tubes on board ballistic missile submarines. These weapons can be launched on short notice and deliver a totally destructive retaliatory nuclear strike. In contrast, liquid-propellant rocket systems require extensive prelaunch preparations and are much more suitable for space launch operations.

Solid rockets generally have the advantage of a higher (propellant) mass fraction. Liquid-propellant rocket systems require fluid feed lines, turbopumps, and tanks, all of which add additional (inert) mass to the vehicle. A well-designed, "ideal" solid-propellant rocket vehicle will have a mass fraction of between 91 and 93 percent (i.e., 0.91 to 0.93). This means that 91 to 93 percent of the rocket's total mass is solid propellant. Large liquid-propellant rocket vehicles generally achieve a mass fraction about 90 percent or less.

The main disadvantage of solid-propellant rockets involves the burning characteristics of the propellants themselves. Solid propellants are generally less energetic than the best liquid propellants. With lower values of specific impulse than liquid propellants, solid propellants deliver less thrust per unit mass of propellant burned. Once ignited, solid-propellant motors burn rapidly. This is another significant disadvantage, because flight-control and range-safety personnel cannot throttle or extinguish an improperly burning solid rocket. In contrast, they can start or stop liquid-propellant engines at will.

Today, military rocketeers use solid propellants for strategic nuclear missiles (e.g., the U.S. Air Force [USAF] Minuteman), for tactical military missiles (e.g., the USAF Sidewinder), for small expendable launch vehicles (e.g., NASA's Scout), and as strap-on solid boosters for a variety of liquid-propellant launch vehicles, including NASA's mostly reusable space shuttle and the expendable USAF Titan IV. Aerospace engineers also use solid-rocket motors in small sounding rockets and in many types of upper-stage vehicles, such as the USAF-sponsored Inertial Upper Stage (IUS) vehicle.

Other Types of Rockets

Besides chemical rockets, we will examine two other important types of rockets: the nuclear-thermal rocket and the electric rocket. For different technical reasons, aerospace engineers restrict the use of these rockets to outer space. Nuclear-radiation safety issues limit the operation of nuclear-thermal rockets to locations beyond Earth's biosphere. Nuclear-thermal rockets provide much higher values of specific impulse than chemical rockets. As a result, they are excellent candidates for priority (rapid) interplanetary missions in which shorter flight time and larger payload capacity prove essential for success. The inherently low thrust output of electric rockets confines them exclusively to outer-space propulsion roles. Electric rockets are especially well suited for nonpriority (leisurely) payload-delivery applications where constant, gentle thrusting over long periods is acceptable.

Nuclear-Thermal Rocket

The nuclear-thermal rocket employs a nuclear-fission reactor to heat propellant. A nuclear reactor is a device that can start, maintain, and control a nuclear-fission chain reaction. There is no chemical combustion in a nuclear rocket. Instead, the energy needed to heat the propellant comes from the controlled fission (splitting) of nuclei of uranium-235 atoms. Just like chemical rockets, the nuclear rocket operates according to Newton's third law of motion. Very hot hydrogen gas, accelerating out a nozzle, creates a reactionary thrust that drives the vehicle forward.

To help us understand why a nuclear rocket works better than a chemical rocket, we must write the specific impulse equation (equation 4-6) in a slightly different form. Aerospace engineers also use the following equation to describe the performance of a rocket engine:

$$I_{sp} = AC_f \sqrt{(T_c/M_w)} \qquad (4\text{-}7)$$

The specific impulse (I_{sp}) still has the same meaning as before. The symbol A is a performance-factor constant, related to the thermophysical properties of the propellant. The symbol C_f is another constant, called the thrust coefficient. Engineers obtain a numerical value for C_f from the design parameters of the rocket's nozzle. We do not have to consider these two terms any further. Our interest concerns the two terms under the square-root symbol ($\sqrt{}$), namely, T_c (the chamber temperature in degrees Kelvin) and M_w (the molecular weight of the exhaust gases).

If we look at equation 4-7 again, we can relate the specific impulse (and therefore the performance) of a rocket engine to the combustion-chamber temperature and the molecular weight of the expelled gases. Again, the nuclear-thermal rocket simply heats a single propellant (usually cryogenically stored hydrogen) to a very high (chamber) temperature. This process occurs without chemical combustion as the propellant flows through heat-transfer channels in the nuclear reactor's core. The high-temperature gas then exits to space through the rocket's nozzle. The higher we can make the chamber temperature and the lower the molecular weight of the expelled gases, the higher the specific impulse of our rocket engine. Very high-temperature hydrogen, with its low molecular weight of two, is ideal for propulsion purposes.

A solid-core-reactor nuclear rocket can heat hydrogen up to about 2,700 K. The term *solid-core reactor* means that the reactor core (nuclear fuel and structure) gets very hot as the rocket operates, but remains in solid form due to the flow of propellant. Since hydrogen has a molecular weight of two, the theoretical specific-impulse value for this nuclear rocket is about 9,600 meters per second. This is more than twice the specific-impulse value for the best chemical rocket (namely, about 4,300 m/s for LH_2 and LO_2 combustion). Aerospace engineers recognize that for the same amount of total propellant mass expelled from the nozzle, the nuclear rocket gives more than twice the total thrust.

Between 1956 and 1973, American nuclear and aerospace engineers teamed up in the ROVER (ROcket VEhicle Reactor) program to develop a variety of successful nuclear-thermal rocket designs. Using captive-firing equipment, a test stand that prevents the rocket from moving, they tested complete nuclear rocket-engine assemblies at the Nuclear Rocket Development Station (NRDS) on the Nevada Test Site (NTS) in southern Nevada. During a typical downward-firing static test, the surrounding desert basin became an inferno as extremely hot hydrogen gas, after exiting the inverted rocket engine, spontaneously ignited upon contact with the air and burned (with atmospheric oxygen) to form water.

In the 1960s, aerospace engineers wanted to use nuclear rockets to send human explorers on faster missions to Mars. They reasoned that the quicker (for example, 36 versus 54 months) round-trip mission to Mars made possible by more powerful nuclear rockets would relieve stress on the life-support system and provide a more favorable psychological environment for the isolated crew. These NASA planners considered Mars as the next logical human destination after the Apollo lunar landings. In their general mission strategy, Project Apollo's giant Saturn V chemical rocket would lift the nuclear rocket into orbit. The astronauts would then board the nuclear rocket in low Earth orbit and depart for Mars by firing its powerful engine. However, changing American space priorities led to the cancellation of the nuclear-rocket program in 1973, just before ROVER program engineers could flight-test a nuclear rocket in space.

Projecting nuclear-reactor technologies to reasonable twenty-first-century technology limits, aerospace and nuclear engineers arrive at molten-core and gaseous-core reactor designs. The term *molten-core reactor* means that the nuclear-reactor fuel is in molten (liquid) form. Similarly, the term *gaseous-core reactor* means that the nuclear-reactor fuel is in gaseous form. Studies suggest that a molten-core-reactor design would make the hydrogen propellant reach a chamber temperature of 5,000 K, and a gaseous-core-reactor design would heat hydrogen to a chamber temperature of 20,000 K. Such liquid and gaseous nuclear-fuel forms overcome the temperature limits encountered with solid-core-reactor designs. Unlike chemical-rocket combustion chambers, the core of the nuclear rocket's reactor is not energy output limited, but heat transfer limited. (Energy output limited means that chemical reactions have only a very small amount of energy release per atomic or molecular reaction, on the order of a few electron volts of energy per chemical reaction. In nuclear-fission reactions, about 200 million electron volts are released per nuclear reaction, so there is no real limit to the amount of thermal energy that can be released in a confined volume in a nuclear rocket. The problem is getting all this energy out before everything melts or vaporizes, that is, the heat-transfer limit.) By going to more exotic liquid and gaseous reactor-core designs, engineers believe that they can overcome such heat-transfer limitations. They know that the hotter the hydrogen propellant gets, the better the performance of the nuclear-thermal rocket. Engineers estimate the theoretical specific impulse for such advanced-design nuclear-thermal rockets as 25,500 meters per second (molten-core reactor) and 66,000 meters per second (gaseous-core reactor), respectively. By the close of the twenty-first century, these advanced nuclear-rocket engines could open the outer solar system to extensive scientific investigation and human exploration.

Electric Rockets

The electric rocket generates thrust by using electric power to accelerate an ionized propellant to a very high exhaust velocity. The basic electric rocket consists of three major parts: some type of electric thruster to accelerate the ionized propellant, a source of electric power, and an appropriate propellant (like mercury, cesium, xenon, or argon) that ionizes easily. The acceleration of charged atomic particles requires a great amount of electricity. Aerospace engineers can use either a compact, space-qualified nuclear reactor or a large array of solar cells to provide the constant supply of electric power needed to operate this type of low-thrust rocket. Nuclear-electric propulsion (NEP) systems can operate anywhere in the solar system, while solar-electric propulsion (SEP) systems are most efficient on missions within the orbit of Mars. Beyond Mars, the inverse square law significantly limits the amount of solar energy available for collection and conversion into electric power.

There are three general classes of electric rocket engine: electrothermal, electromagnetic, and electrostatic. The basic *electrothermal rocket* uses electric power to raise a propellant, like ammonia, to a high temperature. The rocket then expands the gaseous, high-temperature propellant through a nozzle to generate thrust. Propellant heating occurs when it flows through an electric arc. While the "arc-jet" electric engine produces exhaust velocities higher than those achieved by chemical-combustion rockets, the dissociation (breaking apart) of the propellant molecules places an upper limit on just how much energy we can add to it with this technique. Because of this physical limit and electric-arc-induced material erosion, aerospace engineers usually restrict the use of the arc-jet engine to large-spacecraft stationkeeping and orbital-transfer-vehicle propulsion.

The *electromagnetic engine* heats a propellant (such as argon) to create a plasma. This engine then uses intense electromagnetic fields to generate forward thrust by accelerating the plasma rearward. Sometimes called a magnetoplasmadynamic (MPD) engine, the device can operate in either a steady state or a pulsed mode. Aerospace engineers consider the one-megawatt electric-class, steady-state MPD engine an attractive propulsion option for a reusable orbital transfer vehicle that operates within cislunar space.

The *electrostatic rocket engine* (or ion engine) uses its ionizer to pull electrons off atoms of a propellant, like cesium or mercury. The engine then uses an electrostatic field to accelerate the newly created, positively charged propellant ions to very high exhaust velocity. By firing these high-speed "atomic bullets" rearward, the ion engine generates a forward thrust. A neutralizer injects many electrons into the departing ion beam. This beam of electrons allows the spacecraft to remain electrically neutral and is necessary for successful thruster operation. Aerospace engineers consider modern, high-exhaust-velocity (typically 100,000 meters per second) ion thrusters that use argon or xenon propellant quite appropriate for propulsion duties on interplanetary and deep-space missions.

Starting with Robert Goddard, space visionaries have recognized the important role electric propulsion plays in the exploration of space. Electric propulsion systems are best in high-performance missions that start in a low-gravity field, such as that found in Earth orbit or lunar orbit. In comparison to their chemical-rocket cousins (which are high-thrust, short-duration burn devices), electric rockets are inherently low-thrust, long-duration-operation devices. As a result, electric rocket engines are very high-specific-impulse devices with propellant efficiencies that range between 2 and 10 times those achieved by traditional chemical rockets. While traveling through interplanetary space, electric rockets work continuously for long periods, smoothly changing a spacecraft's trajectory. For deep-space missions to the outer planets, electric rockets provide a gentle, continuous acceleration that eventually yields a shorter trip time and more scientific payload capacity than can be achieved by an equivalent-mass chemical rocket fired from low Earth orbit.

CONTEMPORARY MISSILES AND LAUNCH VEHICLES

An intercontinental ballistic missile is a rocket vehicle designed to ascend into near-Earth space following a curved (ballistic) trajectory that eventually takes it down to a target thousands of kilometers away. A launch vehicle, on the other hand, is an expendable or reusable rocket vehicle that lifts a payload into an orbit around Earth or places it on an Earth-escape trajectory into interplanetary space. Aerospace engineers often refer

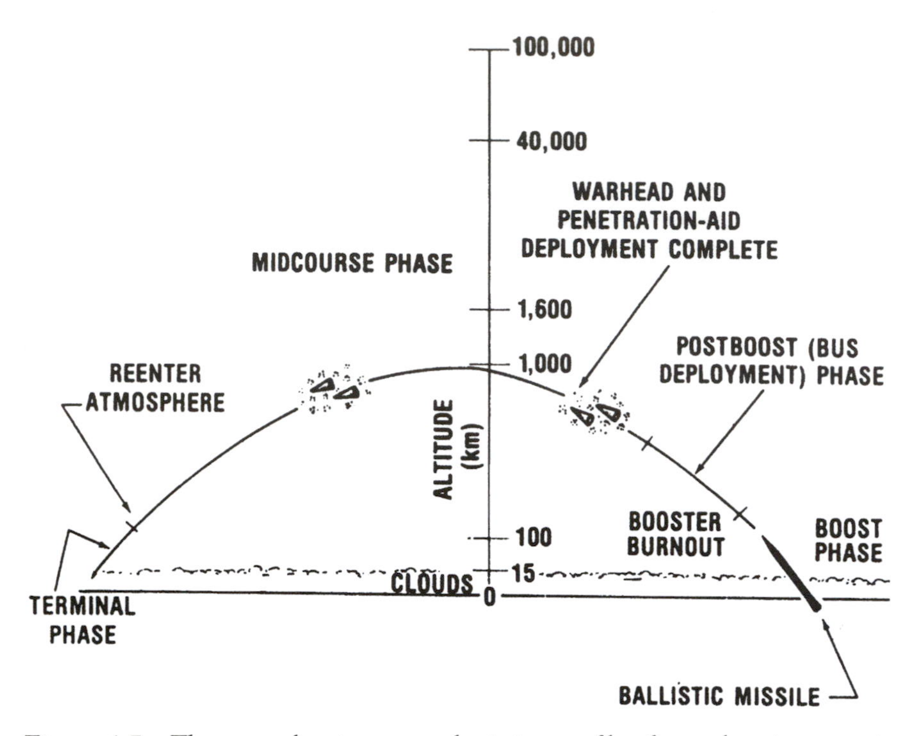

Figure 4.7 The general trajectory and mission profile of a modern intercontinental ballistic missile (ICBM). Image courtesy of U.S. Department of Defense.

to these rockets as *boosters* or *space lift vehicles*. Many expendable launch vehicles (ELVs) used in the twentieth century were modified versions of large chemical-propellant rockets created as part of the Cold War strategic missile race. Within the American space program, these "dual-purpose" rockets included the Redstone, Jupiter, Thor, Atlas, and Titan vehicles.

Intercontinental Ballistic Missiles (ICBMs)

Figure 4.7 shows a typical mission profile for a strategic, nuclear-warhead-carrying intercontinental ballistic missile (ICBM). On receipt of an authenticated launch order (the "go code") from the National Command Authority (NCA), the military crew fires the solid-propellant missile from either an underground silo or a ballistic missile submarine. As it travels toward the target, the powerful booster exhausts its propellant supply. An upper-stage ("bus") vehicle separates from the expended booster and carries the payload of nuclear-armed reentry vehicles (RVs) on a ballistic (unpowered) trajectory. A strategic nuclear missile can carry several reentry vehicles. At just the right time in this ballistic trajectory, the postboost bus vehicle deploys each reentry vehicle. To confuse any enemy missile defense system, the postboost bus vehicle could also dispense various types of penetration-aid devices that confuse or defeat an enemy's missile defense system. Each warhead-carrying reentry vehicle then follows a separate ballistic trajectory to its preassigned target. Somewhere between 20 and 30 minutes after ICBM launch, each reentry vehicle, traveling at hypersonic velocity, descends on its target, and a powerful nuclear detonation occurs.

The U.S. Air Force Minuteman III is an example of contemporary American ICBM technology. It is a three-stage, solid-propellant intercontinental ballistic missile that guides itself to the target by an all-inertial guidance and control system. Aerospace engineers developed and deployed the first version of this ICBM in the 1960s. The Minuteman missile is an extraordinary technical achievement. The missile and its innovative underground-silo basing concept provided significant advances beyond the relatively slow-reacting, liquid-fueled ICBMs of the previous generation of strategic military missiles such as the Atlas and the Titan (in the late 1950s and early 1960s). From the very beginning, Minuteman became a quick-reacting, highly survivable component of America's nuclear deterrent force. At the start of the twenty-first century, U.S. Air Force personnel controlled more than 500 Minuteman III missiles, each deployed and dispersed in hardened underground silos throughout the western United States.

Expendable Launch Vehicles (ELVs)

Most rockets launched to put payloads into space are expendable launch vehicles (ELVs). Aerospace engineers design the ELV as a "throwaway" rocket. The ELV often includes several liquid-propellant rocket stages, assisted by a cluster of solid-propellant booster rockets that fire simultaneously with the first liquid-propellant stage. Figure 4.8 shows a powerful Titan IV ELV that has a liquid-propellant main (core) rocket engine, assisted by two giant, strap-on solid propellant rockets. Aerospace engineers do not design the disposable rocket stages, solid or liquid, for recovery and reuse. As the ELV ascends to space, it discards expended propulsion stages, and the jettisoned hardware falls back to Earth to harmlessly impact in a remote ocean or land area.

This "throwaway" rocket philosophy started when aerospace engineers first adapted early strategic military missiles for service as space launch vehicles. Military planners treated the rocket as a totally expendable piece of weaponry and designed the first long-range ballistic missile systems accordingly. When aerospace engineers converted any powerful new military

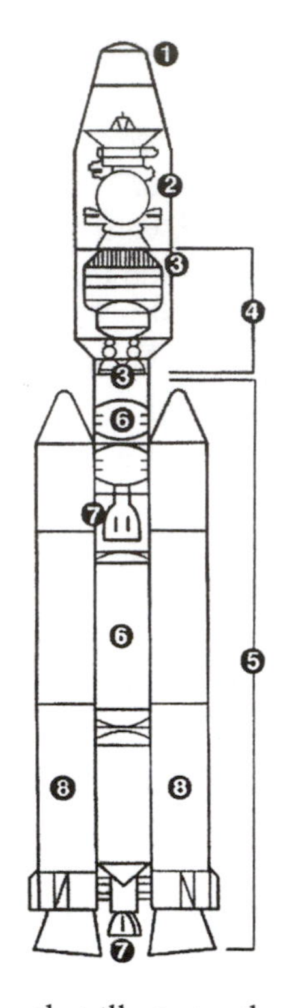

Figure 4.8 A cutaway drawing that illustrates the major components of the powerful Titan IVB/Centaur rocket system that launched the *Cassini* spacecraft on its seven-year journey to Saturn in October 1997. The components include (1) payload pairing; (2) *Cassini* spacecraft (payload); (3) guidance and navigation systems (including computers and gyroscopes); (4) Centaur rocket (upper stage); (5) Titan IVB rocket (lower stage); (6) liquid-propellant rockets (stage one and stage two of Titan IVB); (7) rocket engines for stage one and stage two; and (8) solid-propellant rocket motors. Image courtesy of NASA.

missile for space launch duty, its disposable design came along with the advanced rocket-technology package.

Similarly, the intense political pressures to beat the Soviet Union to the Moon encouraged Wernher von Braun and his team of rocketeers at NASA's Marshall Space Flight Center to develop their powerful new Saturn rockets as a family of throwaway rockets. During the Moon race of the 1960s, there simply was not enough time for NASA engineers to carefully research and demonstrate how to reduce launch costs by recovering and refurbishing rocket hardware. The throwaway-booster approach was again expedient, but this time it proved very expensive. Look at the photo on page 44. The giant Saturn V vehicle stood 111 meters tall, but only the very small top portion of this colossal vehicle, containing the Apollo Command and Service Module, went to the Moon. Everything else got tossed away. All three stages of the Saturn V vehicle (officially called the S-IC Stage, S-II Stage, and S-IV Stage) used liquid oxygen (LO_2) as the oxidizer. NASA engineers selected kerosene as the first-stage (S-IC Stage) fuel and liquid hydrogen (LH_2) as the fuel for the second and third stages. The ground quite literally shook as this mighty vehicle slowly rumbled up from Launch Complex 39 at the Kennedy Space Center. At liftoff, its first-stage engines produced a combined thrust of 34.5 million newtons. Unlike the space shuttle and many modern ELVs, there were no strap-on solid-rocket boosters to help the Saturn rocket get off the pad. Because of their timely and successful design, the Saturn family of rockets allowed American astronauts to successfully walk on the Moon—a truly great accomplishment. Unfortunately, much like their military rocket cousins, the Saturn rockets were very expensive and accomplished little in reducing the cost of getting into space.

This handed-down "throwaway" rocket philosophy is one of the major reasons why putting objects into space is still very expensive. At the beginning of the twenty-first century, it cost between $5,000 and $20,000 per kilogram (or more) to place mass, any mass, into low Earth orbit. Of course, the final cost of putting a payload into space depends on many factors, including the efficiency of the rocket vehicle, the location of the launch site, and the payload's final destination in space. Many times, for example, aerospace engineers must attach a special, upper propulsive stage to help the payload reach that destination. Imagine how much an airline ticket for a nonstop flight from New York to San Francisco would cost if the airline discarded the entire passenger aircraft after each flight. This is why aerospace engineers work so hard to reduce even a little of the inert (nonuseful) mass from an expendable rocket vehicle or its payload. One of the prime space-technology objectives for the United States (and other spacefaring nations) in the twenty-first century is to reduce the cost of accessing space. Much less expensive expendable launch vehicles and fully reusable launch vehicles are prime technical goals. If the cost of placing a kilogram of mass into low Earth orbit can be reduced to just $100 per kilogram by the year 2020, we will witness an incredible expansion in the use of space for innovative commercial activities, detailed scientific investigations, and human habitation.

Propulsive Upper Stages

Quite often, getting a payload into low Earth orbit (LEO) is only part of the overall propulsion effort. Some payloads need an upper-stage vehicle, a rocket that boosts them to a higher-altitude operational orbit or onto an interplanetary Earth-escape trajectory. Aerospace engineers use an upper-stage vehicle to accomplish this task. This special vehicle rides into LEO attached to the payload. Once deployed from the launch vehicle, the upper-stage rocket engine then fires to send the payload to its final space destination.

Engineers design upper-stage vehicles with either liquid-propellant or solid-propellant rocket engines. Upper-stage rocket engines have to survive the rigors of launch from Earth's surface and then remain functional for some period in the space environment following deployment from the booster. Aerospace engineers employ different types of upper-stage vehicles with expendable launch vehicles and the space shuttle. However, NASA human spaceflight safety regulations require that all upper-stage vehicles deployed from the space shuttle contain only approved solid-propellant-rocket upper-stage vehicles because of the fear of accidental igniting of liquid propellants. All upper-stage vehicles now in service are expendable, one-time-use-only systems.

One important early American upper-stage vehicle was the Agena. The U.S. Air Force originally developed this pioneering liquid-propellant vehicle in the late 1950s to support a variety of surveillance-satellite projects. Later design versions supported many important NASA missions to Mars and Venus as well as the Gemini Project. The Agena had a very special design feature: its rocket engine could restart in space. Frequent launch-vehicle configurations included the Thor-Agena and Atlas-Agena. The Thor was a successful intermediate-range ballistic missile (IRBM) developed in the 1950s by the U.S. Air Force. Its design evolved into the very successful Delta rocket-vehicle family. Similarly, the Atlas was the first operational American intercontinental ballistic missile, developed by the U.S. Air Force in the 1950s. Its design also evolved into an important family of space launch vehicles.

The Centaur is a versatile and powerful rocket, originally developed by the U.S. Department of Defense in the late 1950s to serve as a high-performance upper-stage vehicle for the Atlas missile. The vehicle's liquid rocket uses the energetic cryogenic-propellant combination of liquid hydrogen and liquid oxygen. As a historic note, Centaur was the first American rocket to successfully burn liquid hydrogen as its fuel. Aerospace engineers combine the Centaur with Atlas or Titan expendable boosters. (The powerful Titan rocket family emerged from U.S. Air Force ICBM efforts that started in 1955.) The Centaur is actually wider than the main (core) liquid-propellant engine of the Titan launch vehicle. Therefore, when Centaur sits on top of a modern Titan launch vehicle, it creates an unusual sight that aerospace engineers call the "hammerhead" configuration.

With an eye toward future space propulsion needs, the U.S. Air Force developed the Inertial Upper Stage (IUS) vehicle in the 1970s. Aerospace mission planners fly the IUS with either NASA's space shuttle or the Air Force's family of expendable Titan launch vehicles. IUS has supported and continues to support many important military and NASA missions. The standard two-stage IUS configuration measures approximately 2.9 meters in diameter by 4.9 meters in length and comes with two individual solid-rocket motors that accommodate the in-space propulsion needs of an attached payload. The first-stage motor of the IUS contains about 9,700 kilograms of solid propellant and generates a thrust of approximately 185,000 newtons. The second-stage solid-propellant motor contains about 2,720 kilograms of propellant and generates a thrust of approximately 78,300 newtons. An extendable nozzle exit cone on the second-stage rocket motor increases performance.

Space Transportation System (STS) (Space Shuttle)

The official name for NASA's space shuttle program is the U.S. Space Transportation System (STS). The space shuttle is a mostly reusable, delta-winged aerospace vehicle that launches into space like a rocket, but then returns to Earth by gliding through the atmosphere and landing on a runway like an airplane. The space shuttle is the world's first and only operational reusable aerospace vehicle. As the name "aerospace" implies, this vehicle operates in Earth's atmosphere and in outer space. The shuttle provides routine access to space for its human crew and many types of payloads. It can also retrieve satellites from Earth orbit and repair them or bring them back to Earth for repair, refurbishment, and reuse. In late 1998, the shuttle fleet began another major task, the construction and logistical support of the *International Space Station (ISS)*.

Unlike expendable rockets, aerospace engineers have designed all the major components of the space shuttle flight system except the large external tank for refurbishment and reuse. That is why we call it a "mostly" reusable launch vehicle. The space shuttle flight system has three main components: the delta-winged orbiter vehicle (OV), the giant "disposable" external tank (ET), and two large solid-rocket boosters (SRBs) (see photo).

The orbiter is the crew- and payload-carrying portion of the STS system. About the same size and mass as a medium-sized commercial jet aircraft, the orbiter contains a pressurized crew compartment, a cavernous

The space shuttle *Atlantis* lifts off from Pad 39-B at NASA's Kennedy Space Center to start the U.S. Department of Defense–dedicated STS-27 mission (December 2, 1988). The mostly "reusable" space shuttle flight system has three main components: the delta-winged orbiter vehicle, a giant disposable external tank, and two large solid-rocket boosters that are recovered and refurbished. Photograph courtesy of NASA.

cargo bay (18.3 meters long and 4.57 meters in diameter), and three main liquid-propellant engines mounted on its aft end. The vehicle itself is 37 meters long and 17 meters high and has a wingspan of 24 meters. Since each of the three operational vehicles in the fleet (*Atlantis*, *Discovery*, and *Endeavour*) varies slightly in construction, the orbiter vehicle has an empty mass that lies between 76,000 and 79,000 kilograms.

Each of the orbiter's three main engines, the space shuttle main engines (SSMEs), uses cryogenic propellants to generate a thrust of approximately 1.7 million newtons at sea level. The liquid-hydrogen/liquid-oxygen–fueled engine is a high-performance, reusable design that can function at various levels of thrust from 65 percent to 109 percent of the nominal maximum rating. The astronaut crew (commander and pilot) can throttle the orbiter's main engines over a wide range of thrust levels. At liftoff and initial ascent, they make sure that these engines function together at a high thrust level. Then they reduce the thrust level during the final portions of vehicle ascent to limit the acceleration level to three *g* or less. (Remember, one *g* represents the acceleration due to gravity at sea level on Earth, approximately 9.8 meters per second squared [m/s^2]). This more gentle launch-acceleration environment opens spaceflight to a wider group of people than was possible in the early days of space flight. Then astronauts had to be able to survive 6 g. Engineers designed the SSMEs with gimbals so that by swiveling the engines, one can provide pitch, yaw, and roll control during the shuttle vehicle's boost phase (look back at Figure 4.2).

The delta-winged orbiter vehicle does not carry any of the cryogenic propellant to feed the SSMEs. Instead, these propellants come from a huge, disposable external tank (ET), which serves as the orbiter's "gas tank." The propellant tank is 47 meters long and 8.4 meters in diameter. At launch, it has a total mass (including propellant load) of about 760,000 kilograms. The ET contains two inner propellant tanks: the forward (smaller) interior tank holds a maximum of 0.54 million liters of liquid oxygen (LO_2) (oxidizer), while the aft (larger) interior tank contains a maximum of 1.46 million liters of liquid hydrogen (LH_2) (fuel). An intertank region separates the two and provides structural support. Approximately 8.5 minutes into the flight, with almost all its propellant consumed, the shuttle crew jettisons the external tank, and it falls away to burn up in the upper atmosphere. Any surviving pieces of ET debris splash down harmlessly in remote ocean areas. The external tank is the only major part of the space shuttle system that personnel do not recover, refurbish, and reuse.

The shuttle's solid-rocket boosters (SRBs) operate in parallel with the orbiter's main liquid-propellant engines to provide additional thrust at liftoff. Each SRB is 45.4 meters high and 3.7 meters in diameter and has a mass of approximately 590,000 kilograms. The solid propellant consists of a mixture of powdered aluminum (fuel), ammonium perchlorate (oxidizer), and a trace of iron oxide to control the propellant's burning rate. A polymer binder holds this solid-propellant mixture together. Each booster produces a thrust of about 13.8 million newtons for the first few seconds after ignition. The SRB thrust then declines gradually for the remainder of its two-minute burn. The tapered-thrust design prevents overstressing of the shuttle flight vehicle. When the two solid boosters burn with the orbiter's three main engines, the shuttle vehicle generates a total liftoff thrust of about 32.5 million newtons. Approximately two minutes after liftoff, the solid boosters run out of propellant, and the shuttle crew jettisons them. Special ships recover the spent boosters and return them to land, where aerospace personnel refurbish and reload them for another flight.

Each orbiter vehicle also has two smaller orbital maneuvering system (OMS) engines that operate only in space. The OMS engines, located in external pods on each side of the aft fuselage near the orbiter's tail, provide thrust for orbit insertion, orbit change, altitude change, rendezvous operations, and deorbit maneuvers. The OMS rockets use a hypergolic-propellant combination of nitrogen tetroxide (N_2O_4) and monomethyl-hydrazine (MMH). Onboard storage tanks, nestled in each OMS pod, supply these hypergolic propellants.

NASA launches the shuttle vehicles from Complex 39 at the Kennedy Space Center in Florida. Depending on the requirements of a particular mission, the space shuttle can carry up to 22,700 kilograms of payload into low Earth orbit (LEO). An assembled and fully fueled shuttle flight vehicle has a typical liftoff mass of approximately two million kilograms.

Every shuttle flight has its own specific crew, activity requirements, and mission profile. To help us understand how the shuttle works, however, we can explore the general sequence of events and activities that happen in a nominal mission. A few seconds before the final commitment to launch, the shuttle's three main liquid-propellant engines come to life and reach full power. Although the shuttle vehicle tries to move, it cannot because giant bolts hold it securely to the launch pad. If any problems show up in the behavior of any of the three main engines, a computer automatically shuts them all down. Under this circumstance, the solid-rocket boosters are not sent the ignition signal. Instead, mission controllers scrub the planned flight and begin backing out of the countdown. As part of this backout, the astronauts quickly exit the shuttle vehicle, and flight personnel secure all hazardous systems by placing them in a "safe" condition. Engineers then search for, find, and resolve the problem so the shuttle can fly another day.

But this is a great day for flying into space, so we will assume that the three main engines reach full power without incident. When the countdown reaches zero, the solid-propellant rockets ignite, and the hold-down bolts, severed by small explosive charges, release the vehicle. The shuttle leaps toward space on a brilliant pillar of fire.

The solid-rocket boosters continue to burn along with the three main liquid engines until the shuttle vehicle reaches an altitude of about 45 kilometers and a speed of about 5,000 kilometers per hour. At this point in the flight (about two minutes after liftoff), the crew jettisons the expended solid boosters, which then parachute back to Earth. The shuttle's three main engines, fed by the giant external tank, continue providing thrust for another six minutes. At MECO (main-engine cutoff), the engines stop firing, and the empty external tank separates and falls back to Earth. By now (about eight minutes into the flight), the orbiter vehicle is almost in orbit. The crew fires the orbiter's OMS engines to generate the final thrust necessary to reach the desired Earth orbit for the particular mission. The crew might briefly fire the OMS engines several times to make precise adjustments in the vehicle's "working" orbit. For example, a second OMS burn will circularize the orbit. While in space, the crew fires the OMS engines to raise or adjust the orbiter's altitude to satisfy the needs of a particular mission.

Shuttle flights can last from a few days to more than a week, and shuttle missions frequently include deploying satellites. Since the orbiter only operates in low Earth orbit (generally between 250 and 400 kilometers altitude), a shuttle-deployed satellite often has an upper-stage propulsion unit attached. Once the deployed satellite reaches a safe distance from the orbiter, the upper-stage unit fires and propels the spacecraft to its operational location in space. Sometimes the deployed spacecraft flies to geostationary orbit; other times it departs low Earth orbit on an interplanetary trajectory.

To return to Earth, the astronaut crew reverse-fires the OMS engines. This retrograde burn reduces the vehicle's orbital velocity and allows the orbiter to reenter the upper regions of Earth's atmosphere. However, unlike a guided missile's reentry vehicle or earlier U.S. crewed spacecraft (Mercury, Gemini and Apollo), the winged orbiter does not follow a simple ballistic trajectory to the ground. Instead, it behaves like a giant glider capable of maneuvering to the right or left of its entry path by as much as 2,000 kilometers. The vehicle's heat shield protects it from the consequences of aerodynamic heating. Its special, reusable insulation survives temperatures of up to 1533 K (1260° C) and sheds heat so readily that one side remains cool enough to hold in your bare hands, while the other side glows red hot. After gliding down through the atmosphere in a series of energy-dissipating maneuvers, the orbiter touches down like an airplane on the runway at the Kennedy Space Center in Florida (primary landing site) or at Edwards Air Force Base in California (the alternate landing site if weather conditions are unfavorable for landing at the primary site). Once it rolls to a stop, a fleet of servicing vehicles welcomes the orbiter and its crew home and begins preparing the shuttle for its next journey into space.

Reusable Launch Vehicle (RLV)

The reusable launch vehicle (RLV), currently only in the conceptual state, is an aerospace vehicle that includes functional designs and fully reusable components to provide low-cost access to space. The RLV employs twenty-first-century space technology and innovative payload- and vehicle-processing techniques to achieve airline-type operations in delivering payloads into low Earth orbit (LEO). Aerospace experts believe that they can construct an operational RLV, but recognize that complete reusability represents a significant engineering challenge. Unlike expendable launch-vehicle components that must function properly just once, all the parts of an RLV must resist deterioration and survive multiple launches and reentries without requiring extensive refurbishment or replacement between flights.

Payload hauling with NASA's space shuttle, the world's only partially reusable aerospace vehicle, has proven very expensive and requires long turnaround times between flights. This occurs because of the high cost of maintaining the orbiter fleet and the need to extensively inspect and refurbish each "reusable" shuttle vehicle between flights.

We will use another airline analogy to clarify the situation. Consider a nonstop coast-to-coast flight from New York to San Francisco. How much would the ticket cost if the airline had to pull the engines off the jet when it landed in San Francisco and completely rebuild them before the jet could fly back to New York? Now, imagine that the airline also has to remove all the passenger seats and replace them with other seats, remove every restroom and galley, repaint the outer surface of the aircraft, and replace all the tires. The airline would need to perform this detailed overhaul activity after each flight. Under this make-believe model, the operational support process for "routine" air travel becomes very time consuming, labor intensive, and expensive. Our imaginary airline might operate under this arrangement if it could find a very wealthy customer who tolerated the high cost of a coast-to-coast ride. However, most potential customers simply could not afford to participate in air travel and enjoy its benefits.

Unfortunately, this is quite similar to what occurs each time the orbiter lands and begins preparation for its next journey. The space shuttle represents outstanding advances in space technology, but its operational costs are anything but routine and inexpensive (as originally proposed).

It is very expensive to send payloads into orbit with either expendable launch vehicles or the space shuttle. In the year 2003, for example, launch costs ranged between $5,000 and $20,000 per kilogram (or more), with ELV-delivered payloads at the low end of this range and shuttle-delivered payloads dominating the high end. The specific 'LEO delivery bill' depends on many factors, including the efficiency of the launch vehicle, the location of its launch site, the type of payload, and its final orbital destination. That is why aerospace engineers are responding to the challenge of creating a practical reusable launch vehicle (RLV) early in the twenty-first century.

The RLV can place payloads into orbit more cheaply. From the start, aerospace engineers base the vehicle's performance on fully reusable components that combine the latest advances in space technology with uncomplicated, functional designs. A philosophy of airline-type operation also helps reduce the cost of sending payloads into space. The RLV features innovative operational techniques, such as automated vehicle checkout and efficient payload processing. Smart materials, embedded sensors, and intelligent microprocessors all play a major role. This streamlined approach to RLV operation should also eliminate the expense associated with maintaining a large launch complex.

The first generation of RLVs promises to deliver payloads into low Earth orbit at a projected cost that is at least 50 percent less than the lowest current cost of accessing space. Aerospace engineers further project that the second generation of RLVs (available perhaps by the year 2025) should reduce this cost by another 50 percent at a minimum. If these projections hold, by the year 2025, we could ship cargo into low Earth orbit for less than $250 per kilogram. This favorable transportation cost should greatly expand the commercial application of space technology and encourage a new generation of space entrepreneurs.

There are three single-stage-to-orbit (SSTO) reusable launch-vehicle designs that provide payload capacity to meet most government and commercial space-transportation requirements. Each candidate configuration has unique technical obstacles, but also offers distinct advantages for reducing the cost of accessing space. The candidate configurations are the vertical-takeoff/vertical-landing "conical" vehicle, the vertical-takeoff/horizontal-landing winged-body vehicle, and the vertical-takeoff/horizontal-landing lifting body vehicle. The conical configuration provides a simple aerodynamic shape with a low-mass airframe that does not need massive wings. However, the conical vehicle must restart its rocket engines after reentry to accomplish a safe vertical landing and offers a limited payload volume. The second RLV candidate, the winged-body SSTO, provides a simple fuel-tank design and easy maneuvering during reentry. However, this design has a limited payload volume and a high landing speed. The final candidate, the lifting-body SSTO, offers low reentry temperatures, a low landing speed, and a low-mass design. However, it does require a more complicated airframe.

LAUNCHING A ROCKET

Getting into Space

When we want to place a spacecraft into orbit, the launch vehicle we choose has to successfully accomplish two important tasks: vertical ascent and horizontal (tangential) acceleration to orbital speed. First, the rocket vehicle must provide enough thrust to lift itself and its payload while ascending vertically up through the atmosphere. Second, once the rocket vehicle (or its final stage) reaches an appropriate altitude above Earth's surface and atmosphere, the rocket vehicle must pitch over (tilt) and provide the spacecraft a sideways nudge that is sufficient to keep it "falling" around Earth. Aerospace engineers call this height the altitude of orbit insertion. If the rocket's horizontal (sideways) orbit-insertion thrust is not strong enough, the spacecraft will eventually fall back to Earth and burn up in the atmosphere. The force of gravity is quite relentless and unforgiving. If the insertion burn (our sideways nudge) is too much, the spacecraft escapes Earth's gravity completely and heads off on an interplanetary trajectory. Finally, if this horizontal (tangential) thrust is just right, the spacecraft achieves a velocity that keeps it "falling" around Earth in a closed path. We call this special velocity the *orbital velocity* (for the particular altitude), and we call the spacecraft's closed path an *orbit*. This orbit is either circular or elliptical, depending upon its eccentricity, or ovalness. An orbit with an eccentricity of zero ($e = 0$) forms a circle, while an orbit with a high eccentricity (say $e = 0.9$) represents a long, thin ellipse. We will discuss the physics of orbiting objects shortly.

Aerospace engineers and space lawyers consider that outer space begins (for operational and legal purposes) at an altitude between 100 and 200 kilometers. Below this altitude range, Earth's residual atmosphere produces significant aerodynamic drag on an orbiting object. As a result, any spacecraft traveling around Earth at an altitude below about 150 kilometers remains in orbit only if it can provide additional thrust to overcome the retarding influence of atmospheric drag. Otherwise, the spacecraft will continue to slow down and lose altitude. In this orbit-decay process, the object (still traveling at very high speed) eventually encounters the denser regions of Earth's atmosphere, where it burns up due to severe aerodynamic heating. Objects jettisoned by a multistage launch vehicle late in the powered portion of its ascent (such as a spent second stage, clamps, and other interstage hardware) sometimes undergo this unstable orbit behavior. We will learn more about this when we discuss the issue of space debris in chapter 6.

Aerospace personnel launch a large rocket vertically so the vehicle can travel the minimum distance necessary through the denser portions of the lower atmosphere at progressively increasing speeds that are still much slower than orbital velocity. This prevents breakup of the rocket and its payload due to excessive aerodynamic force and frictional heating.

However, we cannot put a satellite into orbit by simply launching it upward in a vertical direction. When a rocket reaches outer space, it must pitch over (tip) and provide the necessary horizontal (tangential) speed to the spacecraft. It is this horizontal (sideways) velocity that places the spacecraft in a stable, low-altitude Earth orbit.

Why not put a satellite into orbit by firing it horizontally at very high velocity from a very tall mountain? The great English scientist Sir Isaac Newton examined this question in the seventeenth century. Much like the illustration in Figure 4.9, Newton's notes contained a sketch that described the possible trajectories of a cannonball fired horizontally from a very tall mountain at different speeds. Although there are many engineering limitations with this suggestion (for example, atmospheric friction and excessively high "launch" accelerations that would destroy most payloads), the concept is quite useful in exploring "launch physics." If we look back to chapter 3, we find that Jules Verne, the famous French writer, suggested a similar "gun-launch-to-space" approach to send his passengers on their fictional voyage around the Moon. We will also use the cannon-on-a-tower concept to help explain why orbiting objects "fall" around Earth.

Let us perform another thought experiment. Imagine that we construct the world's tallest tower—a magnificent structure that rises about 350 kilometers above Earth's surface (see Figure 4.9). We now securely attach a large cannon on the top of our tower and fire a cannonball in the horizontal direction. This horizontal direction is perpendicular to a radial line drawn from the center of Earth. On our first shot from the 350-kilometer-high tower, the cannonball leaves with a horizontal velocity of about 1 kilometer per second. The ball tries to travel in a straight line, but the pull of Earth's gravity bends it back, and it crashes on the ground at a place we will call Point A. (For this simple example, we neglect all atmospheric influences on the cannonball's ballistic trajectory.)

We try again, adding more powder to the cannon. Now, the ball flies out with a horizontal velocity of 4.5 kilometers per second. It travels farther; still, Earth's gravity tugs away, and down it comes at point B. Finally, we load the cannon with a really high-energy charge. Our third shot sends the ball zooming out horizontally at a velocity of about 7.8 kilometers per second. Gravity still bends the ball's trajectory. However, this time, because of its altitude and tangential speed, the ball keeps falling around Earth in a complete circle. Our cannonball is in orbit. Of course, we must now lower the tower a bit to avoid the embarrassment of getting hit by our own projectile as it comes around.

This thought experiment also introduces another very important physical idea. Because the inertial trajectory of an orbiting spacecraft compensates for the force of Earth's gravity, the orbiting spacecraft and all its contents

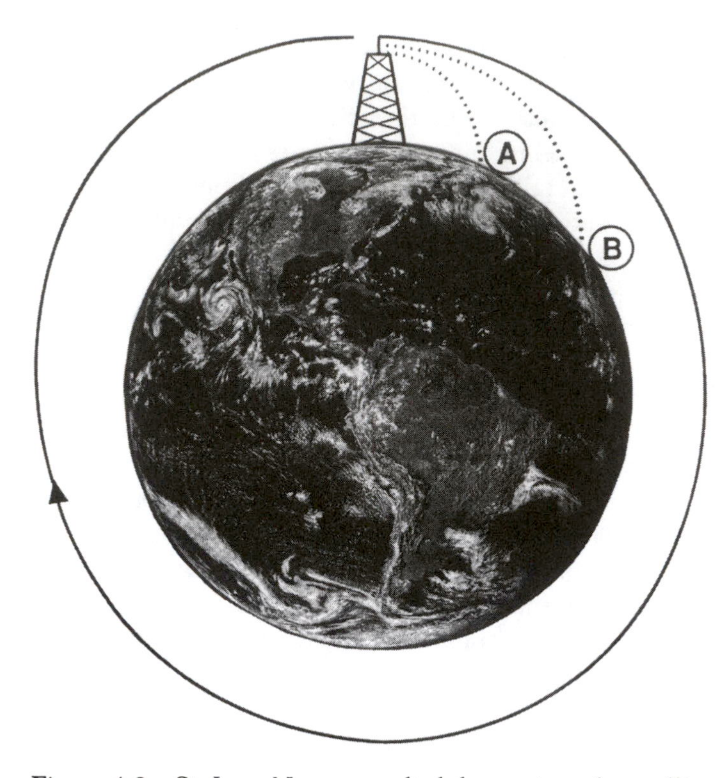

Figure 4.9 Sir Isaac Newton studied the motion of a satellite around its parent body by first imagining the behavior of a cannonball fired from a cannon in the horizontal direction, when that cannon was placed and securely attached to a very tall tower or mountain. As the muzzle velocity of the cannonball increased, it would fall to Earth farther and farther away from the tower until it reached a sufficient velocity (called the orbital velocity) that it could actually travel all the way around Earth without ever falling to the ground. NASA image modified by author.

(including astronauts, test animals, plants, and equipment) experience a continuous state of *free fall.* Scientists say that the (radially inward) gravitational pull of Earth is offset (balanced) by the (radially outward) centrifugal force related to the curved motion of the spacecraft. In this state of free fall, all objects inside the spacecraft appear "weightless." Aerospace engineers sometimes call this condition "zero gravity." If we are not careful, the term *zero gravity* can be a bit misleading. Our planet's gravity is very much present and continuously influencing the path of the orbiting spacecraft and its contents as the spacecraft falls around Earth. For example, at the altitude NASA's space shuttle generally operates in space, gravity's pull is approximately 94 percent of the pull experienced at Earth's surface.

Einstein's principle of equivalence tells us that the physical behavior inside a system in free fall is identical to that inside a system far removed from matter that could exert a gravitational influence. With Einstein's help, we can use the terms *weightlessness* and *zero gravity* when we describe conditions in an orbiting spacecraft.

People sometimes ask what the difference between mass and weight is. Why do we say "weightlessness" and not "masslessness"? Mass is the physical amount of substance in an object—it has the same value everywhere. Weight is the product of an object's mass and the local acceleration of gravity. Weight is a force we calculate with Newton's second law of motion. Therefore, you would weigh about one-sixth as much on the surface of the Moon as you do on Earth, but your mass would still be the same in both locations.

We cannot completely achieve a "zero-gravity" environment in an Earth-orbiting spacecraft. It is really an ideal situation. The venting of gases from the space vehicle, the minute drag imposed by a very thin residual terrestrial atmosphere, and even crew motions create nearly imperceptible forces on people and objects alike. Scientists call this collection of tiny forces in a free-falling spacecraft *microgravity*. In a microgravity environment, astronauts and their equipment float around almost, but not entirely, weightless. A typical microgravity value in an orbiting spacecraft is about 10^{-5} g (that is, 0.00001 of the value of the acceleration of gravity at Earth's surface). This may not seem significant, but too high a level of microgravity (say 10^{-3} or 10^{-4} g) can upset delicate materials-processing experiments done on board an orbiting spacecraft. As we will discover in chapter 6, contamination of its microgravity environment is a major issue for the *International Space Station* (*ISS*).

The Launch Site

A launch site is the specific place from which we send a rocket or an aerospace vehicle into space. For safety reasons, aerospace engineers select a remote location surrounded by large amounts of undeveloped land so the rockets can fly without passing over inhabited areas. They also prefer a site with pleasant, mild weather conditions. A site near the equator is especially valuable, since rockets departing that location in an easterly direction (90° launch azimuth) receive the maximum "natural" velocity boost from Earth's west-to-east rotation. Good transportation (land, sea, and air) to the site will avoid excessive shipping costs and prevent delays in the delivery of the rocket vehicles, their payloads, propellants, and other bulk materials.

We use the term *launch complex* to describe the complete collection of launch sites (pads), control center, support facilities, and equipment used to launch rockets or aerospace vehicles from a given geographic area. A launch complex, like NASA's Kennedy Space Center, might also include a landing strip to allow the return of an aerospace vehicle from its space mission. In this case, the complex serves as both a doorway to space from the surface of the planet and a port of entry from space to the planet's surface. We call this facility a *spaceport*.

Aerospace personnel perform many different operations at a launch complex. They receive and assemble launch vehicles, integrate payloads with their rocket vehicles, test and check out the combined payload/flight-vehicle configuration, fuel the launch vehicle at the pad, coordinate weather, range-safety, and tracking activities, and conduct countdown and launch operations. For reusable aerospace vehicles, the complex also supports landing operations and vehicle refurbishment. The successful operation of a spaceport requires the services of many qualified people who possess a variety of technical skills. Some needed skills include design engineering, safety and security, quality control, hazardous-materials handling, cryogenic-fluids management, maintenance, logistics, computer operations, communications, and documentation. Payloads involving live animals and plants need special handling and facilities, both before and after flight. In the space-station era, the spaceport functions as the prime terrestrial transportation node that connects orbiting research facilities with research scientists and space entrepreneurs on Earth.

At the start of the twenty-first century, there are several major launch complexes on Earth, but only the adjoining Cape Canaveral/Kennedy Space Center complex in Florida qualifies as a spaceport. The U.S. Range includes launch facilities at both Cape Canaveral Air Force Station and NASA's Kennedy Space Center. Located on the central east coast of Florida, this sprawling complex supports both expendable and reusable (shuttle) launch vehicles. These vehicles fly primarily to equatorial (low-inclination) Earth orbit and interplanetary destinations. The United States also operates the Western Range at Vandenberg Air Force Base on the west central coast of California. This complex sends payloads into high-inclination (polar) orbits around Earth. The European Space Agency (ESA) operates a major expendable-vehicle (Ariane) launch complex in Kourou, French Guiana, on the northeast coast of South America. The Russian Federation maintains a major expendable-vehicle launch complex, called the Baikonur Cosmodrome, in the Republic of Kazakhstan. Russian cosmonauts ride expendable launch vehicles from Baikonur to access space.

Launch Activities and Operations

When aerospace engineers launch a rocket, they do so by performing a detailed process called the *countdown*. This is a step-by-step, carefully scheduled set of procedures that ultimately leads to the ignition of the rocket's engines. During the countdown, aerospace personnel bring the rocket vehicle to the launch site and load it with payload and propellants. Using launch-center computers that communicate with sensors on board the rocket, they monitor all of the important systems on the launch vehicle and its payload. They need to monitor both the rocket and the payload because it makes no sense to put a broken object into space. Launch personnel also carefully watch the weather and wait for the proper *launch window*. The launch window is that precise time interval during which aerospace personnel can launch a rocket so that the payload can reach the proper orbital destination. It is primarily controlled by celestial (orbital) mechanics, that is, where various objects are now and are projected to be when we want to get to them. Bad weather can have a secondary influence on the launch window, but it is not a major factor. "Hitting the launch window" is a very demanding and challenging part of aerospace operations. For example, spacecraft rendezvous missions in low Earth orbit and interplanetary missions usually have very narrow launch windows—sometimes just minutes per day. The launch crew must fire the rocket during this window, or else the mission fails. Get the launch vehicle ready too soon, and its cryogenic propellants might boil off. Get the vehicle ready a little too late, and expensive delays and "recycled" countdowns result. If the flight involves a human crew, the astronauts will usually board the vehicle as late in the countdown as practical. One space shuttle mission missed its first launch window because ground support personnel could not properly close the hatch (door) on the orbiter after the astronauts went on board.

Weather at the launch site plays a major, but uncontrollable, role in the countdown. Unfavorable environmental conditions, such as high winds, thunderstorms, and the threat of lightning strikes, will delay a launch. The presence of unauthorized persons who suddenly enter the safety "keep-out" area at and near the launch site can also interfere with a countdown. NASA had to delay a space shuttle mission because a foolish person flew his small private plane into clearly identified restricted ("no-fly") airspace. He did this so his family could get a better view of the launch. At least, that is what he told federal authorities when they arrested him and revoked his private pilot's license. There truly is no trouble-free launch.

During American launch operations, aerospace personnel use *T-time* to reference specific times (plus or minus) to the zero time (launch time) that occurs at the end of the countdown. For example, early in a countdown we might hear: "The bird reached the pad at T minus 30 days." Translated from aerospace language, this means that 30 days before the scheduled launch date, the rocket vehicle arrived at its assigned launch site. Getting closer to launch time, we will hear such announcements as "T minus 20 seconds and counting." This means that we are precisely 20 seconds away from rocket-engine ignition and the countdown is continuing smoothly.

Sometimes, however, we might hear: "T minus 20 seconds and *holding*." This means that launch personnel have suspended the countdown process. Aerospace workers use the word *hold* to identify a halt in the normal sequence of events during a countdown. The step-by-step process now stops while launch personnel investigate and remove the problem. When launch personnel resolve the problem (often called a *glitch*), they usually resume the countdown from the point where they left off.

Sometimes, however, solving one problem creates conditions that could lead to another problem. Under such circumstances, launch personnel might backtrack in the countdown process to make sure that no new glitches have popped up. We might hear the launch director announce a recycling of the countdown from T minus 20 seconds to T minus 80 seconds, or whatever. All launch-control personnel would then go back to their list of detailed instructions and resume the countdown by performing the activities called for at T minus 80 seconds.

Launch personnel insert "planned holds" in a countdown sequence. They do this so the launch-support computers can run automatic checks on the rocket vehicle moments before engine ignition. Other times, an "unscheduled hold" occurs because a problem suddenly appears during the countdown. For example, launch-site weather conditions might quickly go from good to marginal. A "weather hold" gives launch personnel time to fully examine the deteriorating environmental conditions before continuing the countdown. If the launch director and his/her staff resolve the sudden problem, they release the hold, and the countdown clock ticks down to zero.

At zero time (T equal to zero), the launch director sends the ignition signal to the rocket's engines, and they roar to life. As the vehicle rises from the pad, the countdown clock enters the positive (after-liftoff) part of T-time. For example, at approximately T plus two minutes, the space shuttle jettisons its expended solid-rocket boosters. Once a space vehicle achieves orbit, flight-control personnel switch to another convenient reference time called *mission elapsed time* (MET). This is the time from the beginning of a mission, usually taken as the moment of liftoff.

If a problem that defies immediate resolution happens during the countdown, the launch director will postpone, or *scrub*, the launch for that particular day. Depending on the severity of the problem and the complexity of the launch vehicle, another countdown could start within a day, or it may be a week or more before the launch crew attempts to send the rocket into space. While the launch director and his/her staff make every effort to get a fully fueled rocket vehicle off the pad on time, large rockets are very temperamental, dangerous, and expensive devices. Their payloads are even more expensive. Astronauts and cosmonauts recognize and accept the fact that riding rockets into space is a high-risk profession, but no prudent launch director or mission commander will unnecessarily risk a vehicle, its crew, or its payload. The countdown process, with its sometimes disappointing holds and scrubs, helps avoid unnecessary loss of life and property. Launching a large, modern rocket (with or without human crew on board) is not yet a "routine" activity, like the operation of a commercial jet aircraft.

Sometimes, despite careful preparation and checkout, a rocket misbehaves when an important piece of equipment suddenly fails. Aerospace personnel then have to *abort* (cut short or cancel) the mission. Launch-vehicle aborts can occur on the pad or in flight. In one type of on-the-pad abort, the engines of a fully fueled rocket fail to ignite. The rocket now is a very dangerous device that could explode without warning. Launch personnel must exercise extreme caution in backing out of the countdown and securing the vehicle.

Have you ever attempted to fly a small model rocket that did not respond after you pressed the fire switch? Now imagine a 30- or 50-meter-tall container of highly explosive propellants in that very same hang-fire condition. Hopefully, you did not just run up to your misbehaving model rocket. There are well-established safety guidelines for handling on-the-pad, hang-fire aborts, and it is essential to observe them. Every launch director knows such procedures. In fact, launch personnel with premature gray hair have probably worked through several hang-fire pad aborts. If the rocket vehicle has a human crew, emergency crew evacuation is the first and most important activity. Other activities involve patiently securing all electrical systems and circuits and carefully off-loading any liquid propellants. To the greatest extent possible, workers conduct these backout activities from a safe, remote location—perhaps assisted by teleoperated robots.

Sometimes a rocket's engines ignite but do not develop sufficient thrust. The vehicle struggles to get a few meters into the air, lingers suspended over the pad, and then settles back on the ground, disappearing in a huge explosion. This is an on-pad (or near-pad) explosive abort. The violent explosion showers dangerous debris and shrapnel all over the launch-site area. This is clearly a lethal environment for anyone caught in the vicinity. At a well-designed rocket range, launch personnel supervise the final stages of the countdown from a safe, distant control center. (Early rocketeers called this protected launch-control center a "blockhouse.") Except perhaps for a loss of professional pride, launch personnel should experience no injury due to a rocket explosion on the pad or while the malfunctioning vehicle attempts to fly.

Once a rocket clears the pad, any serious malfunction creates an in-flight abort of the vehicle. Every operational rocket range has a command destruct officer whose primary duty is to closely watch the flight trajectory of an ascending rocket vehicle. If the rocket begins to veer off its planned course for whatever reason, this officer prepares to send the command destruct signal. Once the trajectory of the misbehaving rocket touches the *destruct line* (an imaginary boundary line clearly defined for each launch), the command goes out for the errant vehicle to self-destruct. The command destruct signal is an encrypted (coded) electromagnetic signal. When sent, the special signal activates the self-destruct explosive system that every missile and rocket flow on a range must carry.

This is an integral part of the range-safety process. The shape and extent (footprint) of the destruct line varies with the type of rocket being flown and the location of the launch site. For example, population centers that have developed near a launch site will severely limit the direction a rocket can leave, and range-safety officers must calculate where any debris will fall if the rocket explodes. Command destruct of an erratic rocket prevents it from endangering people and property outside the boundaries of the rocket range. Safety experts also use the destruct line to protect personnel and support facilities within the launch complex because some rockets have turned back and tried to blast the launch-control complex.

Sometimes as a launch vehicle rises toward outer space, its second- or third-stage rocket malfunctions. In this case, there is usually no immediate danger to people or property off the rocket range. The range's *launch-azimuth* restrictions prevent downrange in-flight aborts from showering debris on inhabited regions of Earth. The launch azimuth is the initial compass heading of a powered rocket vehicle at launch. For example, aerospace safety considerations require that all rockets launched from Cape Canaveral fly on trajectories with a launch azimuth between 35 degrees and 120 degrees. This means that a rocket can leave this range by flying only in a northeasterly, easterly, or slightly southeasterly direction. Flying due east (90° launch azimuth) is actually quite favorable, since the rocket vehicle picks up the full ("free") velocity increment provided by the natural west-to-east spin of Earth.

The range's launch-azimuth restrictions also guarantee that planned stage impact hazards associated with the normal jettisoning of expended booster stages, fairings, and hardware stay within acceptable levels. A modern rocket range generally extends for thousands of kilometers away from the launch site. The idea is to minimize any impact hazard from normal launch-vehicle debris as intentionally discarded equipment falls back to Earth. Planned impact zones in uninhabited, broad ocean areas prove quite suitable. Downrange tracking stations help personnel monitor the entire ascent trajectory of a launch vehicle. Should an upper stage fail in some unusual way, creating danger to people half a world away, personnel at a downrange station can also send a command destruct signal to the flight vehicle.

Unlike expendable launch vehicles, the space shuttle orbiter is a winged aerospace vehicle with several crew-interactive abort alternatives. However, these alternatives require separation of the solid-rocket boosters and the external tank. There are two basic types of ascent abort modes for space shuttle missions: intact aborts and contingency aborts. NASA uses the intact-abort procedures to provide a safe return of the orbiter vehicle and its crew to a planned landing site should a main engine fail or a significant cabin leak occur during ascent. NASA also created contingency-abort procedures to promote crew survival following more severe failures, when the crew cannot perform an intact-abort procedure. A contingency abort usually would result in an orbiter vehicle ditch operation.

The four types of space shuttle intact aborts are abort-to-orbit (ATO), abort-once-around (AOA), transatlantic-landing (TAL), and return-to-launch-site (RTLS). In the abort-to-orbit procedure, the crew flies the orbiter vehicle into a temporary orbit that is at a lower orbit than the planned mission orbit. The crew and launch-control personnel then take time to evaluate the circumstances and decide whether to execute a deorbit maneuver or continue the planned mission with acceptable modifications.

In the abort-once-around procedure, the crew flies the malfunctioning shuttle into orbit. They go around Earth just once and immediately execute a deorbit maneuver that provides a normal entry and landing. The transatlantic-landing abort procedure allows the crew to make an intact landing on the other side of the Atlantic Ocean at an emergency landing site in either Spain, Morocco, or Gambia (West Africa). This abort mode uses a ballistic trajectory that does not require the crew to fire the orbital maneuvering system (OMS). Finally, the return-to-launch-site abort mode lets the crew fly the shuttle downrange to dissipate propellant. Then, while under power, they turn the vehicle around, jettison the external tank, and glide back for a landing at the Kennedy Space Center. The type of failure (such as the loss of one main engine) and the time during the ascent when the failure occurs determine which abort mode the shuttle astronauts select. Throughout the shuttle's powered ascent, NASA's ground personnel continuously interact with the flight crew and carefully monitor the status of the vehicle. Should a major vehicle problem appear, the launch director would quickly know what abort actions are available at that time. The fatal *Challenger* explosion (January 28, 1986) occurred suddenly during shuttle vehicle ascent while the solid-rocket boosters were still burning. Unfortunately, there was no relevant abort sequence for this catastrophic malfunction. The crew of seven perished.

If the payload contains a nuclear power supply, the United States imposes additional launch safety procedures. For radioisotope sources, aerospace nuclear safety policy requires that the power supply contain the radioactive material under all anticipated flight circumstances, including launch-pad accidents and ascent malfunctions that return the device to Earth. Aerospace safety rules for a payload with a space nuclear reactor require that the reactor's design keep it subcritical (not capable of supporting a sustained chain reaction) under all launch conditions. Flight controllers will make the reactor critical only after the payload reaches a safe orbit.

SPACECRAFT

A spacecraft is basically a platform that travels in orbit around a planet or through interplanetary space. A launch vehicle, sometimes in combination with an appropriate upper-stage transfer vehicle, puts the spacecraft into outer space so it can accomplish its mission. Engineers usually design a spacecraft to function as an automated (robotic) machine. For spacecraft supporting human spaceflight, however, the engineers include a special pressurized compartment and life-support equipment to keep the astronauts and any other living creatures alive and comfortable while the platform operates in space.

A set of functional subsystems lets the spacecraft find its way, operate, and survive in the outer-space environment. These subsystems, which are common to most spacecraft, include structural, thermal control, data handling and storage (including the spacecraft clock), telecommunications (including telemetry packaging and coding), navigation, attitude and articulation control (stabilization), and power.

The spacecraft usually contains a mission-oriented payload that is its raison d'être. This payload, when properly supported by the spacecraft's functional subsystems, executes the mission-oriented tasks. One spacecraft payload might consist of a collection of electro-optical sensors that remotely collect data about planet Earth. Another might involve a group of instruments that directly sample the solar wind as the spacecraft travels through interplanetary space. The "working payload" of a communications satellite might be the specialized cluster of electronic equipment that helps relay and broadcast electromagnetic signals at radio frequencies.

Because of the wide variety of interesting missions that spacecraft perform, they are designed for specific types of missions. For example, engineers design certain spacecraft (called meteorological satellites) to orbit around Earth and make frequent environmental measurements that help scientists forecast the weather and study climate changes. Other spacecraft (called military surveillance satellites) help defense leaders protect the United States and its allies by vigilantly watching the globe for the telltale signs of a hostile missile attack. Scientific spacecraft travel great distances through interplanetary space to reach other planets. Once there, they use an assortment of remote sensing instruments to conduct detailed investigations of these alien worlds. Sometimes, a far-traveling robot explorer lands on another planet's surface and then scurries about performing in situ (in-place) scientific experiments.

Aerospace engineers must design spacecraft to tolerate several distinctly different physical environments. On Earth, the spacecraft travels by land, sea, or air to the launch site. While the spacecraft is being integrated (joined) to its launch vehicle, exposure to dust, dirt, moisture, and toxic materials could damage sensitive spacecraft components. Launch personnel take special precautions to keep these sensitive spacecraft parts as clean and dirt free as possible. That is why they normally work in dust-free clean rooms and wear special clothing (called bunny suits) during a spacecraft's prelaunch processing and testing. Biological contamination by hitchhiking terrestrial microorganisms represents another design concern for spacecraft that will land on another potentially life-bearing planet, like Mars.

The ride into space on a pillar of fire exposes the spacecraft to a severe vibration and mechanical stress environment. Aerospace engineers compensate for this with reinforced structural designs, high-strength (but relatively low-mass) materials, and the use wherever possible of tapered-thrust rocket engines that can limit stresses during certain critical phases of the ascending flight. If the spacecraft survives the shake, rattle, and roll rocket ride into orbit, it must then function for reasonable amounts of time in another very hostile environment. Aerospace engineers characterize operation in outer space as continuous exposure to the following conditions: very hard (but not total) vacuum, very low (but not zero) levels of gravity (microgravity), ionizing radiation, micrometeoroids, space debris, and severe thermal gradients (temperature differences).

Spacecraft engineers must consider the influence of hard vacuum on all electrical, mechanical, and thermal subsystems. Before selecting a material for use on a spacecraft, the engineer must thoroughly understand how that material will behave in hard vacuum. We know that in a vacuum, most materials will outgas (release volatile surface materials) to some extent. When a spacecraft first arrives in space, extensive outgassing of its equipment can cause problems. For example, the outgassing process might eventually coat optical sensors and other special surfaces with an unwanted layer of condensed material.

Similarly, the microgravity environment associated with spaceflight provides interesting engineering challenges, including fluid storage and handling, equipment stowage, and astronaut hygiene (e.g., the design of a microgravity toilet or shower). The natural space radiation environment (trapped particles, solar flares, and cosmic rays) represents a significant problem for astronauts and certain electronic systems. The radiation-induced upset or failure of important electronic subsystems degrades a spacecraft's ability to perform its mission. Ionizing radiation, when absorbed in sufficient doses, is also lethal to humans. For example, a large solar flare can endanger the lives of astronauts, especially if they are caught traveling through interplanetary space when the flare occurs. A high-speed collision with a micrometeoroid can damage a spacecraft, puncture and compromise a pressurized habitat, and rip open a space suit. The impact of a piece of space debris produces similar consequences.

Finally, during orbital or interplanetary flight, a spacecraft experiences varied and extreme thermal environments. The only way heat transfers through a vacuum is by thermal radiation. A spacecraft receives thermal radiation from the Sun (at about 5,800 K) and rejects thermal energy to deep space (at about 3 K). These extreme source and sink temperatures can create severe thermal stresses on the spacecraft.

If the aerospace engineer successfully overcomes all of these space-environment challenges, the spacecraft design problem may still not be over. Our spacecraft, after riding into orbit and traveling through interplanetary space, might also have to land and operate on the surface of an alien world—a world with its own peculiar set of environmental conditions. Spacecraft are definitely marvelous examples of space technology.

Spacecraft Design Trade-offs

When aerospace engineers design a spacecraft, they face a large number of space-technology trade-offs. They often have to make difficult decisions involving the use of new (perhaps non-flight-tested) technology versus existing (flight-proven) technology. New technology will let the spacecraft do much more, but "older" technology works. How much risk is the spacecraft's sponsor willing to tolerate for additional performance gains during a particular mission?

Mission requirements (what the spacecraft must accomplish), budget, and schedule drive the design trade-off process. A program manager sometimes learns that for another $10 million, the spacecraft can accomplish an additional experiment. Is the additional "science data" worth the expense? If the engineer allocates some of the spacecraft's mass budget for additional attitude-control-system propellant, then the mission will have a longer operating life in orbit. However, the mission's sponsor might want to assign that "extra mass" to a new scientific instrument—one that can significantly increase the technical payoff of the entire mission. It is not easy to make such trade-offs. Each spacecraft ultimately represents a struggled compromise between introducing new technologies, getting the "job" (mission) done, staying on schedule, and keeping within budget. The systems engineer serves as the mediator and carefully administers the spacecraft's mass budget. If the spacecraft ends over mass (even a kilogram or so extra on a small spacecraft), the launch vehicle might not be able to send it to where it needs to go, and the mission may not fly.

Did you ever pack for a long vacation on which you could take only one piece of weight-restricted luggage? How did you divide up (allocate) your "mass budget"? If you forgot something (like a toothbrush), you probably "saved the trip" by just purchasing the item at a vacation-spot convenience store. Unfortunately, there are no convenience stores in outer space. We must pack our spacecraft just right at the beginning of its journey and hope that all its equipment functions well. If we are clever in our "packing," we might anticipate the need for "work-arounds" to bypass a component that breaks during the mission. Aerospace engineers use reliable, low-mass equipment and design flexibility to provide some level of redundancy to a spacecraft. During the mission, they can send instructions to a distant, malfunctioning spacecraft and help the platform "fix itself." Sometimes human controller ingenuity and spacecraft redundancy sustain a mission despite the appearance of unexpected glitches; sometimes the unanticipated problems are simply too severe.

Spacecraft requirements do not exist independent of the mission. Once we decide what we want to accomplish on a particular mission in space, we can establish the design parameters for the functional subsystems that the spacecraft needs to support the mission-oriented payload. For example, a solar-cell/storage-battery subsystem generally proves quite adequate for satisfying the electric power needs of an Earth-orbiting spacecraft. However, only a nuclear power source (the radioisotope thermoelectric generator) can now supply adequate electric power to a spacecraft exploring any of the outer planets (Jupiter, Saturn, Uranus, Neptune, or tiny Pluto).

As we previously discussed, one of the biggest constraints on any spacecraft is its total allowable mass. Launch-vehicle payload compartments also put restrictions on the volume and shape of a spacecraft. To comply with these restrictions and still build large spacecraft, engineers sometimes design the spacecraft with components that can bend and fold. After the spacecraft is on orbit, they send commands that carefully extend and deploy the articulating components. Of course, a stuck, partially deployed component can prevent or limit the accomplishment of a mission, so engineers design moving spacecraft components with a great deal of ingenuity and care. Have you ever had an umbrella fail when you tried to open it? Imagine a similar thing happening to a large, umbrella-like antenna on a robot spacecraft that is one million kilometers away from Earth and heading for Jupiter.

Functional Subsystems

A spacecraft's functional subsystems support the mission-oriented payload and allow the spacecraft to operate in space, record data, and communicate back with Earth. Engineers attach all the other spacecraft components on the structural subsystem. They often select conventional airframe materials from the aviation industry. Aluminum is by far the most common spacecraft structural material. A wide variety of aluminum alloys exist and provide the designer with a broad range of physical characteristics, such as strength and machinability. A spacecraft's structural subsystem might also contain magnesium, steel, titanium, beryllium, or fiberglass.

The *thermal-control subsystem* regulates the temperature of a spacecraft and keeps it from getting too hot or too cold. This is a very complex problem because of the severe temperature extremes the spacecraft experiences during a typical space mission. While traveling in space, the spacecraft "sees" the Sun as an extremely hot (approximately 5,800 K) source of thermal radiation. It also "sees" deep space as a very low-temperature (approximately 3 K) heat sink. If the spacecraft operates near Earth or another celestial object, it will also "see" the large planetary object as a source of thermal radiation. For example, Earth is a source of thermal (infrared) radiation at a radiating temperature of about 288 K. Spacecraft surface properties, the influence of sink and source temperatures, and the view geometry (what the spacecraft "sees" of another object) all play a significant role in complex radiation-heat-transfer calculations.

In the vacuum of space, radiation heat transfer is the only mechanism by which the thermal energy (heat) flows in and out of the spacecraft. Engineers use radiation-heat-transfer techniques to achieve an acceptable energy balance for the spacecraft while it operates in space. In some special cases, engineers might also include an emergency, open-loop "heat-dump" technique, based on the expulsion of a cooling fluid. However, this approach to thermal control provides only a temporary, very limited solution.

Spacecraft designers use two general approaches to thermal control: passive and active. Engineers choose from the following passive thermal-control techniques: special paints and surface coatings, radiating fins, Sun shields, insulating blankets, heat pipes, and spacecraft geometry. Thermal conductivity normally controls the flow of heat between adjacent, interior spacecraft components and from these components to the spacecraft's outer surface. Aerospace engineers can also apply active techniques within a spacecraft's thermal-control subsystem. Some of the more common active techniques involve electrically powered heaters and coolers, louvers and shutters, and closed-loop fluid pumping. The engineer's overall objective remains the same—to keep the temperature of the spacecraft and all its sensitive components within acceptable levels throughout the mission.

Engineers place a computer on board a spacecraft to manage its overall activities and keep track of time. This computer interprets commands from Earth; collects, processes, and formats mission data for transmission to Earth; and manages fault-protection systems, designed to protect the overall system if something goes wrong. Spacecraft designers call this computer the spacecraft's *command and data-handling subsystem*. The spacecraft clock is an integral part of this subsystem. The clock, usually a stable electronic circuit in the one-megahertz (MHz) range, meters the passage of time during the life of the spacecraft and regulates nearly all of its activities. Often, spacecraft controllers use a specific clock count to execute a command or to start downlinking (transmitting) data.

A spacecraft's *telecommunications subsystem* links it to Earth by means of radio signals. We call the radio signal we send to a spacecraft the *uplink* and the radio signal sent from the spacecraft the *downlink*. Because a spacecraft has only a limited amount of power available to transmit a radio signal that sometimes must travel a million or even a billion kilometers, aerospace engineers often concentrate all this power into a narrow beam that they transmit toward Earth. The spacecraft's dish-shaped, high-gain antenna (HGA) accomplishes this task. The term *gain* refers to an amplification or increase in signal strength. Engineers sometimes include a low-gain antenna as well to give the spacecraft almost omnidirectional telecommunications coverage.

We call downlink-only communications with a spacecraft *one-way communications*. Mission controllers say that *two-way* telecommunications occur when the spacecraft receives an uplink signal at the same time a downlink signal arrives on Earth. Even the concentrated, high-gain-antenna signals from a distant spacecraft contain very small power levels. To permit telecommunications with very distant spacecraft, we use special, very large (70-meter-diameter) radio receivers here on Earth. These sophisticated radio antennas, such as those in NASA's Deep Space Network (DSN), can detect, track, and record the very low-power signals from spacecraft at the edges of the solar system. NASA's three-station Deep Space Network consists of several highly sensitive radio antennas, strategically positioned about 120 degrees apart on three continents (North America, Europe, and Australia). Therefore, as Earth rotates on its axis, the DSN can keep in continuous contact with a very distant spacecraft. Even at the speed of light, a radio signal from a far-traveling robot explorer at the edge of the solar system will take hours to reach Earth.

Uplink or downlink communications may consist of a pure radio-frequency (RF) tone (called the carrier signal), or we can modify the carrier signal to carry additional information in each direction. Engineers modulate a spacecraft's carrier signal by shifting its phase, frequency, or amplitude, thereby imposing new information in the form of subcarrier signals. We call the respective processes phase modulation (PM), frequency modulation (FM), and amplitude modulation (AM). Engineers call data modulated onto the downlink signal *telemetry*. This term describes the process of making measurements at one point and then transmitting the data to a distant location for evaluation and use. Telemetry data from a spacecraft include its science or mission-related data, as well as the spacecraft's subsystem state-of-health data. Spacecraft controllers also modulate commands (as binary data) on the uplink carrier signal. When we send a burst of commands to a spacecraft, we call the telecommunications process an *upload*. Similarly, when a spacecraft sends a burst of telemetry data to Earth, we call the process a *download*. A modem (*mo*dulator/*dem*odulator device)

A view of the 70-meter-diameter antenna of the Canberra Deep Space Communications Complex, located outside Canberra, Australia. This facility is one of the three complexes that comprise NASA's Deep Space Network (DSN). The other complexes are located in Goldstone, California, and Madrid, Spain. The national flags representing the three DSN sites appear in the foreground of this image. Image courtesy of NASA.

on either the spacecraft or on Earth detects these modulated subcarrier signals and processes such data separately from the main carrier radio signal. We make excellent use of the same technique to communicate with each other here on Earth. The facsimile (fax) machine scans images and text and transmits them as digitized data over a telephone line; a personal computer's audio-frequency modem uses the telephone line to provide access to the Internet and its multimedia collection of information.

The *stabilization subsystem* carefully controls a spacecraft's attitude, or orientation in space. This system is vital for communications and data collection. We want to accurately point the spacecraft's high-gain antenna back to Earth. We also need to precisely point the onboard instruments for accurate data collection and subsequent interpretation and analysis of the collected data. Only a properly pointed spacecraft can execute the precise propulsive maneuvers needed to fine-tune its trajectory and make it arrive at the desired mission location.

Aerospace engineers stabilize a spacecraft by spinning it or by giving it a three-axis stabilization subsystem. With spinning, the gyroscopic action of the rotating spacecraft's mass provides the stabilization mechanism. Engineers will then fire tiny thrusters to provide any necessary changes in the spin-stabilized spacecraft's attitude. Similarly, tiny thrusters gently nudge a three-axis stabilized spacecraft back and forth within a deadband of allowed attitude error. Engineers also use electrically powered reaction wheels, called momentum wheels, to achieve three-axis stability.

Either approach to spacecraft stabilization involves advantages and disadvantages. Spin-stabilized spacecraft provide a continuous sweeping motion. This motion is quite suitable for certain types of field and particle instruments, but complicates pointing electro-optical instruments and high-gain antennas. Engineers often have to design a complex mechanism that "despins" a portion of the spacecraft so instruments can point at a target and the high-gain antenna can lock on Earth. In contrast, a three-axis stabilized platform can precisely point instruments and antennas but must then rotate certain types of science instruments to achieve useful measurements in all directions.

The spacecraft's attitude-control and articulation-control subsystem (AACS) manages all the tasks involved in platform stabilization. It communicates with the navigation and guidance subsystem to make sure the spacecraft is maintaining the desired attitude as a function of mission trajectory. Celestial reference (star trackers) and inertial reference (gyroscopic) navigation data tell the spacecraft where it is and how it is "tilted" with respect to the planned flight path.

The AACS also closely interacts with a spacecraft's propulsive subsystem and makes sure that the spacecraft points in the right direction before a major rocket-engine burn or a sequence of tiny thruster firings takes place. Minor attitude adjustments usually take place automatically, with the spacecraft essentially driving itself through space. However, major trajectory corrections involve direct interaction between the spacecraft and its human controllers, who uplink the appropriate firing commands, make mission-critical decisions concerning large changes in platform attitude, and choose where to point scientific instruments. When the round-trip telecommunications distance between a robot spacecraft and Earth exceeds ten light-minutes, any type of "real-time" supervision by a human operator becomes essentially impossible. Therefore, we must make our far-traveling explorers very smart, very patient, and very fault tolerant.

On some spacecraft, the AACS also controls the articulation of movable components, such as the high-gain antenna, folding solar panels, and scanning electro-optical instruments mounted on a platform. As a robot spacecraft approaches a planetary body, initial (long-distance) imagery data may help scientists target the spacecraft's scanning instruments for maximum scientific return during the close encounter. The spacecraft's operators uplink appropriate commands to the AACS, and the spacecraft responds by positioning itself so the appropriate instruments point at the precise planetary location that scientists want to investigate.

The *power subsystem* satisfies all the electric power needs of the spacecraft. Engineers commonly use a solar-photovoltaic (solar-cell) system in combination with rechargeable batteries to provide a continuous supply of electricity. The spacecraft also has a well-designed, built-in electric utility grid that conditions and distributes power to all onboard consumers.

A solar cell directly converts sunlight (solar energy) into electrical energy by means of the photovoltaic effect. Engineers combine large numbers of solar cells to increase the electric power output. They call a collection of solar cells a *solar array* or a *solar panel*. The solar cell has no moving parts and produces no pollution when generating electricity. However, the ionizing-radiation environment in space can damage solar cells and significantly reduce their useful lifetime. Solar arrays work very well on Earth-orbiting spacecraft and on spacecraft that operate in the inner solar system (within the orbit of Mars). Rechargeable batteries provide electric power when the spacecraft's solar panels cannot view the Sun.

Some spacecraft must operate for years in deep space or in very hostile planetary environments, where a solar-photovoltaic power subsystem proves infeasible. Under these mission circumstances, the aerospace engineer uses a dependable, long-lived nuclear power supply called the radioisotope thermoelectric generator (RTG) to provide continuous electric power to the spacecraft. The RTG converts the decay heat from a radioisotope directly into electricity by means of the thermoelectric effect. The United States uses the radioisotope plutonium 238 as the nuclear fuel in its spacecraft RTGs.

How much electric power does a spacecraft need? Aerospace engineers know from experience that a sophisticated robot spacecraft needs between 300 and 3,000 watts (electric) to properly conduct its mission. Small spacecraft, with less complicated missions, might require only 25 to 100 watts (electric). However, the less power available, the less performance and flexibility the engineers can build into the spacecraft. The *International Space Station* (*ISS*) generates 110 kilowatts (electric) to serve the needs of the crew, their support equipment, and scientific experiments.

Types of Spacecraft

Scientific Spacecraft

Aerospace engineers custom-design scientific spacecraft to meet the widely varying needs of the scientific community. The result is an assortment of interesting platforms that come in all sizes and shapes. Most are robot spacecraft, but during Project Apollo, NASA engineers built a special spacecraft called the lunar excursion module (LEM) to carry human explorers to the surface of the Moon and then return them safely to lunar orbit. For convenience, we will limit our discussion to the characteristics of robot scientific spacecraft. We can group these scientific spacecraft by how they physically perform the mission. Scientific spacecraft include flyby spacecraft, orbiter spacecraft, atmospheric probes, atmospheric balloon packages, lander spacecraft, surface-rover spacecraft, and surface-penetrator spacecraft.

Flyby spacecraft follow a continuous trajectory to the target planetary body, but the planet's gravity does not capture the spacecraft into an orbit around it. Engineers must give the flyby spacecraft an ability to use its onboard instruments while swiftly passing the target—perhaps even compensating for the target's apparent motion in an optical instrument's field of view. Have you ever tried to take a picture from a moving automobile or bus? If the object of interest is far away, you might be able to obtain a picture that does not have much far-field distortion, but what about an object that is nearby as you zip past? Most of us will usually create disappointing blurs. Our smart robot spacecraft has to help its instruments make the right adjustments, or else we get very blurred planetary images.

The flyby spacecraft has several other demanding design constraints. It must be able to transmit data at high rates back to Earth and also be able to store large amounts of data for those periods when its high-gain antenna is not pointing toward Earth. Engineers also have to design the craft so it can cruise through interplanetary space in a quiet, powered-down mode for months or years and then suddenly spring to life for a critical planetary encounter that will last for only a few hours or even minutes. NASA uses the flyby spacecraft during the initial, or reconnaissance, phase of solar-system exploration. At present, only tiny Pluto and its interesting companion moon, Charon, await the initial visit of a flyby spacecraft. In the first three decades of the space age (from about 1962 to 1989), NASA's flyby spacecraft visited all the other major planetary bodies in our solar system. This is a remarkable achievement in the scientific application of space technology.

Engineers design an *orbiter spacecraft* to travel to a distant planet and then orbit around that planet. The spacecraft needs a substantial onboard propulsion capability to decelerate (retrofire) at just the right moment to have the planet's gravity capture it into a useful orbit. Once the spacecraft is in a proper orbit, it begins a detailed observation of the planet. This period of scientific observation normally lasts for a year or more. Because the orbiter circles the target planet, it experiences periods of occultation when the planet shadows the Sun or prevents telecommunications with Earth. Engineers must, therefore, design this type of spacecraft for uninterrupted power production, effective thermal control, and scientific data storage during these periods.

Some space-exploration missions deploy one or more smaller, instrumented spacecraft, called *atmospheric probes*. While the main spacecraft (the "mother ship") is approaching the target planet, the probe separates from it and follows a ballistic trajectory into the planet's atmosphere. As the probe descends, it collects scientific data that it transmits back to the mother ship. The main spacecraft relays these data immediately back to scientists on Earth or else stores them for later transmission. Scientists regard the probe as an expendable scientific instrument whose descent provides a one-time data-collection opportunity. The probe's mission ends when it either crashes into the planet's surface or else encounters excessively high pressures in the dense, lower regions of the planet's atmosphere.

Engineers design an *atmospheric balloon package* to hang from a buoyant gas-filled bag that floats under the influence of the winds in a planet's atmosphere. A spacecraft releases a protective capsule containing the instrument package and an uninflated balloon on a ballistic trajectory into the atmosphere. Parachutes deploy and slow the capsule's descent. At a suitable altitude, the slowly descending capsule releases a self-inflating balloon and its instrument package. Engineers include a battery power supply and transmitter in the instrument package. As the balloon drifts, the instrument package sends scientific data back to the mother ship. The "floating spacecraft" measures the composition, temperature, pressure, and density of the alien world's atmosphere for some extended period. Exploration continues until the balloon fails, the power supply gives out, or the mother ship departs.

A *lander spacecraft* makes a soft landing on another world and survives at least long enough to transmit useful scientific data back to Earth. These data include panoramic images of the landing site, in situ measurements of the local environment, and an examination of soil composition. One of the most successful robot explorers built in the twentieth century was NASA's Viking Lander. In 1976, two of these nuclear-powered spacecraft successfully touched down on Mars and studied the surface of the Red Planet extensively for many months.

A *surface-rover spacecraft* is a robot vehicle that can move away from the landing site and perform experiments. Like the lander spacecraft, the rover makes a soft landing on the surface of a planet. However, once there, it leaves the landing vehicle and begins exploring the surrounding area. Engineers often design a rover to operate in partnership with a stationary lander spacecraft. The smaller rover arrives nestled inside the lander and then deploys from the lander. While the lander serves as the base camp and relays data to and from Earth, the rover scampers about the local area. Mission scientists can explore the alien landscape by teleoperating (remotely controlling over a great distance) the rover. After reviewing panoramic images of the landing site collected by the lander, the scientists will direct the rover to those locations that appear especially interesting. They will also help the rover survive hazards by choosing what looks like the safest route. However, the rover should have some degree of machine autonomy and obstacle avoidance, because at interplanetary distances there is a significant time delay in round-trip telecommunications. For example, depending on the relative positions of Earth and Mars, an average time delay of 10 to 15 minutes occurs when we send a round-trip radio signal between these two worlds. As a result, by the time a human controller "sees" (through the rover's cameras) an unanticipated crevasse, the obedient, remotely controlled mechanical explorer has probably driven over the edge. We can avoid this problem if we give the rover a significant level of artificial intelligence. Our "smart" robot can then travel slowly, but safely, on its own without direct human guidance. When the rover encounters a possible problem or an unusual surface condition, it stops immediately, sends an alert signal, and patiently waits for human instructions. In 1997, NASA's *Mars Pathfinder* mission successfully demonstrated the use of a robot-lander/rover-spacecraft combination. The mission featured semiautonomous surface-rover operation and slow surface movement through cautious teleoperation by human controllers on Earth.

Aerospace engineers design a *surface-penetrator spacecraft* much as they would a well-instrumented, impact-tolerant, steel spear. Approaching the target planet or moon, the carrier spacecraft (mother ship) releases the pointed, slender projectile. The penetrator follows a ballistic trajectory, descends through any planetary atmosphere, and impacts on the planet's surface. The high-velocity impact partially buries the slender spacecraft. The front (forebody) of the penetrator contains impact-resistant instruments designed to investigate the subsurface environment; the back portion (afterbody) remains on the surface, makes surface environmental measurements, and communicates with the mother ship. A flexible, data-carrying cable connects the two sections. We can make a variety of interesting measurements with a well-instrumented penetrator. These measurements include seismic activity, surface meteorology, and surface/subsurface–characterization studies involving heat flow, soil moisture content, and geochemistry.

We might use several penetrators to perform network science on those planets and moons with particularly interesting solid surfaces. The Jovian moon Europa represents one exciting candidate because of the possibility of a liquid-water ocean beneath its smooth icy surface. Mars is another excellent candidate for detailed investigation by a network of penetrator spacecraft. An orbiting mother ship can collect data from each penetrator and relay the findings of the entire network back to Earth.

Earth-Observing Satellites

Earth-observing satellites are robot spacecraft that use remote sensing technology to routinely collect a wide variety of data across the electromagnetic spectrum. These data support national defense, scientific research, environmental monitoring, meteorology, and commercial activities. Aerospace engineers custom-design each Earth-observing satellite (or family of such spacecraft) to serve the needs of a particular user. For example, Earth-orbiting military satellites apply remote sensing techniques to monitor hostile regions, to support friendly forces, and to verify treaties. Aerospace engineers build civilian satellites with similar (but often less sensitive) remote sensing instruments. The growing family of civilian environmental Earth-observing spacecraft has greatly improved meteorology, provided new insights about the complexities of our home planet (Earth system science), and created a library of high-quality multispectral images of Earth's surface. Multispectral images of Earth support many scientific, environmental-monitoring, and innovative commercial applications.

In some cases, the U.S. Department of Defense pioneered a new type of space-platform technology and then converted the military spacecraft to civilian and scientific applications. The very successful family of military weather satellites, known as the Defense Meteorological Satellite Program (DMSP), is an excellent example of military-to-civilian spacecraft-technology transfer. These polar-orbiting military weather satellites take daily close-up visual and infrared images of the cloud-cover conditions for every region of the world. Military weather forecasters use this imagery to detect developing weather patterns throughout the globe. In particular, DMSP helps them identify, locate, and estimate the severity of thunderstorms, hurricanes, and typhoons, since such severe weather conditions can greatly influence military operations.

Remote sensing is the examination of an object, phenomenon, or event without having the sensor in direct contact with the object under study. Information flows from the object (target) to the sensor by means of electromagnetic radiation. Modern sensors use many different portions of the electromagnetic spectrum, not just the narrow band of visible light we see with our eyes.

We can divide all remote sensing instruments (including those we place on a spacecraft) into two general classes: passive sensors and active sensors. Passive sensors observe the sunlight reflected off an object or the characteristic electromagnetic (EM) radiation emitted by that object. The thermal signature of a rocket's exhaust plume is an example of characteristic infrared radiation emitted by a hot object. Active sensors, like an imaging radar system, provide their own (microwave) illumination on the target and then measure the reflected signals from the target. Engineers employ both passive and active remote sensing instruments to create information-rich, high-resolution images of a scene. We call an image that simultaneously examines the same scene in several distinct spectral bands a multispectral image.

Different sensors respond to different bands or regions of the electromagnetic spectrum, but collectively, modern remote sensing instruments on Earth-observing platforms cover the visible portion of the EM spectrum and extend well into its infrared and microwave regions. However, the intervening effects of Earth's atmosphere dictate what spectral bands we can employ when we want to look at Earth's surface from a satellite.

Scientists also place "nonimaging" remote sensing instruments on Earth-orbiting spacecraft. This type of instrument measures the total amount of radiant energy (within a certain portion of the EM spectrum) that appears in its field of view (FOV). A radiometer detects and measures the total radiant energy within a fairly broad region of the EM spectrum. A spectrometer measures incoming radiant energy as a function of both intensity and wavelength.

Passive sensors collect reflected sunlight or object-emitted radiation. An imaging radiometer detects an object's characteristic visible-, near-infrared-, thermal-infrared-, or ultraviolet-radiation signature and then creates an image of the object. When an imaging radiometer looks at Earth from a spacecraft, the intervening atmosphere restricts its view of our planet's surface to certain wavelength bands, called atmospheric windows. (Imaging radiometers on scientific spacecraft observing other planetary bodies may have similar surface-viewing restrictions or may be totally unhampered if there is no appreciable intervening atmosphere.) An atmospheric sounder is an example of a nonimaging remote sensor. It looks down through a column of air and collects the radiant energy (typically at infrared or microwave wavelengths) emitted by certain atmospheric constituents, like water vapor and carbon dioxide. Meteorologists evaluate sounder data from which they then infer atmospheric temperature and humidity values.

Earth-orbiting platforms provide a synoptic (comprehensive) view of our planet's surface, its oceans, and atmosphere totally unhindered by natural or political boundaries. Unlike aircraft flying through the "bumpy" atmosphere, robot spacecraft travel through space without vibration. A vibration-free platform greatly enhances the quality of data collected by remote sensing instruments.

Two general factors influence satellite-based remote sensing: the orbit we select for the space platform and the portions of the electromagnetic spectrum we want our sensors to operate in. In choosing spectral regions, we must remember that Earth's atmosphere and prevailing sunlight conditions influence the performance of certain sensors. Other instruments, like an imaging radar system, operate independent of sunlight conditions and experience minimal interference from Earth's atmosphere. We can create quality radar images of Earth's surface day or night, even when the atmosphere is so full of clouds that we cannot see the ground from space.

Depending on the orbital path we select for our space platform, we can look at regions on Earth continuously or else visit them only briefly, but on a regular, repetitive basis. (We will discuss the physics of satellite motion shortly.) For example, we can make a satellite appear to "stand still" over a point on Earth's equator. A geostationary satellite takes as long to complete an orbit of Earth as it takes our planet to complete one rotation on its own axis. Certain types of military surveillance satellites, weather satellites, and global communications satellites take full advantage of this special property of the geostationary orbit. These platforms succeed in their respective missions because they can continuously view a very large (hemispheric-size) portion of the globe.

However, in remote sensing, distance also makes the target's signature grow weaker. A satellite in geostationary orbit has an altitude of 35,900 kilometers above Earth's equator. At this altitude, some optical instruments become severely limited in their spatial resolution. The spatial resolution of an optical instrument is the size (physical dimensions) of the smallest object it can detect. Different remote sensing data applications (military, scientific, or commercial) require different levels of spatial resolution.

Engineers help solve the spatial-resolution problem by placing imaging instruments on satellite platforms that orbit Earth at much lower altitudes. One very useful low-altitude orbit is the polar orbit. An Earth-observing spacecraft in a polar orbit has an inclination (the number of degrees the orbit inclines away from the equator) of approximately 90 degrees and travels around our planet alternately in north and south directions. Because Earth is rotating on its axis from west to east beneath it, a polar-orbiting spacecraft eventually passes over the entire surface of the planet. If we put an imaging instrument on a low-altitude polar satellite, the instrument will eventually collect high-spatial-resolution data for every location on Earth. However, we must pay a "coverage" penalty. Because of the physics of satellite motion, the observation time over a particular spot on Earth becomes quite limited during each orbital pass. When we select the orbital parameters for our polar satellite, we also determine its revisit time over a particular location. This is usually several days. If we want to observe a special place "up close" and "frequently" from space, we must fly a constellation of satellites (several satellites of the same type). Through orbital mechanics, we then arrange the time of each spacecraft's visit to this interesting location to match our remote surveillance needs. Government agencies use the polar orbit's special properties for military reconnaissance (spy) satellites, weather satellites, and environmental-monitoring satellites.

In the context of scientific environmental monitoring, an *Earth-observing satellite* (EOS) carries a specialized collection of sensors that simultaneously monitor many important environmental variables. Scientists select two types of orbits for these important spacecraft: either geostationary orbit, which provides a continuous hemispheric view of Earth, or a low-altitude polar orbit that systematically provides a closer view of every part of our planet, including its most remote regions, like the Arctic and the Antarctic. We often refer to these spacecraft simply as environmental satellites or green satellites. Weather satellites form a very important subclass of the environmental satellites. They were the first type of civilian Earth-observing spacecraft and still provide important data for day-to-day weather forecasting and tropical-storm warning. Some of the environmental variables studied by modern, highly instrumented Earth-observing spacecraft include the following:

- Cloud properties
- The energy exchange between Earth and space

- The planet's surface temperature (land and sea)
- The structure, composition, and dynamics of the atmosphere, including wind, lightning, and rainfall (precipitation)
- The accumulation and melting (ablation) of snow
- Biological activity on land and in near-surface waters
- The circulation patterns of the world's oceans
- The exchange of energy, momentum, and gases between Earth's surface and its atmosphere
- The structure and motion of sea ice
- The growth, melting, and flow rates of glaciers
- The mineral composition of exposed soils and rocks
- The changes in stress and surface elevation around global faults
- The input of radiant energy and energetic particles to Earth from the Sun

Of course, no one space platform monitors all of these important environmental variables. Instead, engineers construct a variety of Earth-observing satellites. Some, like NASA's *Upper Atmosphere Research Satellite* (UARS), have a very specific, narrowly focused mission. Others, like NASA's *Terra* spacecraft (previously called the *Earth-Observing Satellite (EOS)–AM*), simultaneously collect comprehensive sets of environmental data that allow scientists to thoroughly investigate the coupled interactions of Earth's major natural systems (such as the atmosphere and the hydrosphere) in a way never before possible. The exciting new multidisciplinary field of Earth system science (ESS) exists and flourishes because of Earth-observing spacecraft. Simultaneity of environmental data collection is essential for scientists who study Earth as an integrated system.

The *weather (meteorological) satellite* was the first civilian application of an Earth-observing spacecraft. Today, meteorologists employ two general classes of weather satellite: geostationary and polar, named after their respective operational orbits. Modern weather satellites observe and measure a wide range of atmospheric properties and processes to support weather forecasting and warning. Imaging instruments provide detailed pictures of clouds and cloud motions, as well as measurements of sea-surface temperature. Infrared and microwave sounders provide meteorologists profiles of atmospheric temperature and moisture content as a function of altitude. Other instruments measure ocean currents, sea-surface winds, and the extent of snow and ice cover.

The polar-orbiting *Earth resources satellites* collect high-resolution multispectral images of our planet's surface. These multispectral images represent an important data set for resource managers, urban planners, and environmental scientists. NASA launched the world's first Earth resources satellite, called the *Earth Resources Technology Satellite–A* (*ERTS-A*), on July 23, 1972. Later renamed *Landsat-1*, this pioneering spacecraft established an important Earth-observing tradition that continues today with the high-resolution, multispectral imagery from the *Landsat-7* spacecraft, launched on April 15, 1999. Careful evaluation of these multispectral images helps us understand and monitor natural and people-caused (anthropogenic) changes in Earth's physical, biological, and human environments.

Military satellites also observe Earth. The *reconnaissance (spy) satellite* collects high-resolution images of denied (hostile) areas from space. Satellite-based remote sensing for military purposes started in August 1960, when the U.S. Air Force successfully launched its (then) top secret *Corona XIV* spacecraft. The photoreconnaissance satellite became an integral part of American national defense. During the Corona program (1960 to 1972), the quality of reconnaissance-satellite imagery improved from an initial ground resolution of between 7.6 and 12.2 meters to a final ground resolution of about 1.8 meters. In other words, the satellites could pick out something the size of a small car. The capabilities of current photoreconnaissance satellites are classified, but by comparison, current civilian systems have a resolution of about 1.0 meter.

The U.S. military also employs *surveillance spacecraft*. The early warning satellite detects and reports the launch of an enemy ballistic missile attack. Surveillance spacecraft in geostationary orbit have special infrared sensors that continuously scan Earth's surface (land and ocean) for the hot exhaust-plume signatures characteristic of ballistic missile launches. The U.S. Air Force started development of infrared-sensor surveillance spacecraft in 1966. Since 1970, a family of Defense Support Program (DSP) missile surveillance satellites has formed the main element of the U.S. early warning program. Providing 24-hour worldwide surveillance from geostationary orbit, these remote sensing sentinels stand ready to alert national authorities about any intercontinental ballistic missile (ICBM) attack.

Other Earth-Orbiting Spacecraft

Free from interference by Earth's atmosphere, *orbiting astronomical observatories* can collect incoming signals within the electromagnetic spectrum. These space observatories are robotic platforms that carry advanced, highly sensitive instruments. Engineers design each spacecraft to investigate a particular region of the EM spectrum. For example, NASA's Hubble Space Telescope (HST) (1990–present) is the most complex and sensitive orbiting optical telescope ever made. This observatory is approximately the size of a railroad car. This powerful orbiting observatory lets scientists view the visible universe out to distances (and therefore back to times) never before obtained. HST imagery has revolutionized optical astronomy and made many significant contributions to astrophysics and cosmology.

Similarly, NASA's Compton Gamma Ray Observatory (CGRO) (1991–2000) provided the most comprehensive look at the universe in the gamma-ray portion of the EM spectrum. The newest of NASA's great observatories, the Chandra X-Ray Observatory (CXRO) (1999–present), is the most sophisticated X-ray observatory ever built. Earth's atmosphere absorbs X-ray and gamma-ray signals before they can reach astronomers on our planet's surface. Orbiting observatories, such as the CGRO and CXRO, represent the only way scientists can perform a sustained and comprehensive study of the universe within the most energetic portions of the EM spectrum. Gamma-ray astronomy reveals the explosive, high-energy processes associated with such astrophysical phenomena as supernovas, exploding galaxies, quasars, pulsars, and black holes. X-ray emissions are also associated with very energetic, violent processes occurring in the universe.

Earth-orbiting observatories study the universe in other specialized regions of the EM spectrum, including microwave, infrared (IR), ultraviolet (UV), and extreme ultraviolet (EUV). NASA's *Cosmic Background Explorer* (*COBE*) (1989–1990) detected the cosmic microwave background (at about 2.7 degrees Kelvin), providing scientific evidence of the radiation remnants of the big-bang explosion that started the current universe.

A *communications satellite* is an Earth-orbiting platform that relays signals between two (or more) stations on Earth. There are two general classes of communications satellites: active and passive. The passive communications satellite simply reflects radio signals from one Earth station to another. In 1960, NASA launched *Echo 1*, a giant, inflatable balloon that became the world's first passive communications relay satellite. The active communications satellite receives, regulates, and retransmits radio-frequency signals between stations on Earth, at sea, or in the air. The active communications satellite is the type used to support the global communications infrastructure.

The commercial communications-satellite industry started with the successful launch of *Early Bird* (*INTELSAT 1*) into geostationary orbit on April 6, 1965. Today, many advanced-design active communications satellites use geostationary orbit to provide a wide variety of information-transfer services to a global marketplace. As Arthur C. Clarke predicted in 1945, there are many technical advantages of placing large, multifunctional communications satellites in a fixed (geostationary) position above Earth. For example, just three strategically placed spacecraft establish a global communications network.

We can use other orbits around Earth to satisfy special communications needs. The Russian government pioneered the use of a novel, highly elliptic 12-hour orbit, called the Molniya orbit, in 1965. A satellite in a Molniya orbit (named after the Russian family of communications satellites) spends the majority of its time above the horizon in view of the high northern latitudes and very little of its time over southern latitudes. Two such spacecraft can provide continuous communications service to the sprawling, northern-latitude regions of the Russian Federation. Similarly, companies seeking to create a global cellular communications network can establish an appropriate constellation of many (perhaps 50 to 70) small satellites. This "swarm" of radio-frequency-linked spacecraft travels around Earth in polar orbit at an altitude of about 650 kilometers. Interaction with just one spacecraft results in access to the networked constellation. Citizens of the twenty-first century have an insatiable demand for information-transfer services. Through modern space technology, the communications satellite plays a major role in connecting the world and creating a global village.

A *navigation satellite* orbits Earth in a well-known position and broadcasts a precise radio-frequency signal. With the proper equipment, users receiving these special signals from several (at least four) navigation satellites in the operational constellation can calculate their location, the time, and their relative velocity (if they are moving) anywhere on Earth, at sea, or in the air. We measure the distance from any particular navigation satellite in the constellation in terms of "transit time," the time it takes the satellite's signal to reach us. To measure the true transit time of a signal from a navigation satellite to our receiver, the clock in the satellite and the clock in our radio-signal receiver need to be precisely synchronized. Two key technologies made possible the navigation satellite: ultrastable spacecraft clocks and very stable space platforms in precisely known, predictable orbits.

The U.S. Air Force developed the Global Positioning System (GPS) for the Department of Defense to serve the navigation needs of American military forces throughout the world. For example, during Operation Desert Storm (the 1991 Gulf War), allied troops relied heavily on GPS data to navigate precisely through the featureless regions of the Saudi Arabian desert. When the air force developed GPS, civilian use was only a sec-

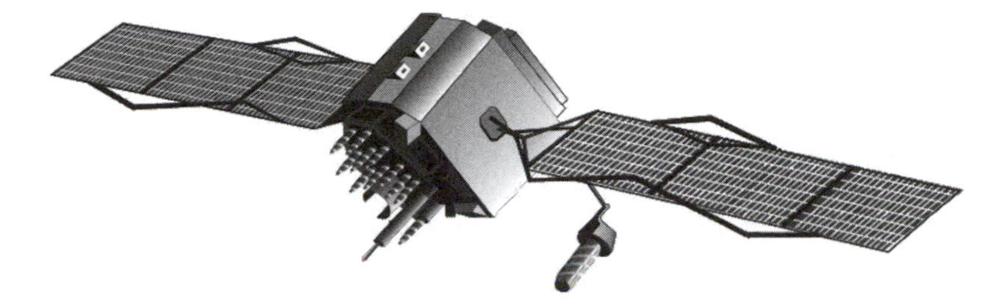

Figure 4.10 A detailed drawing showing the current Global Positioning System (GPS) satellite developed and operated by the U.S. Air Force. Illustration courtesy of U.S. Air Force.

ondary consideration. Nevertheless, civilian applications of satellite-based navigation have become a billion-dollar global industry (see Figure 4.10).

The baseline GPS satellite constellation consists of 24 identical spacecraft, deployed in circular 20,350-kilometer-altitude orbits around Earth. U.S. Air Force satellite controllers operate these spacecraft in six orbital planes inclined at 55° to the equatorial plane. Each orbital plane contains four satellites, distributed in a somewhat unequal fashion. The complex constellation allows a user anywhere in the world (on land, at sea, or in the air) to simultaneously view at least four GPS satellites. Each spacecraft broadcasts radio-ranging signals and a navigation message. With the right receiver, we can use the signals from four satellites (at a minimum) to calculate our location and (if we are moving) our velocity.

The U.S. government operates GPS in two basic modes: the authorized Department of Defense–user mode provides an encrypted radio signal that generates extremely precise location data to support American military operations. The civilian mode involves a degraded radio signal that produces less precise location data. The dual-signal strategy prevents forces hostile to the United States from taking full advantage of GPS navigation data in time of war. The signals provided by GPS are so accurate that for authorized (military) users, time can be calculated to 100 nanoseconds accuracy, horizontal position to 22 meters accuracy, and vertical position to 27.2 meters accuracy. Civilian users throughout the world enjoy use of the "degraded" GPS signal data without charge or restriction. With most civilian receivers, the degraded signals result in a time accuracy of 340 nanoseconds, a horizontal-position accuracy of 100 meters, and a vertical-position accuracy of 156 meters.

Civilian and military engineers have developed a wide variety of GPS receivers for use in aircraft, ships, and land vehicles, as well as for handheld applications by individuals in the field. Satellite-based navigation is one of the fast-growing segments of the information-technology industry.

PHYSICS OF ORBITING BODIES

Earlier in this chapter, we discussed how we launched a satellite into orbit with a rocket. We also used the analogy of a cannon firing a projectile from a very tall tower (see Figure 4.9). Let us now resume that discussion, but this time let us assume that the cannonball leaves the cannon with a velocity that slightly exceeds the minimum horizontal (tangential) velocity needed to place the projectile in a circular orbit around Earth. What happens to the path of the projectile in this case? The cannonball makes a stretched-out circle, called an ellipse, in its path around Earth. In this elliptical path, sometimes the cannonball (or satellite) is nearer to Earth than at other times. We call the point at which our satellite's orbital path comes nearest to Earth the *perigee* of the orbit. The point at which the satellite is farthest away from Earth we call the *apogee* of the orbit (see Figure 4.11). (For objects orbiting the Sun, we use the term *aphelion* to describe the point in the orbit farthest from the Sun and *perihelion* for the point nearest the Sun.) If we continue to increase the horizontal velocity of the cannonball, it will eventually reach a velocity that allows it to completely escape Earth's gravitational attraction. The projectile (or satellite) then travels on a path called a hyperbola and never returns to Earth.

We must understand the physics of how objects move in space if we wish to launch, control, and track a spacecraft and to predict the interplanetary motion of natural and human-made objects. Engineers and scientists call this branch of mathematical physics orbital mechanics.

We begin our brief discussion of orbital mechanics by introducing some very important ideas and terms. An *orbit* is the closed path in space along which a smaller object (called the secondary) moves around a much larger object (called the primary). When viewed from space, a single orbit is a complete path around the primary. However, sometimes the primary also rotates on its own axis. A single orbit is different from a revolution. We define a revolution as the condition when an orbiting object passes over the longitude (or latitude) on the primary from which it started. For example, the space shuttle *Atlantis* completes a revolution of Earth whenever it passes over approximately 80 degrees west longitude (the longitude of its launch site at the Kennedy Space Center). However, while *Atlantis* orbits from west to east around the globe, Earth itself is also rotating from west to east. The time *Atlantis* needs to make one revolution is actually longer than its orbital period. This happens because *Atlantis* has to catch

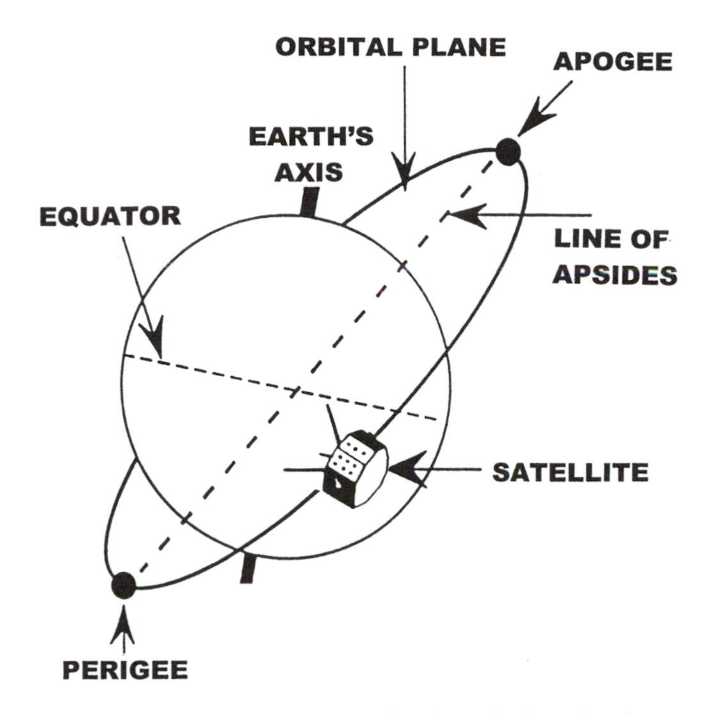

Figure 4.11 Various terms used to describe the orbital motion of a satellite around Earth. NASA image modified by author.

up to the west-to-east movement of the reference longitude line. When a satellite travels in a polar orbit, we say that it completes a period of revolution whenever it passes over the latitude from which it started.

Another important idea we encounter in orbital mechanics is that of the *orbital plane*. We can visualize the orbital plane if we imagine a giant flat plate that cuts Earth in half and around whose outer edge the satellite travels (see Figure 4.11). The *inclination* is the number of degrees the orbit of a satellite is inclined away from the equator. The inclination tells us how far north and south a particular satellite will travel during its orbit around Earth. A satellite in an equatorial orbit has zero inclination. In contrast, a satellite in polar orbit has an inclination of 90 degrees. The *line of apsides* is the straight line passing through the center of the primary (here Earth) that connects the apogee and perigee points of an orbit (see Figure 4.11).

Several basic scientific principles describe the fundamental motions of both celestial objects and human-made spacecraft. One is Newton's law of gravitation; the others are Kepler's laws of planetary motion. Specifically, Newton's law of gravitation tells us that two bodies attract each other in proportion to the product of their masses and inversely as the square of the distance between them. From Newton's law of gravitation, we know that a satellite needs more velocity to stay in a low-altitude orbit than in a high-altitude orbit. For example, if our spacecraft orbits Earth at an altitude of 250 kilometers, it needs an orbital speed of about 7.8 kilometers per second. In contrast, the Moon is about 442,200 kilometers from Earth and orbits our planet at a speed of about 1 kilometer per second. Of course, if we want to lift a spacecraft from Earth's surface to a high-altitude (versus a low-altitude) orbit, we need to expend a great deal more energy.

Any spacecraft launched into orbit moves according to the same laws of motion that govern the movement of the planets around the Sun. In the seventeenth century, Johannes Kepler (1571–1630) formulated three basic laws to describe planetary motion. Kepler's first law tells us that each planet revolves around the Sun in an orbit that is an ellipse, with the Sun as its focus (or primary body). We call Kepler's second law the law of equal areas. This law says that the line from the center of the Sun to the center of a planet (the radius vector) sweeps out equal areas in equal times. It also relates to the motion of a satellite in orbit around Earth. Kepler's third law tells us that the square of a planet's orbital period is equal to the cube of its mean distance from the Sun. We can extend this statement to spacecraft orbiting Earth by saying that a spacecraft's orbital period increases with its mean distance from the planet. In other words, a high-altitude satellite takes longer to go around Earth than one orbiting at a lower altitude.

About a century after Kepler formulated the laws of planetary motion, Newton published the *Principia*, a monumental work containing his law of gravitation and his three laws of motion. Newton's laws provided the mathematical basis for a complete physical understanding of the motion of planets and satellites. Today, engineers and scientists use six parameters (called an object's *orbital elements* or Keplerian elements) to describe its position and path. We call these six elements the semimajor axis of the elliptical orbit (a), the eccentricity (e), the inclination (i) of the orbit plane with respect to the central body's equator, the right ascension of the ascending node (Ω), the argument of perigee (ω), and the true anomaly (θ).

Any detailed discussion of how spacecraft operators use the orbital elements to specify a spacecraft's location and path exceeds the scope and level of this book. Our objective is simply to introduce the basic physical principles of orbital motion and to recognize that calculating the precise orbital motion of a spacecraft depends on complicated mathematical procedures.

Spacecraft controllers encounter many interesting orbital mechanics problems. For example, when they want to transfer a spacecraft from one orbital altitude to another orbital altitude in the same orbital plane, they can do so with a minimum expenditure of energy by using the Hohmann transfer orbit. Sometimes, however, they must do so as quickly as possible. Then they select an energy-intensive, fast-transfer option. They can also change the inclination of a spacecraft's orbit, change the location of its perigee, and even rendezvous with another spacecraft that is in a different orbital plane. To do so, they must expend propulsive energy by providing the right amount of thrust in the right direction at precisely the right time. Sophisticated computer programs help them complete the detailed calculations needed to place a spacecraft in the precise orbital location for a particular mission.

Sending a spacecraft to another planet is a somewhat more challenging task. First, our spacecraft most have enough velocity (the escape velocity) to leave the control of Earth's gravity. The spacecraft then travels through interplanetary space, where it is under the control of the Sun's gravity. Finally, our spacecraft approaches the target planet, whose gravitational attraction influences the end phase of the mission. For the flight between the two planets, we can take advantage of the slower, but more energy-efficient, Hohmann transfer orbit technique. Otherwise, we can select a more rapid, fast-transfer approach that uses a great deal more propulsive energy.

We cannot just decide to leave Earth and head for another planet. Since both planets are moving around the Sun, there are interplanetary transfer windows when it is possible to launch such missions. For some planets, like Venus and Mars, these launch opportunities occur often—about every 1.6 years for Venus and every 2.2 years for Mars. For the more distant, outer planets, these launch opportunities are much less frequent. For example, once every 176 years the giant outer planets (Jupiter, Saturn, Uranus, and Neptune) align themselves in such a pattern that a spacecraft launched from Earth to Jupiter at just the right time could also visit the other three planets on the same mission, using a technique called *gravity assist*. In 1977, NASA scientists took advantage of this rare celestial alignment and sent the *Voyager 2* spacecraft on its unique "grand tour." With an incredible feat of interplanetary navigation, the far-traveling robot spacecraft visited Jupiter (1979), Saturn (1981), Uranus (1986), and Neptune (1989).

We now consider what happens to our spacecraft's trajectory when it encounters the target planet after crossing interplanetary space (see Figure 4.12). We call the first possible trajectory an *impact trajectory*. As shown in Figure 4.12(A), our spacecraft crashes into the planet or makes a hard-impact landing. The second type of trajectory is the *orbital capture trajectory*.

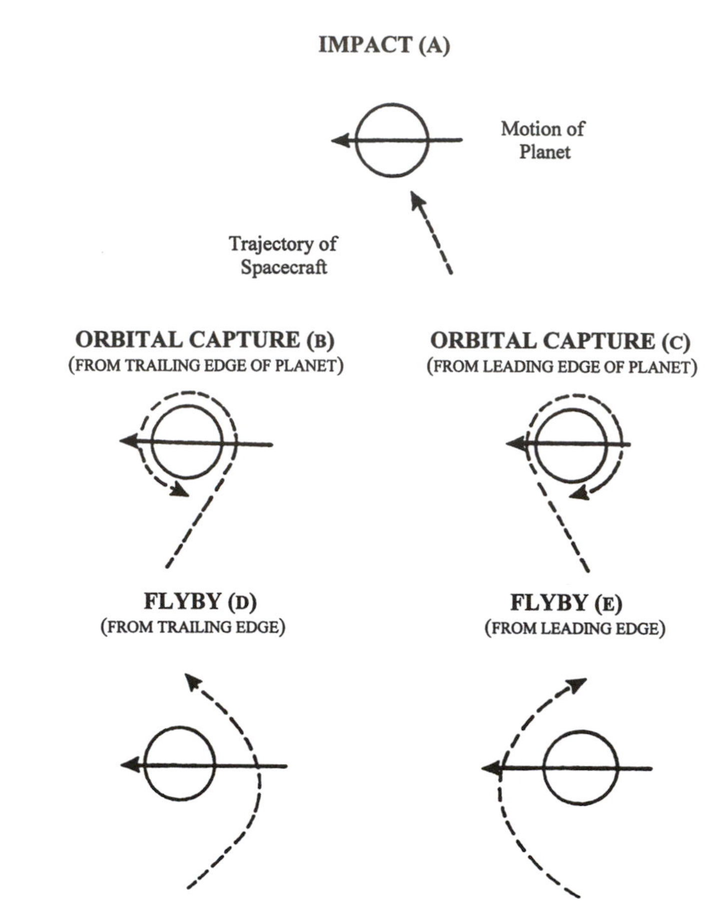

Figure 4.12 Spacecraft planetary-encounter scenarios. NASA artwork modified by author.

The gravitational field of the target planet captures our spacecraft, and it enters orbit around the planet. Depending upon its speed, arrival altitude, and other factors, our spacecraft can enter this captured orbit from either the trailing edge (see Figure 4.12[B]) or the leading edge of the planet (see Figure 4.12[C]). The third type of planetary trajectory is the *flyby trajectory*. Our spacecraft remains far enough away from the planet to avoid gravitational capture, but passes close enough to have the planet's gravity strongly affect its flight path. In this case, our spacecraft experiences an increase in speed if it approaches from the trailing side of the planet (see Figure 4.12[D]) and a decrease in speed if it approaches from the leading side (see Figure 4.12[E]). In addition to changes in speed, the direction of the spacecraft's motion also changes.

The trailing-edge flyby encounter represents the basic gravity-assist maneuver. The increase in the spacecraft's speed comes from an infinitesimal decrease in speed of the target planet. In effect, our spacecraft gets pulled along by the target planet while it performs a flyby trajectory from the trailing side of the planet.

This brief discussion is a greatly simplified treatment of complex encounter phenomena. For example, if the target planet has an atmosphere, this atmosphere could play a significant role (helpful or detrimental) in the gravitational-capture process. While traveling through interplanetary space, our spacecraft might perform several trajectory-adjusting rocket-engine burns. Finally, when it arrives near the target planet, our spacecraft might reverse-fire (retrofire) its onboard rocket engine to assist in the gravitational-capture process, or else briefly fire a thruster to shape and strengthen the gravity-assist, flyby maneuver.

SPACECRAFT FOR HUMAN SPACEFLIGHT

When aerospace engineers design a spacecraft for human passengers, they must devise a vehicle that can protect a person from the temperature extremes, vacuum conditions, and radiation hazards of outer space. They must also keep conditions inside the spacecraft cool and comfortable during the intense aerodynamic heating processes that accompany high-speed reentry.

Human spaceflight is basically a BYOB (bring your own biosphere) activity. The spacecraft's life-support subsystem must maintain life from launch to landing. Depending on the particular mission, the flight environment can include ascent and descent through Earth's atmosphere, extended travel in outer space, and activities on the surface of another world (for example, excursions across the lunar or Martian surface). The life-support subsystem must reliably provide the astronauts with all their daily needs of clean air, potable water, food, and reliable waste removal. A properly functioning life-support subsystem ensures healthy and productive lives while human beings ride inside the spacecraft.

The aerospace engineer designs a modern, human-rated spacecraft with a pressurized cabin or module so the crew has sufficient room to work and live in "shirtsleeve" comfort. When a spacecraft provides this, the human occupants must wear bulky, individual space suits only to perform an extravehicular activity.

Engineers must also design the craft's pressurized cabin/module with sufficient interior space so each crew person has enough room to work, eat, relax, and sleep. A viewport is another important design feature. We know from the space shuttle experience that astronauts often spend their off-duty time engaged in "recreational" viewing of Earth from orbit. Space-station astronauts will do likewise.

Human-factors engineering and psychology play very important roles in designing spacecraft for long-term occupancy. The absence of "personal space" can lead to tension and disagreements with an isolated crew. A well-designed space-station habitation module keeps noisy, group activities in the active zone and acknowledges the need for individual privacy and silence in the quiet zone. Have you ever had the misfortune of sitting in the middle seat between two people who constantly talked on a long-distance airplane flight? Now imagine that instead of just four hours of anguish, you have one full year in Earth orbit to "enjoy" such continuously noisy social interaction. The basic pressurized habitation module could also serve as an important space-technology building block for a lunar-surface base or a Mars expedition vehicle.

Life-support-subsystem technologies deal with three major functions: water reclamation, air revitalization, and waste management. Water reclamation must satisfy the potable-water needs of each crew member. Air revitalization maintains a breathable, comfortable atmosphere of the proper gas composition, temperature, humidity, and pressure. Sea-level pressure and an 80 percent nitrogen and 20 percent oxygen composition create a suitable, familiar atmosphere for the spacecraft's pressurized cabin/module. The air-handling system must also control airborne contaminants of all types. Finally, the waste-management system has to efficiently handle, treat, and dispose of human by-products, both liquid and solid. Sanitary engineering is a challenging problem in the microgravity environment of an orbiting space platform. A constantly malfunctioning toilet will stress even the best-trained crew.

Traditionally, engineers use a "once-through" (open-loop) approach to life-support-subsystem operation. In this approach, they store all the air, food, and water needed by the crew on board the spacecraft at the start of the mission. For extended missions in Earth orbit, they might resupply some of these consumables through the frequent delivery of logistics modules. Astronauts store waste products and trash in appropriate containers and return them to Earth. To keep the amount of stored liquid waste manageable, they jettison liquid waste products into space (for example, an occasional dirty-water dump from the orbiting spacecraft). Because of the growing space-debris problem, they cannot dump solid "trash" into orbit.

The once-through approach may continue to remain acceptable for short-duration missions with small crews. However, as the duration of human-crewed missions grows and the number of crew members in particular spacecraft increases, this approach becomes extremely expensive and impractical.

For long-duration missions to Mars, for example, engineers are considering using a regenerative life-support system (RLS). This system would recycle air and water, important life-sustaining consumables. During an interplanetary expedition to Mars, resupply of these consumables from Earth would be impractical. Instead, the space vehicle's RLS would recover and recycle some (if not most) of the daily human crew needs for water and air. Of course, when we recycle air and water in a small closed system, we must be confident that our recycling system is both reliable and safe.

Later in the twenty-first century, a large space base or orbiting settlement with a thousand or more inhabitants would perform full-scale recycling and treatment of air, water, and waste products. Save for the input of solar energy, the system should approach total closure; that is, everything needed to sustain life would be found inside the settlement, including an energy source. We call a mostly closed or completely regenerative life-support subsystem a *controlled ecological life-support subsystem* (CELSS).

LIVING AND WORKING IN SPACE

Microgravity represents an intriguing experience for space travelers. They can float with ease through a space cabin. They can push off one wall of a space station and drift effortlessly to the other side. They can perform slow-motion somersaults and handsprings and can move "heavy" objects that now feel weightless.

However, life in microgravity is not necessarily easier than life on Earth. Aerospace engineers have to redesign many ordinary things for use under microgravity conditions. For example, a beverage in an open container clings to the container's inner or outer walls. If we suddenly move or shake an open container, a glob of beverage will slide out. If someone or something disturbs this runaway glob of liquid, the crew cabin is filled with hundreds of tiny free-floating droplets. These free-floating liquid droplets are not simply an inconvenience. They can annoy crew members and represent a definite hazard to equipment, especially sensitive electronic devices and computers. Air flows constantly through the crew cabin to collect floating droplets and particles before they cause damage.

Liquids are a problem in space. When we are thirsty here on Earth, we think nothing of pouring water (or another beverage) from a container into a glass. In an Earth-orbiting space shuttle, astronauts cannot do this. To serve water in microgravity, astronauts must use a specially designed dispenser that they turn on or off by squeezing and releasing a trigger. If they want to drink another beverage, such as orange juice, they carefully insert a plastic straw into a sealed container. When they stop sipping, they must clamp the straw shut. Otherwise, the fluid can creep up and out of the straw by capillary action.

Microgravity living also calls for special considerations when we eat foods. Foods that crumble come in bite-size pieces. This prevents leftover crumbs from floating around the pressurized cabin. Gravies, sauces, and dressings have a viscosity (stickiness) that usually prevents them from lifting off food trays and floating away. Engineers equip the space food trays with magnets, clamps, and double-adhesive tape to hold metal, plastic, and other utensils. Astronauts can use forks and spoons in orbit, but they must learn to eat without sudden starts and stops so their lunch does not float away.

The kinds of food the astronauts eat are not mysterious concoctions but foods prepared here on Earth. Each crew member selects his/her menu from many choices, designed to supply the crew with all the recommended dietary allowances of vitamins and minerals necessary to perform in the environment of space. Space-station crew members can select 30-day flight menus. The crew stores the food they have chosen in the galley on board the space station. There are eight general categories of space food: rehydratable food (e.g., oatmeal), thermostabilized food (e.g., fruit cup), intermediate-moisture food (e.g., dried peaches), natural-form food (e.g., nuts), irradiated food (e.g., beef steak), frozen food (e.g., quiche), fresh food (e.g., banana), and refrigerated food (e.g., cream cheese). Pleasant-tasting, nutritious food is important not just for health maintenance but also for psychological well-being. Space shuttle food is a great improvement over the "food tubes" used during the early American human spaceflight program. Space-station food service builds upon this tradition, but expands the available menu to provide more variety and to cater to international tastes.

Personal hygiene is equally challenging in microgravity. For example, the shuttle does not have a shower facility, so astronauts must take sponge baths. Habitation module engineers designed a special waste-water-removal system using a flow of air to send wastewater from bathing down a drain that leads to a sealed tank. In the future, space-station crews might enjoy the comfort of an Earthlike shower facility.

One of the most common questions asked astronauts is how they go to the bathroom in space. Waste elimination in microgravity represents a very challenging problem. Aerospace engineers designed a special space toilet so shuttle astronauts could approximate the normal sanitary procedures they perform here on Earth. However, in space, the astronaut must use a seat belt and foot restraints to keep from drifting. The space toilet flushes away the waste products using a flow of air and a mechanical "chopper-type" device.

The shuttle's waste-collection system has controls that an astronaut adjusts to set the system for various operational modes, including urine collection only, combined urine and feces collection, and emesis (vomit) collection. The complete microgravity toilet system consists of a commode (or waste collector) to handle solid wastes and a urinal assembly to handle fluids. A similar space toilet supports the needs of space-station crew members.

Both male and female astronauts use the shuttle's urinal. The astronaut must hold the urinal while standing or sitting on the commode with the urinal mounted to the waste-collection system. Since the urinal has a contoured cup with a spring assembly, it provides a good seal with the female crew member's body. During urination, a flow of air creates a pressure differential that draws the urine off into a fan separator/storage tank.

The space shuttle's microgravity commode collects both feces and emesis. When properly functioning, it has a capacity for storing the equivalent of 210 person-days of vacuum-dried feces and toilet tissue. Astronauts can use the shuttle's commode simultaneously with the urinal. To operate the waste collector during defecation, the astronaut positions himself or herself on the commode seat. Handholds, foot restraints, and waist restraints help the individual maintain a good seal with the seat. The crew member uses this equipment like a normal terrestrial toilet, including tissue wipes. The astronaut disposes of the used tissues in the commode. The space toilet then vacuum-dries everything stored in the waste collector—feces, tissues, and the contents of fecal and emesis bags.

Shaving also presents problems in microgravity, especially if an astronaut lets his whiskers end up floating around the cabin. These free-floating whiskers could damage delicate equipment (especially electronic circuits and optical instruments) or else irritate the eyes and lungs of space travelers. One solution is to use a safety razor and shaving cream or gel. The whiskers will adhere to the cream until they are wiped off with a disposable towel. Another approach is to use an electric razor with a built-in vacuum device that sucks away and stores the cut whiskers.

For long-duration space missions, other personal hygiene tasks that might require some special procedure or device include nail trimming and hair cutting. Aerospace engineers have also developed special devices for female astronauts to support personal hygiene requirements associated with the menstrual cycle.

When we live in a microgravity environment, things we often take for granted on the surface of Earth are absent. For example, furniture must be bolted in place, or else it will simply float around the cabin. Tether lines, belts, adhesive anchors, and handholds allow us to move around and to keep ourselves and other objects in place. While working, we learn quickly that we cannot just put a tool down because, in microgravity, there is no down and the object simply floats away.

Sleeping in microgravity represents another interesting experience. Astronauts can sleep either horizontally or vertically while in orbit. Their fireproof sleeping bags attach to rigid padded boards for support, but the astronauts themselves quite literally sleep "floating in air."

Working in microgravity also requires the use of many special tools such as torqueless wrenches, handholds, and foot restraints. Handholds and foot restraints keep a space worker in place at a workstation or research bench. These devices help balance or neutralize reaction forces. If these were not available, an astronaut might find him/herself helplessly rotating around a "work piece" or the workstation.

For human beings, exposure to microgravity causes a variety of physiological changes. For example, space travelers appear to have "smaller eyes," because their faces have become puffy. They also get rosy cheeks and distended veins in their foreheads and necks. They may even be a little bit taller than they are on Earth because their body masses no longer "weigh down" their spines. Leg muscles shrink, and anthropometric (measurable postural) changes also occur. Astronauts often move in a slight crouch, with head and arms forward.

Upon initial entry into microgravity, many space travelers, including veteran astronauts, suffer from a temporary condition resembling motion sickness. On Earth, our brains have learned how to process the combined signals from our eyes, ears, and the nerves in our skin to give us information about where our body is in relation to the 'one-g' world around us. In the space environment, the sight, hearing, and tactile (touch) signals do not match as they do on Earth—primarily because in microgravity there is now no "up" or "down" that a person's brain can use as a reference. In orbit, astronauts can no longer feel the floor beneath their feet nor feel the chair beneath them when they "sit down." This sudden input of confusing signals to the brain causes many astronauts and cosmonauts to experience the temporary condition called *space motion sickness* or *space adaptation syndrome*. Most space travelers overcome this discomforting experience in less than a day, although a few astronauts have lingered in this unpleasant condition for several days. In addition, when astronauts enter microgravity conditions, their sinuses often become congested, leading to a condition similar to a cold.

Many of these microgravity-induced physiological effects appear to be caused by fluid shifts from the lower to the upper portions of the body. So much fluid goes to the head that the brain may be fooled into thinking that the body has too much water. This can also result in an increased production of urine.

Extended stays in microgravity tend to shrink the heart, decrease production of red blood cells, and increase production of white blood cells. A process called resorption also occurs. This is the leaching of vital minerals and other chemicals (such as calcium, phosphorus, potassium, and nitrogen) from the bones and muscles into the body fluids that are then expelled as urine. Such mineral and chemical losses can have harmful physiological and psychological effects. In addition, prolonged exposure to a microgravity environment can cause bone loss and a reduced rate of bone-tissue formation.

While a relatively brief stay (say from 7 to 70 days) in microgravity may prove a nondetrimental experience for most space travelers, long-duration (i.e., one- to five-year) missions could require the use of "artificial gravity." Artificial gravity (i.e., gravity effects created through the slow rotation of the living modules of a spacecraft) should help future space explorers avoid any serious health effects that might arise from very prolonged exposure to a microgravity environment. While cruising to Mars, for example, we can also use this artificial-gravity environment to help condition the explorers for activities on the Martian surface, where they will once again experience the force of a planet's gravity. (The acceleration of gravity on the surface of Mars is 3.73 meters per second squared [m/s^2]—about 38 percent of the acceleration of gravity on the surface of Earth.) Without such conditioning, a Mars explorer might not be able to walk on the planet after traveling several hundred days through interplanetary space in a totally weightless condition.

In the future, very large space settlements will also use "artificial gravity" to provide a more Earthlike home and to avoid any serious health effects that might arise from essentially permanent exposure to a microgravity environment. Of course, engineers could design these large space habitats to provide their inhabitants the very exciting possibility of life in a multiple-gravity-level world, with a variety of different modules or zones that simulate gravity conditions ranging from microgravity up to normal terrestrial gravity.

American astronaut Shannon Lucid exercises on a treadmill set up inside the Russian *Mir* space station as it orbited Earth on March 28, 1996. Regular exercise programs help astronauts and cosmonauts combat some of the undesirable physiological consequences of long-term exposure to the microgravity environment of an orbiting space vehicle. Image courtesy of NASA.

HUMAN SPACECRAFT FOR A MARS EXPEDITION

Beyond the Earth-Moon system, Mars is the only practical target for human exploration and settlement in the early decades of the twenty-first century. Mars also provides the opportunity for demonstrating the practicality of in situ resource utilization (ISRU). With ISRU, the Red Planet could provide air for the astronauts to breathe and fuel for their surface rovers and ascent rocket vehicle (from the Martian surface to the orbiting interplanetary spacecraft). For this reason, ISRU has become a major strategy in expedition scenarios. In one expedition scenario, we would deliver the Mars ascent vehicle (MAV) (the rocket vehicle used for crew departure from the surface of the planet), critical supplies, an unoccupied habitat, and an ISRU extraction facility to the surface of the Red Planet before the human crew ever leaves Earth.

Of course, planning a human mission to Mars is a very complex undertaking. We must evaluate many factors before a team of human explorers can depart for the Red Planet with acceptable levels of risk and a reasonable hope of returning safely to Earth. Our crew will embark on an interplanetary voyage that takes between 500 and 1,000 days (depending on the particular strategy selected). In designing the expedition spacecraft, we must carefully consider the overall objectives of the expedition, the choice of the interplanetary transit trajectories, the desired stay time on the surface of Mars; the primary surface site, the required resources and equipment (whether we haul everything from Earth or use Martian resources for resupply), and crew health and safety throughout the extended journey. Due to the nature of interplanetary travel, there is no quick return to Earth, or even the possibility of emergency help from Earth, should the unexpected happen. Once our crew departs from the Earth-Moon system and heads for Mars, they and their spacecraft must be totally self-sufficient and flexible enough to adapt to new situations.

A crewed expedition to Mars early in the twenty-first century represents the first voyage through interplanetary space by human beings. One popular mission scenario uses a nuclear-electric-propelled spacecraft. This Mars-expedition spacecraft would support a crew of five on their 950-day (2.6-year) interplanetary journey to the Red Planet and return to Earth. The spacecraft also carries a Mars lander vehicle. Twin-megawatt-class, advanced-design space nuclear reactors power the spacecraft's nuclear-electric propulsion (NEP) system. Closed air and water life-support systems and artificial gravity sustain the crew throughout the extended flight.

We would assemble the Mars-expedition vehicle in low Earth orbit (LEO) near the *International Space Station*. Without a crew on board, mission controllers would turn on the electric propulsion system and let the spacecraft gently spiral out from low Earth orbit to geostationary Earth orbit (GEO). At GEO, the crew boards the spacecraft, and the expedition begins an outward-spiraling journey to Mars (about 510 days). (Remember, electric rockets produce a low, but continuous thrust, so electrically propelled vehicles take trajectories that spiral in and out of planetary gravity fields.) After the outbound interplanetary journey, the spacecraft takes another 39 days to perform a capture spiral maneuver around Mars, ending up in a circular, 3,000-kilometer-altitude orbit above the planet. While the expedition spacecraft orbits Mars, the crew engages in a 100-day reconnaissance mission, including a 30-day surface-exploration excursion by three of the five crew members. A 23-day-duration Mars departure spiral starts the electrically propelled vehicle back to Earth. Mars-to-Earth transfer takes about 229 days under optimum interplanetary coasting conditions. Once within the Earth-Moon system, the spacecraft executes a 16-day capture spiral to geosynchronous orbit around Earth. From GEO, the crew would transfer to a special quarantine (if necessary) and debriefing facility on the space station before returning to Earth's surface. There are other interesting Mars-expedition spacecraft designs and proposed mission scenarios. This particular scenario helps us understand the use of advanced space technologies and the complexities of human travel through interplanetary space.

The commitment to a human expedition to Mars, perhaps as early as the second decade of the twenty-first century, is clearly a very ambitious undertaking. In addition to many well-demonstrated improvements in space technology, this mission also requires a political and social commitment that extends for several decades. Based on extrapolations of space technology at the beginning of twenty-first-century space technology, we can say that a safe, reliable, and successful human expedition to Mars will be very expensive—perhaps $100 billion to $1 trillion when we add up all the associated costs. One nation, or several nations in a cooperative venture, must be willing to make a lasting statement about the value of human space exploration in our future civilization. A successful crewed mission to Mars establishes a new frontier both scientifically and philosophically. (We will discuss this point more in chapter 7.) To generate the maximum overall benefit from this expedition, the sponsoring society should view it as the precursor to permanent human settlement of the Red Planet. We should not attempt the expedition as a one-time space-technology adventure with no further purpose.

Chapter 5

Impact

> Man must rise above the Earth—to the top of the atmosphere and beyond—for only thus will he fully understand the world in which he lives.
>
> Socrates, ancient Greek philosopher, fifth century B.C.E.

Recognizing that we are now immersed in a swiftly flowing stream of social and technical change, this chapter describes how space technology enables us to better understand the universe, our home planet, and ourselves. The chapter also discusses the dramatic impact that space technology has on national security and on the global information infrastructure. Finally, the chapter presents several speculative, yet exciting, thoughts concerning the impact of space technology on the human spirit and philosophy in the third millennium.

When the former Soviet Union launched *Sputnik 1* on October 4, 1957, we entered the space age and broke through the last two great physical barriers that hindered our exploration of the universe: gravity and air (that is, Earth's atmosphere). With powerful rockets we could now escape from the relentless pull of our home planet's gravity—that embracing force within which all previous human history had taken place. We could leave the cradle of Earth and reach other worlds.

Through space technology, smart robot exploring machines visited all the major planets except tiny, distant Pluto, investigated them at close range, and in some cases even landed on their surfaces. Human explorers also voyaged into outer space, including the 12 men who walked on the surface of the Moon and gathered interesting rock specimens from this alien world for detailed analysis on Earth. In the short span of just four decades we learned more about the intriguing celestial objects within our solar system than we had in all previous human history. For millions of nonscientists, the marvels of space technology provided spectacular close-up images of previously inaccessible alien worlds, making them almost as familiar as our own Moon.

The arrival of the space age also tumbled the other remaining barrier to our detailed investigation of the universe: the blurring, dimming, or total blocking of the information-rich natural signals from celestial objects and cosmic phenomena caused by Earth's intervening atmosphere. Prior to the space age, scientists attempting to look outward from the ground were severely limited in what and how far they could "see." Their situation was similar to that of a diver at the bottom of a murky, shallow lake or sea trying to study the detailed features on the Moon's surface.

However, through space technology, orbiting instruments and human eyes could now observe the universe directly in all its incredible immensity, variety, beauty, and violence. No longer limited by the protective, yet obstructive, influence of Earth's atmosphere, space-based instruments and observatories provided a dramatic new vision of the universe—a vision that extended across all portions of the electromagnetic spectrum. This new vision created an overflow of intriguing scientific data that keeps toppling previously cherished hypotheses about the physics of the universe. Not only are we learning that the universe is a very strange place, but it is proving to be a much stranger place than anyone dared to imagine. Fresh insights are also reviving long-standing philosophical questions about our cosmic origins and our ultimate destiny in the universe.

The military exploitation of space technology created a powerful, two-edged sword. On one cutting edge lay new offensive weapon systems—primarily nuclear-tipped intercontinental ballistic missiles—that can deliver incredible levels of destruction throughout the planet in less than an hour. Spawned in the Cold War, the threat of "instant" global destruction forced major changes in strategic thinking and induced high levels of geopolitical instability.

On the other edge of this sword lie the space-based reconnaissance and surveillance systems that provide unparalleled levels of critical military information—information needed by national leaders to make stabilizing decisions in very turbulent political times and to avoid nuclear Armageddon. Today, the expanded acquisition and flow of national security information through space-based military systems remains essential in maintaining global stability. When diplomacy and common sense fail, however, battle-space-information supremacy (that is, the ability to make an enemy's position and action on the battlefield totally transparent) significantly enhances the application of military force and promotes a swifter, successful conclusion to any armed conflict.

Space technology provides us with an incredible, politically unobstructed view of our home planet. With the development of reliable Earth-orbiting spacecraft, this "high-ground" synoptic view quickly stimulated the rise of many important military, civilian, and commercial satellite-based information services. It also enabled the rise of Earth system science (ESS), an important new multidisciplinary field that seeks a comprehensive understanding of the terrestrial biosphere and supports intelligent stewardship of its resources.

Perhaps one of the most important impacts of space technology is the creation of the many invisible lines of satellite-based information that are rapidly replacing or supplementing the more traditional lines of communications, such as roads, sea lanes, and cables. Television, weather, voice, images, navigation, and other important "wireless" data streams pour down on Earth from the hundreds of orbiting spacecraft operated by military, civil, and commercial entities. Information-bearing signals also travel up to space from any location on Earth—no matter how remote—and interact with some portion of this global, space-based information network. In a very real sense, these Earth-orbiting spacecraft perform functions similar to those of terrestrial public utilities by dependably providing information-technology services. However, unlike ground-based utilities that generally service only a neighborhood, city, or small region of the planet, satellite-based information-technology services can simultaneously accommodate millions of users in every corner of the world.

Recognizing this circumstance, we can treat the diverse collection of Earth-orbiting information systems as a "global utility" that serves the almost insatiable information needs of the "global village" below. In just a few decades, satellite-based information-technology utilities have become an integral part of national security. They now support effective management of the environment, promote economic growth and vitality, and enhance citizen well-being (including entertainment and safety).

In a totally uncontrolled free-market global economy (driven to profitability without rules of ethical human behavior), access to and use of modern information streams can create a great "digital divide" separating empowered individuals and nations from those without such services. However, with space technology and a cooperative political environment, previously disadvantaged people in remote areas of developing countries can quickly vault across this potentially oppressive digital divide and begin enjoying the advantages of satellite-delivered education, telemedicine, news, economic information, weather forecasting, natural-hazards warning, and entertainment.

Finally, space technology provides access to a unique physical and psychological frontier in which human beings can exercise their creative energies and stretch their imaginations. In the first half of the twentieth century, visions of interplanetary travel propelled the creative spirits of certain individuals, like Goddard, Tsiolkovsky, and Oberth. In the twenty-first century, space technology extends the unique opportunity of solar-system exploration and settlement to another generation of pioneering humans. While space probes did not find any evidence of intelligent "Martians," space technology can deliver intelligent beings to the Red Planet in the twenty-first century. In a very real sense, because of space technology, we have met the Martians, and they are us.

MEETING THE UNIVERSE FACE-TO-FACE

Beginning to understand and appreciate the universe in all its vastness, complexity, and splendor is a most challenging task. Curiosity has helped creative people over the centuries pursue meaningful answers to questions such as the following: How did the universe come to be? What is it made of? What fundamental forces rule its behavior? Why is the universe just the way it is and not otherwise? Finally, what will ultimately become of the universe (and us)?

In seeking knowledge, scientists search for laws that not only describe the universe, but predict behavior within it—all the way from the workings of the smallest atom to the complex dynamics of the largest galaxies. However, knowing such basic laws is not enough, because (as scientists often discover) simple rules do not necessarily produce simple outcomes. Just as the English alphabet with its set of 26 letters and handful of grammatical rules can fill large libraries with exciting ideas, stories, meanings, and intellectual possibilities, so too the fundamental laws of physics give rise to complex tangles of diverse phenomena throughout the universe.

For example, cosmologists, exobiologists, planetary scientists, and modern philosophers now joust (each from a slightly different professional perspective) with this challenging puzzle: Is the evolution of conscious intelligence a normal step in the overall process by which matter and energy evolve in time and space? Or is conscious intelligence, as found here on Earth, a very rare and special phenomenon?

In many unique and special ways, modern space technology is helping scientists better understand just how the basic physical laws translate into the rich diversity we observe in nature, not just here on planet Earth but throughout the solar system and far beyond its limits to the very edge of the observable universe. The exploration of space provides scientists many different opportunities and intriguing new worlds on which to observe first-hand just how the basic (and presumably universal) laws of physics translate into the rich diversity found in nature.

Does life start on suitable planets beyond Earth whenever it can? Is it possible for intelligent life to emerge elsewhere in the Milky Way galaxy? Has it? If so, what will be the social and cultural impact on Earth, should we discover (through space technology) that we are not alone in this vast universe? Prior to the space age, these "twilight technical questions" had to remain in the realm of science fiction. Any scientist who openly attempted to answer them often experienced professional derision, public ridicule, or both. Today, the preliminary results from four decades of space exploration now encourage scientists to cautiously revisit such speculations. In particular, recent data make Mars and Europa (a major Jovian moon) especially attractive candidates for a more focused search for life beyond Earth.

How much has space technology improved our overall knowledge of the universe? One way to examine the impact is to look back at the prevailing view of the universe just before the start of the space age in 1957 with the launch of *Sputnik I*. This approach will help us appreciate how much space-technology-generated knowledge everyone now takes for granted.

Before 1957, it was popular to think of Venus as literally a twin of Earth. People believed that since the planet's diameter, density, and gravity were only slightly less than Earth's, Venus must be similar to Earth—especially since it had an obvious atmosphere and was just a little closer to the Sun. Consequently, during the first half of the twentieth century, visions of Venus as a "prehistoric Earth" appeared frequently in the science as well as the science-fiction literature. Popular fictional stories endowed Venus with large oceans, lush tropical forests, giant reptiles, and even primitive humans. However, visits by numerous American and Russian spacecraft since the 1960s have shattered all prior romantic fantasies of a neighboring prehistoric world. Except for a few physical similarities, such as size and gravity, Earth and Venus proved to be two very different worlds. As space-exploration data reveal, the surface temperature of Venus approaches 500 degrees Celsius, its atmospheric pressure is almost 100 times that of Earth, the planet has no surface water, and its dense, hostile atmosphere contains clouds of sulfuric acid and an overabundance (about 96 percent) of carbon dioxide—conditions clearly representative of a runaway greenhouse of disastrous proportions.

Mars lies at the center of astronomical interest and speculation, as it has since ancient times. In the late nineteenth century, scientists and astronomers used the best available Earth-based telescopes to study the Red Planet and reported what appeared to be straight lines crisscrossing its surface. Some astronomers, like Percival Lowell (1855–1916), enthusiastically, but incorrectly, concluded that an intelligent race of Martians had constructed a large system of irrigation canals to support life on a "dying planet." In 1938, actor Orson Welles broadcast a radio drama based on H.G. Wells's classic science-fiction story *The War of the Worlds*—the fictional account of the invasion of Earth by menacing creatures from Mars. The radio broadcast was so realistic that many people actually believed that Earth was being invaded by Mars, creating a near panic in some parts of the United States.

Unfortunately, an armada of sophisticated robot spacecraft (including flybys, orbiters, landers, and a minirover) shattered the pre-space-age romantic myth of a race of ancient Martians and their system of giant canals to deliver water from the planet's polar regions. The spacecraft data revealed Mars as a "halfway" world. Part of the surface of the Red Planet is ancient, like the surfaces of the Moon and Mercury; part is more evolved and Earthlike. Physical features resembling riverbeds, canyons, gorges, shorelines, and even islands suggest to planetary scientists that large rivers and perhaps even shallow seas once existed on an ancient Mars.

In addition to the popular misconceptions about Mars and Venus, in 1957, scientists treated the Sun as a basically stable, steadily shining star. Mercury was just a blur about whose surface features we had no information. No one knew about the bands of trapped nuclear radiation (the Van Allen belts) that encircled Earth. The principal moons of Jupiter (Io, Europa, Callisto, and Ganymede) were still vague points of light, hardly better seen and understood than when Galileo first detected them in 1610. Among the planets, only Saturn was known to have rings. Today, we know that all four Jovian planets (Jupiter, Saturn, Uranus, and Neptune) have ring systems, and close-up satellite imagery has revealed the marvelous complexity and beauty of the Saturnian ring system. In 1957, we did not know that tiny Pluto had a large companion moon (now called Charon). Only Earth was known to have volcanoes, and no one knew what a Moon rock was like.

Before the space age, black holes and neutron stars were just highly speculative ideas in the minds of a few theoretical physicists, and no one dared even imagine strange objects like pulsars and quasars. The "big-bang" theory was just one of several theories about the origin of the universe. Most cosmologists considered it quite unlikely that they would ever acquire the data to clearly favor one theory over another.

Prior to 1957, no human being had ever traveled in outer space. In fact, the perceived hazards of weightlessness and meteorites cast great doubt that a person could survive, much less work, in such an environment. Finally, even the boldest space visionaries did not believe that human beings would visit the Moon before the end of the twentieth century.

Yet, through space technology, scientists placed sophisticated (often unique) observatories into orbit around Earth that looked farther out into the universe and further back in time than was possible with observing instruments located on our planet's surface at the bottom of a murky, intervening atmosphere. In 1990, delicate instruments on NASA's *Cosmic Background Explorer* (COBE) spacecraft peeked all the way back to the dawn of time and recorded the faint cosmic microwave background radiation (at about 2.7 degrees Kelvin)—that cool, lingering remnant of the primordial "big-bang" explosion at the very edge of the observable universe. The important fields of infrared astronomy, X-ray astronomy, gamma-ray astronomy, cosmic-ray astronomy, and ultraviolet astronomy all became possible because scientists could place sophisticated instruments on modern space platforms and meet the universe face-to-face across the entire electromagnetic spectrum. Contemporary astronomical work in other research areas benefited from large, high-resolution optical systems like the Hubble Space Telescope, operating outside Earth's atmosphere.

From the limited historic perspective of the early twenty-first century, space-technology-enabled scientific accomplishments appear to represent some of the most remarkable human achievements of all time. If the discovery of just one "new world" by Christopher Columbus helped stimulate the great release of human creativity during the late Renaissance period (sixteenth century) and the birth of the scientific age (seventeenth century), what will the near-simultaneous "discovery" of several dozen exciting new worlds (that is, the major planets and their intriguing moon systems) do to the human spirit in the twenty-first century?

At this point, we can only confidently forecast that scientists in the twenty-first century will continue to apply advanced space technology in their efforts to observe the birth of the earliest galaxies in the universe, to detect all extrasolar planetary systems in the Sun's neighborhood (that is, out to about 100 light-years or so), to identify any extrasolar planets believed capable of supporting life, and to learn by direct exploration with robot systems and human expeditions whether life began and possibly now exists elsewhere in the solar system. Scientists will eagerly perform these tasks to help the human family discover its ultimate cosmic roots and destiny—that is, who we are, where we came from, where we are going, and

NASA's Hubble Space Telescope provided this detailed image of the majestic dusty spiral galaxy NGC 4414 through a series of observations in 1995 and 1999. This galaxy is about 60 million light-years away. Such incredibly detailed images from orbiting observatories allow scientists to study and explore the universe in ways never before possible. Image courtesy of NASA, the Hubble Heritage Team, STScI, and AURA.

whether we are on this fantastic cosmic journey through a magnificently beautiful and violent universe all by ourselves.

THE TRANSFORMATION OF WARFARE

The Ballistic Missile and a Revolution in Strategic Warfare

In the middle of the twentieth century, space technology transformed international politics. The marriage of two powerful World War II–era weapon systems, the American nuclear bomb and the German V-2 ballistic missile, ultimately produced the single most influential weapon system in the twentieth century, if not all history—the intercontinental ballistic missile (ICBM). The ICBM and its technical sibling, the submarine-launched ballistic missile (SLBM), were the first weapon systems designed to travel into and through space. The arrival of the first generation of such space weapons in the late 1950s completely transformed the nature of strategic warfare.

The ICBM created a fundamental change in national security policy. Before the ICBM, the chief purpose of the U.S. military establishment had been to fight and win wars. Once the operational nuclear weapon-equipped ICBM arrived, both the United States and the former Soviet Union possessed a weapon that could deliver megatons of destruction to any point on the globe with little or no chance of being stopped. From that moment on, the chief purpose of the U.S. military establishment became the avoidance of strategic nuclear warfare. A wholesale, unstoppable exchange of ballistic-missile-delivered nuclear weapons would destroy both adversaries and leave Earth's biosphere in total devastation. There would be no winners, only losers. With the development of the ICBM, for the first time in history, human beings possessed a weapon system that could end civilization in less than a few hours.

Therefore, the first space weapon made the deterrence of nuclear war the centerpiece of national security policy—a policy appropriately called mutual assured destruction (MAD). Military leaders no longer focused on "winning" the next major war; rather, they created a variety of schemes and technologies to help them prevent any large-scale confrontation that could escalate to the use of nuclear weapons. The final course of action in this strategic plan was quite simple. If all else failed and a nuclear war started, each side would inflict lethal damage on the other.

It is interesting to note that the threat of nuclear Armageddon has helped restrain those nations with announced nuclear weapon capabilities (such as the United States, the former Soviet Union, the United Kingdom, France, and the People's Republic of China) from actually using such weapons in resolving lower-scale, regional conflicts. Because of this standoff of unstoppable missile against unstoppable missile, political scientists assert that the ICBM created a revolution in warfare and international politics—a revolution making nuclear warfare between rational actors (nations) impossible.

Today, however, as nuclear weapon and ballistic missile technologies spread to other nations (such as Pakistan and India), the specter of a regional nuclear conflict or even nuclear terrorism haunts the world community. The heat of long-standing, culturally based regional animosities could overcome the decades of self-imposed superpower restraint on the use of nuclear-tipped missiles to settle an armed conflict. This emerging regional missile threat is encouraging some American military leaders to revisit ballistic missile defense technologies, including concepts requiring the deployment of antimissile weapon systems in outer space.

For more than four decades, intercontinental ballistic missiles have served as the backbone of America's strategic nuclear deterrent forces. Throughout the Cold War and up to the present day, deterring nuclear war remains the top U.S. defense priority. Since 1959, strategic-force missileers have served around-the-clock on continuous alert. Buried in underground launch facilities, ICBMs are the most rapid-response strategic force available to the American president. The Minuteman III ICBM, for example, is capable of hitting targets more than 8,000 kilometers away within about 30 minutes with outstanding accuracy (see Figure 5.1). In the post–Cold War political environment of the early twenty-first century, military leaders still regard this ICBM force as America's most credible deterrent against nations that possess, or are in the process of developing, weapons of mass destruction (WMD). Submarine-launched ballistic missiles (SLBMs) represent a complementary (mobile) component of this long-standing nuclear deterrent policy.

As part of an ongoing initiative to transform the U.S. military into a twenty-first-century fighting force the Department of Defense merged the U.S. Space Command (USSPACECOM) with USSTRATCOM on October 1, 2002. The new organization, called the United States Strategic Command (USSTRATCOM), is headquartered at Offutt Air Force Base in Nebraska and serves as the command and control center for U.S. strategic forces. It also controls military space operations, computer network operations, information operations, strategic warning and intelligence assessments, as well as global strategic planning.

Figure 5.1 A drawing showing the modern Minuteman III intercontinental ballistic missile (ICBM). Illustration courtesy of U.S. Air Force.

Space Technology and the Information Revolution in National Security

Reconnaissance satellites, surveillance satellites, and other information-related Earth-orbiting military spacecraft changed the nature of military operations and national security planning forever. Space-based information collection produced enormous impacts on peacekeeping and war fighting and immediately became an integral part of projecting national power and protecting national assets.

For simplicity, reconnaissance systems use their sensors to search for specific types of denied information of value to intelligence analysts. Surveillance systems, on the other hand, use their special instruments to monitor Earth, its atmosphere, and near-Earth space for hostile events (usually military in nature) that threaten the interests of the United States or its allies. Such hostile events include the launching of a surprise ballistic missile attack. Surveillance satellites also support treaty monitoring. In this role, their sensors continuously search from the vantage point of space for telltale signals that indicate a violation of an existing arms-control agreement.

For example, nuclear surveillance satellites can detect the characteristic signals from a nuclear weapon secretly tested in Earth's atmosphere.

Many of these important space-technology advances and applications went unnoticed by the general public. While the reconnaissance satellite transformed the art and practice of technical intelligence collection, it did so under the cloak of secrecy. From its formation in 1960, the National Reconnaissance Office (NRO) operated in a highly classified environment. The premier American spy-satellite organization consisted of U.S. Air Force, Navy, and Central Intelligence Agency (CIA) program offices. It built, launched, and operated intelligence-collecting spacecraft for the duration of the Cold War. Today, because of a presidentially directed declassification of its early missions and official public acknowledgment of its very existence, the NRO has emerged a bit from the shadows of its secret operating environment. Only now can this organization's extremely important role in successfully applying space technology to collect intelligence data (imagery and signals) over denied areas in support of national security be acknowledged, at least partially.

In 1966, the U.S. Air Force began developing an important family of surveillance satellites, now known as the Defense Support Program (DSP). Placed in geostationary orbits at an altitude of 35,900 kilometers above the equator, these early warning satellites use special infrared detectors to continuously scan the planet's surface (land and sea) for the hot exhaust-plume signatures characteristic of ballistic missile launches. Since missile-warning satellites first became operational in the early 1970s, they have provided national leaders an uninterrupted, 24-hour-per-day, worldwide surveillance capability.

In addition to guarding against sudden ballistic missile attacks, these satellites also observe the performance of space launch vehicles and monitor remote regions of Earth for clandestine nuclear weapons tests. Continuously improved versions of the DSP satellite served as the cornerstone of the American early warning program and made feasible the national policy of strategic nuclear deterrence. Surveillance satellites would immediately detect any enemy attempt to launch a surprise ICBM attack in a destructive first strike. Before the attacking missiles could impact their targets, national leaders could order the launch of an equally destructive nuclear counterstrike. This "everybody-loses" nuclear-exchange scenario hardly seems appropriate for a mature planetary civilization. However, during the Cold War, military surveillance satellites provided an important level of sanity within a politically divided world that focused on mutual assured destruction (MAD).

In the post–Cold War era, these silent sentinels still stand guard, always ready to alert national authorities about a hostile ballistic missile attack. Now, their missile-surveillance mission has expanded to include shorter-range missiles, launched by rogue nations during regional conflicts. For example, the DSP demonstrated its effectiveness during the Persian Gulf conflict (1991) when the program's surveillance satellites detected the launch of Iraqi Scud missiles and provided warning to civilian populations and coalition forces in Saudi Arabia and Israel. By 1995, new techniques in processing DSP data provided U.S. and allied theater-level forces improved warning of attack by short-range missiles.

In the twenty-first century, a new generation of space-based infrared surveillance systems will help American military leaders satisfy four critical defense missions: missile warning, missile defense, technical intelligence, and battle-space characterization (including battle-damage assessment). Theater (shorter-range) ballistic missile proliferation is becoming an ever-increasing problem in the twenty-first century. Many nations possess theater missiles, and some of these nations have made their missile technology available for purchase by unstable or regionally aggressive political regimes and terrorist organizations. With advanced space-based surveillance systems, the U.S. Strategic Command will provide theater ballistic missile warning to U.S. forces deployed throughout the world.

Three other classes of military satellite systems significantly influence the use of information in national defense and in closely related civilian applications. These are polar-orbiting military weather satellites, a special constellation of navigation satellites called the Global Positioning System, and a family of advanced-technology military communications satellites. The operation of these versatile military spacecraft represents an important dual use (military and civilian) of space technology.

Since the mid-1960s, for example, the United States has used the low-altitude, polar-orbiting satellites of the Defense Meteorological Satellite Program (DMSP) to routinely provide military and other government agencies with important environmental data, much of it acquired over previously inaccessible or politically denied areas. Of special significance is the primary sensor on DMSP satellites that provides continuous cloud-cover imagery in both the visible and infrared portions of the spectrum. The satellites' other specialized meteorological sensors measure the atmospheric vertical profiles of moisture and temperature. Military weather forecasters use these data to monitor and predict regional and global weather patterns, including the presence of severe thunderstorms, hurricanes, and typhoons.

"Space weather" (the collective influence of solar-geophysical interactions) is also of great importance to modern military operations. Consequently, DMSP satellites carry special sensors that measure charged-particle populations and electromagnetic-field strengths in low Earth orbit (about 830 kilometers in altitude). Military space-weather forecasters use these data to assess the impact of changes in the Earth's ionosphere on long-range radar systems and communications networks. Scientists use these data to monitor global auroral activity and to predict the potential effects of changes in the near-Earth space environment on military satellite operations.

The basic principles of the navigation satellite were discussed previously in chapter 4. The fully operational Global Positioning System (GPS) is a constellation of 24 Earth-orbiting satellites that provide navigation data to military and civilian users all over the world. The Department of Defense originally developed this system to support military operations. Operated by the U.S. Air Force, each GPS satellite in the constellation travels in a circular orbit around Earth at an altitude of 20,350 kilometers. As they circle the globe every 12 hours, the GPS satellites continuously emit special navigation (radio-frequency) signals. With proper equipment, users can simultaneously receive the signals from up to six GPS satellites and then combine these data to determine location, velocity, and time. For the military, GPS enhances the functions of contemporary positioning and navigation equipment, greatly improving combat efficiency, search and rescue operations, mapping, aerial refueling and rendezvous, and the performance of precision-guided munitions. The navigation signals provided by GPS are so accurate that users can establish time within a millionth of a second, velocity within a fraction of a kilometer per hour, and geographic location within a few meters or less. Today, both military and civilian GPS receivers are available for use in aircraft, ships, and land vehicles, as well as in handheld applications by individuals in the field.

Military communications satellites provide secure, reliable command and control information services for American strategic nuclear forces and conventional (nonnuclear) forces anywhere on the globe. The family of Defense Satellite Communications System (DSCS) spacecraft orbit Earth at an altitude of 37,400 kilometers (geostationary orbit). These satellites are jam-resistant, superhigh-frequency systems capable of providing worldwide secure voice and data transmission. The spacecraft in this system support high-priority military communications, such as the exchange of wartime information between defense officials in the Pentagon and battlefield commanders anywhere in the world. Military officials also use

DSCS to transmit time-critical space-operations and early warning data to various systems and users.

The more advanced Milstar Satellite Communications System provides similar secure, jam-resistant, worldwide communications to meet essential wartime information requirements for high-priority users. The sophisticated, multisatellite constellation links national command authorities with a wide variety of military resources, including ships, submarines, aircraft, and ground stations. The fully operational Milstar system involves a constellation of four satellites positioned around the world in geosynchronous orbits (approximately 37,400 kilometers in altitude). Unlike previous military communications satellites, each advanced-technology Milstar satellite represents a very smart switchboard in space that can automatically direct defense-related communications traffic from terminal to terminal anywhere in the world. Since the Milstar satellite actually processes the communications signal and can link with other Milstar satellites through cross-links, this design feature greatly reduces the requirement for ground-controlled switching. In response to user direction, each Milstar satellite establishes, maintains, reconfigures, and then disassembles the required communication circuits. Milstar terminals provide encrypted voice, data, teletype, or facsimile communications. An important feature of the Milstar system is interoperable communications among the many different users of U.S. Army, Navy, and Air Force terminals. Geographically dispersed mobile and fixed control stations provide survivable and enduring operational command and control of the Milstar constellation.

Space Technology and the Creation of a Transparent Battle Space

Space technology has greatly influenced the nature and conduct of modern warfare and is now an inseparable part of national defense. Space systems are the primary source of warning of an impending attack and can fully characterize that attack. Highly capable reconnaissance satellites monitor arms-control agreements and continually provide data that help intelligence analysts assess the world situation and avoid military or political surprises. Collectively, these capabilities represent unprecedented global knowledge and awareness of tactical conditions, creating a transparent battle space.

Today, military planners speak in terms of a battle space that extends beyond the traditional land, sea, and air battlefield and includes the application of various military satellites operating far above in the space environment. The concept of a transparent battle space means a state of information superiority that allows the American commander to "see" everything within the battle space—all enemy activity with complete accountability for all friendly forces. As clearly demonstrated during the Gulf War (1991) and again during antiterrorist actions in Afghanistan (2001–present) and the 2003 Iraq war, the vantage point of space is a significant force multiplier that permits the surgical application of American (and allied) military resources with minimum risk to the friendly forces and with maximum impact on enemy forces.

Since the early 1960s, wherever space applications have proved superior to earthbound methods in satisfying national security information needs, the space systems have quickly dominated. Today, military space systems of all types form an integral part of national defense. Important information services such as global weather forecasting, navigation, reconnaissance, surveillance, and tactical communications now extensively rely on the use of Earth-orbiting satellites. Successful war fighting and peacekeeping in the twenty-first century require information dominance over a battle space and *space control.* As defined within contemporary U.S. military doctrine, space control is the combination of abilities to enter, to deny entry to, and to exploit the region beyond Earth's atmosphere. Space control implies that the United States will maintain *space superiority*—a condition that assures friendly forces the use of the space environment while denying its use to enemy forces.

Since the end of the Cold War, the importance of military space systems has grown exponentially. The United States participates in numerous historic alliances while simultaneously addressing major changes in traditional threats. New dimensions, such as energy, the environment, and economic competition, are emerging as significant components in contemporary American national security strategy. In the turbulent post–Cold War period, many regions of our planet (such as the Balkans, the Middle East, or South Asia) are experiencing significant military, social, and political unrest. The international geopolitical situation shifts daily and generates global requirements for space-based reconnaissance to support American national policy and military operations. Furthermore, continued advances in smart weaponry, space technology, and telecommunications provide near-real-time information support for war fighters, arms-proliferation issues, and counterterrorism efforts. American leaders regard information-related military space systems as "force multipliers"—critical systems that provide peacekeeping and combat units on land, at sea, or in the air with a distinctive "information edge" over potential adversaries anywhere in the world.

As far back as Alexander the Great, every military commander has faced the dilemma of needing to see over enemy lines. Having a unique vantage point to observe the entire battlefield from the highest hill certainly worked in times when weapons technology was limited in range and effectiveness and military forces consisted of relatively small numbers of combatants. However, as the size of the military forces and the capabilities of modern weapons increased, so did a commander's need for improved battlefield awareness.

The space environment now provides a very special "high-ground" vantage point for supporting military operations with data derived from navigation, communications, weather, missile-warning, and surveillance satellites. Through the marvels of high-speed information processing and advanced sensor technology, military space systems give the modern commander a winning edge by providing a digitized, synoptic view of the entire three-dimensional battle space. Since space systems can simultaneously observe the front lines, locations deep behind the front lines, adjacent bodies of water, and all the associated airspace above these regions, they create a transparent battle space.

For example, during Operation Desert Fox (the intensive Allied aerial assault against Iraq prior to the sweeping ground offensive during the 1991 Gulf War), space assets provided the primary, low-risk technical means of assessing the battle damage from strikes deep within Iraqi territory. During Operation Desert Storm (the Allied assault portion of the 1991 Gulf War), both military and commercial space assets belonging to the United States, the United Kingdom, France, and the Russian Federation provided the coalition (allied) forces with communications, navigation, surveillance, intelligence, and early warning. For the first time, satellites also brought live television coverage of the war to home viewers around the world. Using some 60 satellites, coalition forces had access to secure strategic and tactical communications for in-theater as well as into- and out-of-theater operations. These satellites bridged the gap for tactical ultra-high-frequency (UHF) and very high-frequency (VHF) signals that had previously depended upon terrestrial line of sight. Consequently, during the conflict, time-sensitive information could be exchanged between the ground, naval, and air units spread throughout the theater. The use of satellites during the Gulf War provided commanders at all levels with unprecedented communications capabilities and marked the beginning of a new era in warfare.

From a military-conflict perspective, the navigation satellite also came of age in the desert of the Arabian Peninsula. The setting, kilometers and kilometers of sand dunes with very few distinguishable landmarks, proved perfect. The U.S. Air Force's Global Positioning System (GPS) provided real-time, passive navigation updates to virtually every Allied weapon system in-theater. Planes, helicopters, tanks, ships, cruise missiles—even the supply trucks used to deliver food and water to the front—relied on GPS receivers to precisely establish their position, speed, and (for aircraft) altitude. Some military historians now refer to the Gulf War of 1991 as the first "space-applications war"—meaning that for the first time in history, a wide variety of space systems directly contributed to the swift and efficient conclusion of a war. In this conflict, coalition forces liberated Kuwait and soundly defeated all opposing Iraqi forces.

Today, using a similar collection of military space systems monitoring the entire battle space, a tactical-force commander in some regional conflict would quickly detect a column of enemy tanks as it emerged from a hidden underground complex, rapidly apply appropriate forces (perhaps ordering a precision weapon strike) against this new threat, and then immediately assess the damage inflicted on the hostile target. Space-system-supported information dominance over the battle space helps support the swift and efficient end of a local or regional armed conflict. This space-technology advantage helps minimize collateral damage and avoid unnecessary civilian casualties while also reducing the combat risks to friendly forces.

Information products generated by modern spy satellites can provide warning of war, monitor arms-control treaties, track arms shipments, and support international peacekeeping missions. On occasion, the U.S. government has provided military photoreconnaissance-satellite imagery to assist with other urgent issues at home and abroad, including disaster-relief operations and the study of certain critical environmental problems. Properly interpreted and analyzed, these high-resolution images can greatly assist the various relief agencies in providing timely and efficient relief and recovery responses. Independent of these important military reconnaissance systems, a new generation of commercial Earth-observation satellites now makes roughly equivalent high-resolution (one-meter) imagery available to various military, civil, and commercial customers throughout the world, helping to create a "transparent globe."

ORBITING INFORMATION UTILITIES AND THE CREATION OF A TRANSPARENT GLOBE

Toward the Creation of a Transparent Globe

One of the unexpected impacts of NASA's Apollo Project was the heightened level of environmental awareness caused by the many inspir-

This inspirational view of the "rising" Earth greeted the three *Apollo 8* astronauts as they came from behind the Moon after the lunar-orbit-insertion burn on December 29, 1968. Illustration courtesy of NASA.

ing, long-distance pictures of Earth taken by the astronauts during their translunar flight. The *Apollo 8* mission (December 1968) was the first time in history that human beings could look back across the interplanetary void and personally view the entire Earth as a beautiful, complex system bursting with life. As they observed this giant "blue marble" from their spacecraft, the *Apollo 8* astronauts could not help but compare its dynamic and bountiful biosphere with the barren and lifeless lunar landscape below them. Mission commander Frank Borman recalls the powerful, almost spiritual, impact of glimpsing Earth above the lunar landscape: "We were the first humans to see the Earth in its majestic totality, an intensely emotional experience for each of us. And it was the most beautiful, heart-catching sight of my life. And I thought: 'This must be what God sees.' "

For other lunar astronauts, the starkness of the Moon and the living beauty of "Gaia" (the Earth goddess in Greek mythology) stimulated equally profound feelings of wonder and inspiration. *Apollo 11* astronaut Michael Collins said, "As viewed from the Moon, the Earth is the most beautiful object I have ever seen." *Apollo 14* astronaut Edgar Mitchell recalls: "My view of our planet was a glimpse of divinity." The inspiring views of the whole Earth from the Apollo Project also helped millions of people begin to recognize the fragile, interconnected nature of our planet's oceans, clouds, atmosphere, snow-covered polar regions, and great variety of landmasses, some bursting with vegetation and others barren.

Space technology not only stimulated a significant increase in environmental consciousness, but also provided the very efficient technical means with which to properly study and understand Earth as a complex, dynamic system. This portion of the chapter describes the enabling role satellite-collected information plays in Earth system science (ESS). It also explains how the application of space technology leads to a more thorough understanding of critical global-change issues, supports sustainable growth, and enhances environmental security in the twenty-first century.

As we continue to learn more about our home planet, new questions also arise for scientists, drawing them deeper into the previously unappreciated complexities of Earth's climate system. While space technology has allowed us to gain an important new understanding of our changing planet, we still do not know the answers to such important questions as the following: Is the current global-warming trend temporary, or is it the precursor of an accelerated increase in global temperatures? As global temperatures rise, how will this affect weather patterns, food-production systems, and the level of the sea? Are the number and size of clouds increasing, and if so, how will this affect the amount of incoming and reflected sunlight, as well as the thermal energy (heat) emitted from Earth's surface? What are the causes and effects of ozone fluctuations? How will climate change affect human health, natural resources, and the global economy over the next century and beyond?

Before the space age, scientists seldom asked such questions because there was no way to respond to them. Today, modern Earth-observing spacecraft can provide the huge amounts of simultaneously collected data needed to begin answering such questions, as well as many future questions we are not yet smart enough to ask. Unobstructed by physical or political boundaries, Earth-observing satellites are forming the high-resolution, reliable information database that leads to a "transparent globe."

The repetitive, detailed observation of our planet from space is not limited only to military satellites. The first civilian application of satellite-based remote sensing occurred in the 1960s with the development of the early civilian weather satellites—a space-technology milestone that created a revolution in meteorology, climate studies, and severe-weather warning. In the mid-1960s, the U.S. Department of the Interior and NASA developed the first environmental-monitoring satellite. This family of pioneering spacecraft (eventually named Landsat) combined emerging space and remote sensing technologies to produce information-rich, multispectral images of Earth's surface—images that began to serve the information needs of environmental scientists, farmers, ranchers, water-resource managers, and many other individuals both inside and outside of government. Building upon Landsat's technical heritage, civilian (scientific) remote sensing from space now represents a key information pathway to sustainable growth and intelligent stewardship of Earth.

Two successful satellite launches in 1999 dramatically changed the publicly accessible Earth-observation equation and further excited the revolution in home-planet information collection. Since September 1999 and the launch of Space Imaging's *IKONOS* satellite, excellent-quality, high-resolution (one-meter or better) imagery from space has become commercially available from a private firm. Before this pioneering commercial Earth-observation spacecraft was launched, there was a clear distinction between tightly controlled, high-resolution military imaging (photoreconnaissance) satellites and civilian (government-owned) Earth-observation satellites (such as the Landsat family) that collected openly available, but lower-resolution, multispectral images of Earth's surface.

With the Land Remote Sensing Policy Act of 1992 and Presidential Decision Directive 23 (PDD-23) issued in March 1994, the U.S. government blurred this long-standing distinction. The government agreed to license Earth-observation satellites owned by private American companies like Space Imaging—satellites engaged in the commercial enterprise of collecting and selling high-resolution imagery from space to civilian, commercial, and military customers from around the world. The immediate impact of the government's decision was the creation of a civilian/-commercial version of the "transparent globe" that photoreconnaissance satellites had granted military leaders since the early 1960s. This knowledge-frontier environment is now stimulating a wealth of creative civil and commercial applications for high-resolution satellite imagery of Earth, generating many new information markets in the process. A little later in this chapter, we will discuss some very interesting contemporary applications of commercial high-resolution satellite imagery data.

The arrival of another nonmilitary Earth-observing spacecraft also accelerated the growth of the orbiting information revolution. In December 1999, NASA successfully placed the new *Terra* spacecraft into orbit around Earth. *Terra* is a joint Earth-observing project between the United States,

Japan, and Canada. It carries a payload of five state-of-the-art sensors that are simultaneously collecting information about Earth's atmosphere, lands, oceans, and solar-energy balance

NASA successfully launched the *Aqua* spacecraft into polar orbit on May 4, 2002. This new Earth-observing spacecraft—a technical sibling to *Terra*—provides scientists with unprecedented information about our planet's global water cycle. Equipped with six state-of-the-art remote sensing instruments, *Aqua*, as its name implies, has the primary mission of gathering data about the role and movement of water in the Earth system.

Earth System Science

Terra serves as an Earth system science flagship. The simultaneous collection of data by its complementary suite of advanced sensors allows environmental scientists from around the world to understand how Earth's climate and environment function as an integrated system. The spacecraft carries a payload of five state-of-the-art sensors that study the interactions among Earth's atmosphere, land surfaces, oceans, biosphere, and radiant-energy inputs from the Sun. It crosses the equator during the descending node of each orbital revolution at 10:30 A.M. local time. (This special orbit gave the spacecraft its original name of *EOS AM-1*, or the "morning spacecraft.")

All of *Terra*'s remote sensing instruments are passive and depend on object-reflected sunlight or object-emitted thermal radiation. *Terra*'s complement of instruments includes the moderate-resolution imaging spectrometer (MODIS), the advanced spaceborne thermal-emission radiometer (ASTER), the multiangle imaging spectrometer (MISR), the measurement of pollution in the troposphere (MOPITT) system, and the clouds and Earth radiant-energy system (CERES).

The information *Terra* gathers is shared freely with scientists, resource managers, and commercial enterprises around the world. It represents a milestone mission in the study of Earth from space. NASA scientists have suggested this simple analogy to help describe its true impact. If we compare planet Earth to a middle-aged person who has never had a physical health checkup, then within this analogy *Terra* is providing scientists with the best scientific tool for conducting the first comprehensive global examination of our planet.

Satellite-based remote sensing has changed the way scientists view the Earth. Data from space are helping to explain (as never before possible) the limits of a majestic, yet finite, world dominated by water and shielded from outer space by a thin layer of atmosphere. From the unique perspective of outer space, scientists see Earth as an integrated whole (or system) consisting of the land, oceans, atmosphere, and a diverse collection of marvelous living creatures that interact to create and sustain the special life-supporting environment we call the biosphere. The ability to make reliable and repetitive global observations from space enables the study of Earth as a unified system.

This systematic approach to studying Earth from space is unobstructed by physical or political barriers. For the first time in history, scientists can measure and understand how local natural or human-caused activities might produce effects on a regional or even global scale. The full range of phenomena and processes involved in Earth system science extends over spatial scales from millimeters to the circumference of Earth, but important, coupled interactions connect many of these processes and, consequently, bridge widely separated spatial and temporal regions. Once a significant change occurs somewhere in this integrated system, it can propagate through the entire Earth system, resulting in a consequence popularly called *global change*.

Volcanic activity, for example, occurs along intersections of Earth's crustal plates and is driven by mantle convection—a very long-time-scale process that ranges from hundreds of millions to billions of years. Yet the effects of a volcanic eruption are felt locally within hours or days and then, over more extended geographic regions, for months or even years afterward as a result of the deposition of dust and gases in the atmosphere. The Laser Geodynamics Satellite (LAGEOS) program represents another interesting way space technology supports the detailed study of our home planet. It involves a series of spherical satellites, launched by NASA and the Italian Space Agency (Agenzia Spaziale Italiana [ASI]). NASA and ASI dedicated these spacecraft exclusively for satellite laser ranging (SLR) technologies and applications. Each LAGEOS spacecraft is a small, but dense, spherical satellite about 60 centimeters in diameter with a mass of 405 kilograms. This compact, high-density design makes the spacecraft very stable in orbit and allows geophysicists around the world the opportunity to perform very precise ranging measurements.

Scientists use SLR to accurately measure movements of Earth's surface—with a precision of centimeters per year in some locations. For example, data from the *LAGEOS-1* spacecraft (launched into a nearly circular 5,800 kilometer, 110°-inclination orbit in 1976) allowed geophysicists to show that the island of Maui in the state of Hawaii is moving toward Japan at a rate of approximately 7 centimeters per year and away from South America at a rate of 8 centimeters per year.

Although most of the global-change-inducing phenomena result from natural events—events that are currently beyond human control—environmental scientists recognize that modern human beings, in their pursuit of a technology-rich, material-goods-oriented planetary society, also represent a powerful agent for environmental change. For example, we have significantly altered the chemical composition of Earth's atmosphere through both the agricultural and industrial revolutions. Improper agricultural and construction practices have dramatically influenced the erosion of continents and the sedimentation of rivers and shorelines. The production and release of toxic chemicals have altered the health and natural distributions of biotic populations.

In addition, the ever-increasing human need for fresh, potable water has altered the patterns of natural water exchange that take place in the hydrological cycle. This important cycle involves the solar-energy-driven circulation of water from the ocean to clouds, to surface freshwater bodies and to groundwater supplies, and then back again to the ocean. The enhanced evaporation rate of water from large-surface-area, human-made reservoirs when compared to the more moderate evaporation rate of water from wild, unregulated rivers is just one example of how human beings have become agents for global change within the hydrosphere.

As the world population continues to grow (perhaps peaking at about 9 billion in 2050) and as our planetary civilization experiences widespread technological and economic development in the twenty-first century, the role of this planet's most influential and "dangerous" animal species (that is, *Homo sapiens*) as an agent for environmental change will grow exponentially. Space technology provides a very efficient way of obtaining the critical information necessary to protect our planet while still accommodating human needs.

During the last few decades, scientists have used satellite systems to accumulate supporting technical evidence indicating that certain ongoing environmental changes are the result of complex interactions among human-sponsored and natural systems. For example, scientists now treat changes in our planet's climate not only as a function of the cloud populations and wind patterns in the atmosphere, but also as due to the interactive effects of human influences on the chemical composition of the atmosphere, Earth's albedo (reflectivity property), the distribution of water among the atmosphere, hydrosphere, and cryosphere (polar ice), and urban-heat-island effects. The most significant global changes that could influence human well-being and the quality of life on our planet in the twenty-first century include global climate warming (or the "greenhouse effect"), sea-level change, ozone depletion, deforestation, desertification, drought, and loss of biodiversity. Each of these elements of global change represents a complex and significant phenomenon worthy of detailed study in its own right. However, scientists now recognize that they cannot fully understand and properly address any of these individual elements of global change unless they investigate them all collectively in an integrated, multidisciplinary fashion.

With the support of modern space technology, the overall goal of Earth system science is to obtain a scientific understanding of the entire Earth system on a global scale by describing how its component parts and their coupled interactions have evolved, how they function, and how they are expected to continue to evolve on time scales that range from days and years to centuries and hundreds of millennia. The scientific community recognizes that long-term continuous global observations of Earth throughout the twenty-first century are necessary for continued progress in Earth system science. In particular, the intimate connections among Earth's components cannot be fully uncovered and documented without simultaneous, systematic observations conducted both in situ (in place) and remotely from space.

Space observations are essential to the successful study of Earth as a system. Only observations made from orbiting platforms can provide the sheer volume of detailed global synoptic data needed to discriminate among worldwide processes operating on short time scales. In addition, space platforms permit scientists to place a variety of complementary instruments at the same vantage point. Such a single vantage point greatly facilitates the integration of remote sensing data and considerably reduces the problem of blending (or fusing) a variety of measurements made from different sites at different times. Finally, satellite-based remote sensing provides a reliable and repetitive method of viewing difficult-to-access terrain without the hindrance of political boundaries or natural barriers.

Quite possibly the greatest service space technology can provide for the human race in the twenty-first century is to enable a much better understanding of the Earth as a complex system. This new knowledge will allow future generations to conduct enlightened stewardship of their home planet and to enjoy the benefits of a sustainable planetary civilization—a world in which all human beings can contribute their individual creative energies to the common pursuit of wisdom, beauty, and cultural development and not have to spend their lives in a constant and desperate struggle for survival. An idealization? Perhaps. But humanity can never successfully reach for the stars as an intelligent species unless we first learn to manage and respect Spaceship Earth and all its passengers.

Weather Satellites and Natural-Hazards Warning

Until the start of the space age, weather observations were limited to areas relatively close to Earth's surface, with vast gaps over oceans and sparsely populated regions. Having a synoptic (long-range) view of our planet remained the perennial dream of meteorologists. Without sensors on Earth-orbiting satellites, they could view Earth's atmosphere only from within and mostly from below. As late as 1952, a U.S. Weather Bureau pamphlet described the "future" of meteorological forecasting with the following wishful opening statement: "If it were possible for a person to rise by plane or rocket to a height where he could see the entire country from the Atlantic to the Pacific."

For pre-space-age meteorologists, an Earth-orbiting satellite held out the exciting promise of providing the detailed view of Earth they so desperately needed to make more accurate forecasts. Even more important was the prospect that a system of operational weather satellites could help reduce the number of lethal surprises from the atmosphere. Some meteorological visionaries speculated quite correctly that cameras on Earth-orbiting platforms could pick up hurricane-generating disturbances long before these destructive storms matured and threatened life and property. With such weather satellites, meteorologists might also be able to detect dangerous thunderstorms hidden by frontal clouds and provide warning to communities in their path. Finally, "weather eyes" in space offered the promise of greatly improved routine forecasting (three- to seven-day predictions)—a service that would certainly improve the quality of life for most citizens.

These dreams became a reality and the wishes of many meteorologists were fulfilled on April 1, 1960, when NASA launched the world's first satellite capable of imaging clouds from space. The Television Infrared Observation Satellite (TIROS 1) operated in a midlatitude (about 44° inclination) orbit around Earth and quickly proved that satellites could indeed observe terrestrial weather patterns. This successful launch represents the birth of satellite-based meteorology and opened the door to a deeper knowledge of terrestrial weather and the forces that affect it.

In the field of satellite-based meteorology, the terms *weather satellite*, *meteorological satellite*, and *environmental satellite* are often used interchangeably, although the last term (environmental satellite) has acquired special meaning within the field of global-change research and Earth system science. *TIROS 1* carried a television camera and during its 78-day operating lifetime transmitted about 23,000 cloud photographs, more than half of which proved very useful to meteorologists.

Within the U.S. government, *TIROS 1* started a long-term, interagency development effort that produced an outstanding operational (civilian) meteorological-satellite system. In this arrangement, NASA conducted the necessary space-technology research and development efforts, while the U.S. Department of Commerce (through the auspices of the National Oceanic and Atmospheric Administration [NOAA]) would manage and operate the emerging national system of weather satellites. Over the years, the result of that pioneering arrangement provided the United States with the most advanced weather-forecast system in the world. As a purposeful part of the peaceful application of outer space, the United States (through NOAA) makes weather-satellite information available to other federal agencies, to other countries, and to the private sector.

Once proven feasible, the art and science of space-based meteorological observations rapidly evolved and expanded. Scientists soon developed more sophisticated sensors, capable of providing improved environmental data of great assistance in weather forecasting. In 1964, NASA replaced the very successful family of TIROS spacecraft with a series of advanced weather satellites called Nimbus (after the Latin word for cloud). Among their numerous technical advances, the Nimbus family of satellites contributed an especially important improvement in space-based meteorology. They all flew in near-polar, Sun-synchronous orbits, allowing meteorologists to piece their data together into mosaic images of the entire globe. As remarkable as the development of civilian polar-orbiting, low-altitude satellites was in the 1960s, this achievement represented only half of the solution to high-payoff space-based meteorology.

To completely serve the needs of the global weather-forecasting community, geostationary operational weather satellites capable of providing good-quality hemispheric views on a continuous basis were also needed. In 1966, NASA placed the *Applications Technology Satellite (ATS-1)* in geostationary orbit over a Pacific Ocean equatorial point at about 150 degrees west longitude. In December, its spin-scan cloud camera began transmitting essentially continuous photographic coverage of most of the Pacific Basin. For the next few years, this very successful experimental satellite provided synoptic cloud photographs and became an important part of weather analysis and forecast activity for this data-sparse ocean area.

ATS-3, launched by NASA in November 1967, provided a similar impact on meteorology. From its particular geostationary vantage point, the satellite's field of view (FOV) covered much of the North and South Atlantic Ocean area, all of South America, most of North America, and the western edges of Africa and Europe. This experimental spacecraft carried an advanced multicolor spin-scan camera that initially provided red-, green-, and blue-colored outputs until the red channel failed during the first year of operation. For a time, this spacecraft's photographs represented the best full-face images of our planet, until they were replaced by the more spectacular "whole-Earth" images taken during the Apollo lunar missions.

Scientists from NASA and NOAA used both *ATS-1* and *ATS-3* data to pioneer important new weather-analysis techniques. Perhaps even more important from a meteorologist's perspective than the cloud-system and wind-field data available from the geostationary-satellite photographs were the relatively small-scale weather events that atmospheric scientists could now observe on an almost continuous basis. The ATS cameras, repeating their photographs at about 27-minute intervals, showed that geostationary weather satellites could watch a thunderstorm develop from cumulus clouds and improve the early detection of severe weather. ATS data also became a routine part of the information flowing into the National Hurricane Center in Florida. In August 1969, for example, *ATS-3* helped track Hurricane Camille and provided reliable and timely warning for the threatened area along the Gulf Coast of the United States.

The outstanding technical accomplishments of these ATS spacecraft formed the technical foundation for an important family of Geostationary Operational Environmental Satellites (GOES) currently used by NOAA to provide a complete line of forecasting services and severe-weather warning throughout the United States and around the world. When NASA launched *GOES-1* for NOAA in October 1975, the field of high-payoff space-based meteorology became fully operational.

Today, the weather satellite is an indispensable part of modern meteorology and influences everyday life. Most television "weather persons" include a few of the latest satellite cloud images to support their forecasts. Professional meteorologists use weather satellites to observe and measure a wide range of atmospheric properties and processes in their continuing effort to provide ever more accurate and timely forecasting services and severe-weather warnings. Imaging instruments provide detailed visible and near-infrared images of clouds and cloud motions, as well as measurements of sea-surface temperature. Atmospheric sounders collect data in several infrared or microwave spectral bands. When processed, these data provide useful profiles of moisture and temperature as a function of altitude. Radar altimeters, scatterometers, and synthetic aperture radar (SAR) imager systems measure ocean currents, sea-surface winds, and the structure of snow and ice cover.

NOAA's operational environmental-satellite system consists of two basic types of weather satellites: Geostationary Operational Environmental Satellites (GOES) for short-range warning and "now-casting" and Polar-Orbiting Environmental Satellites (POES) for longer-term forecasting. As mentioned previously, both types of weather satellites are necessary for providing a complete global weather-monitoring system.

Geostationary weather satellites provide the kind of continuous monitoring needed for intensive data analysis. Because they stay above a fixed spot on Earth's surface and are far enough away to provide a full-disk view of Earth, the GOES spacecraft provide a constant vigil for the atmospheric "triggers" of severe-weather conditions such as tornadoes, flash floods, hailstorms, and hurricanes. When these dangerous weather conditions develop, the GOES satellites monitor the storms and track their movements. Meteorologists use GOES imagery to estimate rainfall during thunderstorms and hurricanes for flash-flood warnings. They also use weather-satellite imagery to estimate snowfall accumulations and the overall extent of snow cover. These data help meteorologists issue winter storm warnings and spring snow-melt advisories.

The NOAA polar orbiters monitor the entire Earth, tracking atmospheric variables and providing atmospheric data and high-resolution cloud images. These NOAA spacecraft primarily track meteorological patterns that affect the weather and climate of the United States. The satellites provide visible and infrared radiometer data that are used for imaging purposes, radiation measurements, and temperature profiles. Ultraviolet-radiation sensors on each polar-orbiting spacecraft monitor ozone levels in the atmosphere and help scientists keep watch on the "ozone hole" over Antarctica. Each day, these polar-orbiting satellites perform more than 16,000 global measurements, thereby providing valuable information to support forecasting models, especially for remote ocean areas, where conventional data are lacking.

Nowhere have operational weather satellites paid their way more demonstrably and made a greater impact on society than in the early detection and continuous tracking of tropical cyclones—the hurricanes of the Atlantic and the typhoons of the Pacific. Few things in nature can compare to the destructive force of a hurricane. Called the greatest storm on Earth, a hurricane is capable of annihilating coastal areas with sustained winds of 250 kilometers per hour or higher and intense areas of rainfall and a storm surge. In fact, scientists estimate that during its life cycle a major hurricane can expend as much energy as 10,000 nuclear bombs. Before the arrival of the weather satellite, life was extremely difficult and risky for persons who lived in hurricane country. Today, because of the weather satel-

Figure 5.2 A computer-generated image of Hurricane Fran, using data from GOES weather satellites (August 1, 1989). Illustration courtesy of NASA.

lite, meteorologists can provide people who live in at-risk coastal regions timely warning about the pending arrival of a killer storm (see Figure 5.2).

Commercial High-Resolution Satellite Imagery

Totally independent of the U.S. government's or Russian Federation's satellite photoreconnaissance systems, commercial high-resolution (one-meter object size or less) imagery from Earth-observing satellites now make the world highly "transparent" and stimulate an information-application revolution within the great information revolution of the twenty-first century. As more and more people become familiar with the robust information content of commercial high-resolution satellite imagery, the uses of these data are rapidly diffusing into many new business sectors and the personal lifestyles of individuals. Today, high-quality commercial remote sensing from space serves as the primary driving force behind innovative geographic information system (GIS) and mapping activities, replacing (or at least significantly complementing) the use of traditional aerial photographs. For example, high-accuracy image maps provide much more detailed information than conventional line-drawn maps. Contemporary image maps, derived from satellite imagery, often include buildings, automotive vehicles and parking lots, bridges and highway overpasses, vegetation, natural land features (such as a swamp or wetlands), and cultural land features (such as historic monuments or the ruins of an ancient city or fortification).

This portion of the chapter introduces several of the more interesting contemporary applications of commercial high-resolution satellite imagery. However, this list of applications will grow exponentially in the next two decades as creative individuals and companies take the luxury out of commercial high-resolution satellite imagery and transform such imagery into an absolute necessity in hundreds of new and exciting information-services markets. A similar situation occurred in 1876 when Alexander Graham Bell (1847–1922) invented the telephone. Human voice transmission via electronic signals carried by wire penetrated the existing telegraph and cable markets as more nineteenth-century businesspeople discovered the numerous uses and untapped potential of this novel information-transfer "luxury." Within a decade or so, the telephone became an absolute communications necessity both in the workplace and at home.

Quite similarly, individuals, organizations, and businesses that previously had no way of acquiring high-resolution (military-quality) satellite imagery on a timely and reliable basis are now discovering this valuable information resource. The transformation represents a great information power shift from government control and very limited distribution of very powerful data-intense resources to open collection and worldwide distribution of roughly equivalent information resources in a free-market environment.

Contemporary high-resolution satellite-imagery markets exist in numerous commercial, government, or consumer sectors, including agriculture, archaeology, commercial news gathering (media), disaster response, rescue and relief, the environment, mineral exploration, infrastructure development, insurance, local government, national emergencies, law, and utilities. We can highlight only a few of the many interesting applications here.

Agriculture, especially American agriculture, depends on high technology to feed a growing world population. An unstable food supply is a great threat to national security. The world population is now approximately six billion people and is anticipated to reach nine billion by the year 2050. According to the United Nations Food and Agriculture Organization (FAO), about 800 million people currently suffer from hunger and malnutrition, and 24,000 people (mostly children under five years of age) die each day from hunger and hunger-related causes. To help combat this problem, high-resolution satellite imagery is supporting global food-security analysis and crop-damage assessments (from severe weather, natural and human-caused disasters, and insect infestations).

High-resolution satellite-acquired images also nurture a form of information-intense agriculture called *precision agriculture*. Properly analyzed and interpreted, such images lead to very accurate crop forecasts for use by government officials and emergency relief agencies, as well as commodities traders. The overall concept of global food security in the twenty-first century relies on timely and accurate crop-yield estimates, for which Earth-observing spacecraft and their unobstructed view of the entire planet are the key. When certain agricultural regions are in a drought pattern, for example, governments and/or commercial food-production companies can base national food-production plans on more intense yields from other agricultural areas not affected by the drought to meet anticipated domestic and global consumption rates. For the first time in history, human beings have a powerful information resource at their disposal to help in the equitable management of food production on a truly planetary basis.

High-technology farmers can use commercial, high-resolution multispectral imagery from space (now available at about four meters' resolution or better) to quickly observe an emerging problem in their croplands. Once they are alerted, crops can be saved and yields increased through the timely administration of mitigating actions, such as using special nutrients, additional water, or pesticides. For example, plant stress resulting from insufficient watering, nutrient deficiencies, or insect attacks becomes very apparent in high-resolution, multispectral satellite imagery—quite often long before the same threatening condition is obvious to a person making a ground-level physical inspection.

Here are two interesting examples involving specialized cash crops. Vineyard managers in California now use high-resolution satellite imagery to assess vine health, to monitor nutrient status, to schedule harvesting, and to battle against the spread of destructive pests. Oil-palm-plantation managers in Malaysia use high-resolution satellite imagery to count palm trees so they can efficiently monitor the production of palm oil (a major tropical-region crop) and assess the value of a particular plantation. Manual tree counting takes time and often includes an unacceptable level of human error. However, since the crown of an individual palm tree is distinguishable in high-resolution satellite imagery, automated image-analysis techniques can perform a tree census with an accuracy of better than 90 percent of the actual tree population.

High-resolution satellite imagery plays a major role in monitoring environmental changes that are taking place in a particular region. Individuals concerned with environmental stewardship (from government agencies, nongovernmental organizations [NGOs], or private industry) use satellite imagery to establish an accurate environmental baseline in a particular region and then monitor and record changes that occur with respect to this documented (digitized) baseline. When effectively integrated into a modern, computer-based geographic information system (GIS), high-resolution color and multispectral imagery can help planners and regulators assess vegetation-cover health and stress, ensure compliance with land- and water-use regulations (including waste disposal and wetlands preservation), and assess the adequacy of wastewater treatment and stormwater management. Using this approach, undesirable changes in a local ecosystem (such as the pollution of surface water or the destruction of wetlands) can be quickly identified and remedial actions taken before costly and possibly irrevocable damage to the local ecosystem takes place.

Ecotourism is a rapidly growing industry in many of the world's remote and delicately balanced ecosystems. High-resolution satellite imagery allows government officials to work closely with private citizens and business organizations in effectively planning and monitoring the use of some popular, but fragile, ecosystem so that positive economic benefits (such as jobs and financial growth) are enjoyed without endangering or destroying the very environmental conditions that attract visitors.

One particularly interesting area within the ecotourism sector is called archaeological ecotourism—an emerging market, especially in developing countries (like Cambodia and Honduras) with abundant ancient ruins in remote areas. Satellite imagery can help locate undiscovered or lost ruins and ancient structures and then assist government organizations and private business enterprises in developing a strategy for the responsible use of each newly discovered site. In 1992, for example, fairly low-resolution Landsat imagery and data from NASA's space shuttle imaging radar system helped scientists locate the remains of the lost ancient city of Ubar (founded circa 2,800 B.C.E.) on the Arabian Peninsula in a remote part of the modern country of Oman.

Through space technology, suspected, lost, or presently unknown archaeological sites can become interesting, genuine ecotourist attractions, benefiting the local economy and generating a revenue stream that contributes to the scientific study and preservation of the site. Radar imagery cuts through clouds and dense vegetation and helps archaeologists detect the ruins of unexplored jungle cities, temples, and fortifications. High-resolution visible imagery provides adventurers, scientists, and archaeologists with a relatively inexpensive and safe "armchair" alternative for conducting exploratory probes of physically or politically inaccessible regions. Turkey's Mount Ararat—the reputed legendary resting place of Noah's Ark—is just such an example. Because of wartime conditions involving Kurdish guerrillas, the government of Turkey has prohibited tourism and expeditions to the suspected site. However, following in the path of previous military reconnaissance collections, high-resolution commercial satellite imagery tantalizingly shows what could be the partially exposed portions of a postulated human-made structure embedded in the mountain's ice cap. In the twenty-first century, a space-age Indiana Jones might be tempted to trade his (or her) bullwhip and explorer's hat for a portable computer loaded with high-resolution satellite imagery.

Finally, as a result of high-resolution commercial satellite imagery, politically denied areas of the world now are far more transparent to scrutiny by nongovernmental organizations (NGOs) and the news media. For example, independent of any government, the International Atomic Energy Agency (IAEA) can now apply high-resolution satellite imagery to support the objectives of its nuclear nonproliferation and international nuclear safeguards programs. One-meter-resolution imagery allows knowledgeable international "eyes in the sky" to peek in on plutonium production and processing facilities, such as those currently found in India and Pakistan, two nations with a long history of conflict and publicly demonstrated ambitions to develop into regional nuclear weapons states. While analysis of such high-resolution imagery may not always allow IAEA inspectors to make a final determination about suspicious activities at a particular nuclear facility, careful analysis of such imagery will alert agency officials to conduct a special on-site inspection, to perform environmental sampling operations, and to encourage nonproliferation-oriented political pressure through channels within the international diplomatic community.

Similarly, the era of commercial news gathering from space experienced a major milestone in April 2001. Following the April 1 midair collision between a U.S. Navy EP-3 reconnaissance aircraft and a People's Republic of China fighter aircraft, the American plane made an emergency landing at Lingshui military airfield on Hainan Island, China. (The damaged Chinese aircraft and its pilot were lost at sea.) Three days after the collision, the *IKONOS* commercial Earth-observing satellite snapped an image of the damaged U.S. Navy aircraft as it sat at one end of the airfield. The reconnaissance aircraft and its crew were suddenly pawns in a Sino-American political chess match. After a flurry of tense negotiations, the crew was eventually released unharmed, and the plane was dismantled and flown back in pieces to American territory. The special significance here is that newsworthy, high-resolution satellite imagery became available without the assistance of either the American or Chinese governments. Space imaging provided *IKONOS* data to the media for broadcast throughout the world. This act represents the first time a breaking news story taking place in a politically denied area was effectively covered by a high-resolution commercial Earth-observing satellite. From a historic perspective, in April 1986 much lower-resolution satellite imagery (from the U.S. Landsat and French SPOT spacecraft) helped confirm and monitor the Chernobyl nuclear-reactor explosion in the former Soviet Union, despite an initial lack of information from Russian officials.

Commercial high-resolution satellite imagery allows news services to investigate breaking stories, environmental disasters and relief efforts, military clashes, treaty violations, acts of terrorism, and the like from the vantage of outer space without the cooperation of national governments. A transparent globe gives great power and flexibility to the news media in democratic societies, but this empowerment also imposes the need for responsible interpretation and use of satellite-derived imagery data (see chapter 6). Can the twenty-first-century news media properly respond to the unique opportunities and challenges space technology now creates?

WIRELESS SWITCHBOARDS IN THE SKY

Early in the twentieth century, physicists and communications engineers recognized that radio waves, like other electromagnetic waves, propagate along the line of sight; that is, they travel in a straight line and cannot (of themselves) bend around the curvature of Earth. Consequently, a radio or television receiver cannot obtain broadcasts from a transmitter that lies beyond the horizon. The higher the transmitting antenna, the farther the line of sight available for direct-wave, wireless communications. This is the reason why antenna towers found on Earth's surface are so tall. Wireless-communications pioneers like Guglielmo Marconi (1874–1937) also discovered that under certain circumstances they could bounce radio waves of a certain frequency off ionized layers in Earth's atmosphere and thereby achieve long-distance shortwave radio broadcasts. However, the ionosphere as a natural phenomenon is subject to many irregularities and diurnal variations, making its use undependable for any reliable, continuously available wireless-communications system.

In 1945, the technology key to dependable, worldwide wireless communications became obvious to one creative individual—make the antenna tower incredibly tall by putting it on a platform far out in space in a special "fixed" orbit around Earth. That concept, first proposed by British space visionary Arthur C. Clarke, is the essence of the geostationary communications satellite—a space-technology application that transformed the world of wireless communications and helped stimulate today's exciting information revolution.

The communications satellite is an orbiting spacecraft that relays signals between two or more communications stations. In other words, the space platform serves as a very high-altitude switchboard without wires. Aerospace engineers divide such satellites into two general types: the *active communications satellite*, which is a spacecraft that receives, regulates, and retransmits signals between stations, and the *passive communications satellite*, which, like a mirror, simply reflects signals between stations. While NASA's first passive communications satellite, called *Echo 1*, helped demonstrate the use of orbiting platforms in wireless communications, hundreds of active communications satellites serve today's global communications infrastructure. The satellite communications industry (including long-distance and remote-area mobile and cellular telephone services) represents the largest segment of the commercial space-applications industry.

As first envisioned by Clarke, the communications satellite uses space technology to function like an extraterrestrial relay station. However, orbiting at an altitude of just 300 kilometers (the nominal operational altitude for NASA's space shuttle), a communications satellite can view only about 2 percent of Earth's surface at any one time. In this concept, one ground station (or Earth station) transmits a signal to this spacecraft, which then relays that signal down to another ground station hundreds to perhaps thousands of kilometers away from the sending station. Unfortunately, a satellite in low Earth orbit travels quickly around Earth and, therefore, views any particular ground station for only a few minutes. However, the higher the orbital altitude of a spacecraft, the farther it has to travel to circle Earth and the longer it remains in view of an Earth station as it passes overhead. At an altitude of approximately 35,900 kilometers, the object is in a special orbit that takes the spacecraft precisely one day to circle Earth. At this altitude, the satellite can view about 42 percent of Earth's surface at one time and, when placed above Earth's equator, appears to remain fixed at the same point in space to an observer on Earth's surface. Aerospace engineers call this special orbit a *geostationary orbit*, and it is of great importance to the communications-satellite industry.

A key advantage of geostationary orbit for the communications satellite is the fact that a ground (Earth) station has a much easier tracking and antenna-pointing job, because the satellite is always in view at a fixed location in space. In addition, the same geostationary communications satellite that is in view from the United States, for example, also has good line-of-sight viewing from Canada, Mexico, Brazil, Venezuela, Colombia, Argentina, and Chile. Because of this, many different countries want to deploy their own communications satellites in approximately the same part of geostationary orbit above the equator, causing a potential crowding problem or a signal-interference problem.

To help resolve these conflicts, international agreements now allocate operational frequencies, so two communications satellites can occupy the same general area of geostationary space and still provide high-quality, interference-free service to their many different customers. By using special curved antennas, modern communications satellites can also focus their transmission signals into narrower beams aimed at particular regions (or ground stations) back on Earth, thereby limiting interference with signals from other satellites.

The communications satellite has created a "global village" in which news, electronic commerce, sports, entertainment, and personal messages travel around the planet efficiently and economically at the speed of light. Time zones no longer represent physical or social barriers. Breaking news events flash around the world with unprecedented speed, often despite or without the control of government officials who might like to hide an unfavorable behavior or act. Mobile television news crews, equipped with the latest communications-satellite linkup equipment, have demonstrated an uncanny ability to pop up anywhere in the world just as some important event is taking place. Through their efforts and the linkages provided by communications satellites, television viewers around the world often witness significant events (good and bad) in essentially real time. No nation, region, or individual is isolated in the twenty-first century without taking very drastic measures to avoid such forms of global scrutiny.

With the help of communications satellites, information and news now diffuse rapidly through political, geographic, and cultural barriers, providing millions of information-hungry people objective, or at least alternative, versions of a particular story or event. This free flow of information throughout our planet ignites the flame of democracy among many politically oppressed societies and represents a major impact of space technology. Communications satellites are helping the classic proverb "The truth will set you free" take on a special, geopolitical significance in the information revolution of the twenty-first century. Of course, no technology, no matter how powerful or pervasive, can instantly solve social inequities and political animosities that have lingered for centuries. However, the rapid and free flow of information throughout the global village is a necessary social condition for creating an informed human family, capable of achieving enlightened stewardship of planet Earth in the twenty-first century.

Communications satellites and sports broadcasting have successfully formed another natural, highly profitable combination. The first sporting event (a major-league baseball game) ever broadcast via satellite happened on August 9, 1975. In that game, the Texas Rangers beat the Milwaukee Brewers by a score of 4 to 1, and the broadcast served as a pioneering information and space-technology experiment. Success quickly followed success. On September 30, 1975, the most important sporting event in establishing the communications satellite as a means of delivering sports entertainment to audiences around the world took place. On that date, broadcast live via satellite from Manila in the Philippines, Muhammad Ali met Joe Frazier in the world heavyweight boxing championship match, often dubbed "the Thrilla in Manila." This sporting event demonstrated the fundamental role of communications satellites in delivering sports entertainment packages both internationally and into rural areas throughout the United States through satellite-delivered pay-television services. The rest is entertainment-industry history.

The worldwide demand for satellite-based sports broadcasting now appears insatiable, at least for the next few decades. For example, because of communications satellites, an estimated 750 million people in 140 countries viewed Super Bowl XXXII on January 24, 1998—an immensely popular American football game in which the Denver Broncos defeated the Green Bay Packers by a score of 31 to 24. Sports have universal appeal, and the viewer market is worldwide and growing. When people in developing nations get their first television set, for example, one of the first categories of programming they demand is sports. The Olympic Games, the soccer World Cup, the World Series of baseball, and many other sporting events draw millions of viewers from every corner of the globe.

Today, to further satisfy growing demands for personal communications services (e.g., cellular phones), information and aerospace companies are exploring the use of constellations of smaller satellites in low Earth orbit. The cellular phone has become an integral part of information-rich living everywhere in the world. A nomad living in a remote portion of the Arabian Peninsula, a missionary in the Brazilian jungle, and an Eskimo hunter on an ice floe in Alaska all have at least one thing in common—with a cellular telephone they can be part of a "virtual electronic tribe" bonded together by invisible lines of wireless communications. Satellite-linked communications also played a dramatic role during the tragic terrorist attacks that took place in the United States on September 11, 2001. Brave passengers on four hijacked commercial jet aircraft sent loved ones tender final messages and attempted to help authorities by describing the terrorists through a variety of wireless communications pathways, including satellite-linked air-phone networks.

The direct-broadcast satellite (DBS) represents another applied space-technology innovation. The DBS receives broadcast signals (that is, television signals) from points of origin on Earth and then amplifies and retransmits these signals to individual end users throughout some wide area or specific region. For example, many households in the United States now receive their television programs directly from space. A DBS can deliver 100 or more channels of television programming to a private home or business through small (less than 0.5 meter in diameter), inconspicuous rooftop satellite dishes.

The information revolution supported by the communications satellite is just beginning. Through satellite technology, rural regions in developed or developing countries can enjoy almost instant access to worldwide communications services, including television, voice, facsimile, and data transmissions. The exponentially growing impact of the communications satellite on health care (e.g., telemedicine), the workplace (e.g., telecommuting), education (e.g., distant learning), electronic commerce (e.g., 24-hour-per-day participation in the global marketplace), and banking (e.g., rapid "digital" monetary transfers across international boundaries) is producing major social and economic changes within our global civilization. For further proof, just look at the international calling section of a current telephone directory. Such exotic and formerly remote locations as Ascension Island (area code 247), San Marino (area code 378), Greenland (area code 299), French Polynesia (area code 689), and Antarctica (area code 672) are now just a communications-satellite link away.

The Communications Satellite: A Catalyst for Democracy and Social Improvement

One of the major impacts of space technology is this marvelous transformation of planet Earth into an informed, socially interactive global village. As information begins to leak and then flood into politically oppressed populations along invisible lines from space, despotic governments are encountering an increasingly difficult time in denying freedom to their citizens and in justifying senseless acts of aggression against their neighbors. One of the most dramatic examples of how the communications satellite has become the tool and symbol of modern democracy came out of the streets of Kabul, Afghanistan, in November 2001. Just hours after the oppressive Taliban government fled from the capital city, satellite-uplinked news broadcasts showed jubilant Afghan citizens walking through the streets of Kabul, carrying their cherished (but long-hidden) satellite dishes. Now liberated, the first thing these long-suffering and information-denied people sought to restore was the inflow of unfiltered information and entertainment from the rest of the human race.

The electronic switchboard in space has now become the modern tyrant's most fearsome enemy. With their unique ability to shower countless invisible lines of information gently and continuously on the free and the oppressed alike, communications satellites serve as extraterrestrial beacons of freedom, supporting the inalienable rights of human beings everywhere to pursue life, liberty, and individual happiness.

Guardian Angels on High: Search and Rescue Satellites

Space technology also helps protect and save human lives. Many modern satellites (especially weather satellites) now come equipped with a search and rescue satellite-aided tracking (SARSAT) system that functions on an international basis to help locate people who are lost and who have appropriate emergency transmitters. (The Russians use "COSPAS"—an acronym in the Russian language that stands for "Space Systems for the Search of Vessels in Distress.") COSPAS-SARSAT–equipped weather satellites can immediately receive and relay a distress signal, significantly increasing the probability of a prompt, successful rescue mission. Sponsored and initiated in 1982 by Canada, France, Russia, and the United States, the satellite-based system aims to reduce the time required to alert rescue authorities whenever a distress situation occurs. The rapid detection and location of a downed aircraft, a sinking ship, or a lost or injured individual are of paramount importance to survivors and rescue personnel.

By the close of 2001, almost 13,000 persons worldwide and more than 4,000 persons in the United States had been rescued by this satellite-based system. For example, on March 2, 2001, the COSPAS-SARSAT system detected a 406-megahertz (MHz) distress signal from the Atlantic Ocean east of Cape Hatteras, North Carolina. (In the COSPAS-SARSAT system, registered rescue beacons that transmit between 406.0 and 406.1 MHz send digitally encoded information that includes a beacon identification for accessing a user registration database.) The sailing vessel *High Noon* was taking on water while in transit from the West Indies to England. Once alerted by the satellite rescue system, the U.S. Coast Guard in Norfolk, Virginia, diverted the merchant vessel *Putney Bridge* to the area of the distress signal, and the merchant ship rescued the two persons who had abandoned their sinking ship.

PHILOSOPHICAL IMPACT OF SPACE TECHNOLOGY

The major philosophical impact of space technology is the off-planet expansion of human consciousness. For the very first time since life emerged on Earth, the use of space technology allows conscious intelligence to leave its terrestrial cradle and meet the universe face-to-face. Space technology offers human beings an *open-world philosophy*. An open-world philosophy considers expansion off the planet into the solar system and beyond as a logical extension of the great evolutionary unfolding of consciousness that started here some 350 million years ago when living creatures left the seas of an ancient Earth and crawled upon the land for the first time. In contrast, a *closed-world philosophy* rejects space as a major pathway for the sociotechnical development of the human race and limits future human activities to a single planet.

Space technology provides the tools to find out whether life, including possibly intelligent life, is a common part of cosmic evolution within the universe. The scenario of cosmic evolution postulates that there is an overarching synthesis or purpose for the long series of alterations of matter and energy that started with the big-bang explosion. Over time, matter formed and then slowly evolved into the Milky Way galaxy, the solar system, planet Earth with its biosphere capable of supporting life, and finally intelligent human life.

The German space travel visionary Hermann Oberth (1894–1989) suggested a possible destiny for the human race when he responded to a reporter's probing question, "Why space travel?" Oberth replied: "To make available for life every place where life is possible. To make inhabitable all worlds as yet uninhabitable, and all life purposeful." In the deepest philosophical sense, space technology offers human beings the universe as both a destination and a destiny.

Chapter 6

Issues

> It is difficult to say what is impossible, for the dream of yesterday is the hope of today and the reality of tomorrow.
>
> Robert H. Goddard

The application of space technology has dramatically altered the course of history. In the previous chapter, we examined some of the very exciting ways that space technology is changing the lives of every human being on Earth. Within some societies, the impact of these changes is already exceptionally large. Within other societies, the impact is less immediate and more subtle. But everywhere on this planet, space technology now exerts a permeating influence that provides unique opportunities for continued technical, social, and intellectual development in the twenty-first century and beyond. However, important technical and social issues frequently accompany the rapid introduction of any powerful new technology. In this chapter, we examine several of the major issues associated with the application of space technology. For ease of discussion, we divide the space-technology-related issues into three broad categories: technical issues, volatile issues, and horizon (or twilight) issues.

Technical issues currently limit the application of space technology, but attempts at resolution generally do not produce strong opposition—that is, there are no pro or con positions associated with the particular issue. We examine three dominant issues within this category: reducing the cost of accessing space, the hazards of human spaceflight, and the growing problem of space debris. The objectives in resolving these particular issues are quite clear and essentially uncontested. What is not necessarily clear, however, is precisely how best to resolve each issue from a technical, economic, or operational perspective. The pathway leading to an optimum technical solution is the subject of much professional debate within the aerospace industry and usually requires a combination of improvements in existing technology, sustained funding, risk taking, and failure tolerance.

Volatile issues emerge from the current use of space technology and produce very strong pro or con positions. We discuss three such space-technology-related volatile issues in this chapter: space nuclear power, sovereignty and personal privacy in the era of high-resolution commercial satellite imagery, and the use of weapons in space. This portion of the chapter includes sufficient technical background to appreciate the basic principles involved with each issue and the major pro and con positions. Seldom, if ever, does the resolution of volatile issues spontaneously generate a "for-the-common-good" consensus. Rather, their resolution usually requires an official government-endorsed decision, driven by prevailing technical, economic, political, and national policies and exercised within the limits of international space law. Even after an official decision is made concerning a volatile issue, strong disagreement lingers within democratic societies.

The third broad category of space-technology-related issues is the horizon (or twilight) issues, so named because they currently reside on the technical horizon in the distant future or in the speculative, twilight world between science fiction and science fact. This chapter examines three such issues: extraterrestrial contamination, responding to an alien message, and ownership and use of space resources. Horizon issues usually encompass a speculative technical, social, or political consequence that could arise from the hypothetical extrapolation of space technology. A full discussion of a horizon issue not only requires an appreciation of projected developments in space technology, but also must consider possible modifications of international space law and various national space policies.

For example, can a commercial company extract and sell lunar-water ice for a profit? Since current international space law states that no nation can exert sovereignty over other celestial bodies, does the harvesting of lunar resources for profit by a company incorporated in a particular nation on Earth imply that that nation is now exercising sovereignty over the Moon? Discussion of such twilight issues involves speculative extrapolation of space technology, as well as scenario construction involving institutional responses to projected future events.

TECHNICAL ISSUES

Reducing the Cost of Accessing Space

Impeding any greatly expanded application of space technology is the critical technical issue of affordable and reliable access to space. The excessively high cost of space transportation (about $5,000 to $20,000 per kilogram of mass delivered into low Earth orbit in 2003), coupled with continuing reliability and safety concerns, places a tight grip on any significant expansion of human activities beyond the boundaries of Earth. Dramatic improvements in launch-vehicle technology must take place to make space transportation more affordable and safer.

The need to reduce the high cost of accessing space is very apparent to aerospace engineers, mission planners, and space entrepreneurs around the world. What is not so clear, however, is how best to tackle the challenge of reducing the prohibitively high cost of delivering people and cargo into low Earth orbit and to orbital destinations beyond near-Earth space. At present, access to space is provided by a variety of expendable launch vehicles (ELVs) and the world's only partially reusable aerospace vehicle, the U.S. Space Transportation System (STS), or space shuttle.

With the exception of the space shuttle, all modern chemical rockets used to place payloads into space fall within the broad category of expendable launch vehicles (ELVs). This term means that all of the rocket vehicle's flight components, such as its engines, propellant tanks, guidance and navigation equipment, and support structure, are simply discarded after just one use. For staged rockets (that is, rocket vehicles with two or more propulsion units stacked vertically, one on top of the other), expended rocket units are jettisoned (discarded) in increments or steps during the ascending flight up through Earth's atmosphere. As each lower stage consumes its supply of chemical propellants, its empty tanks are discarded. In this rocket-vehicle design approach, the upper (or later) propulsion stages function more efficiently, since they have less and less useless mass to carry into orbit.

The "throwaway" launch-vehicle philosophy emerged during the Cold War, when Soviet and American officials pressed their newly developed intercontinental ballistic missiles into additional service as the world's first space launch vehicles. At the time, government officials regarded this rather wasteful approach to space transportation as both politically and technically expedient. However, their decision also proved quite expensive. Decades of throwing away the entire space launch vehicle have created a deeply embedded disposable-rocket mindset within much of the aerospace community, but aerospace-industry people also complain bitterly about the great expense and recognize the critical need for innovative engineering solutions to reduce launch costs.

The following hypothetical example, already mentioned in chapter 4, should help illustrate the significance of this problem. Imagine the price of a commercial airline ticket for a person to fly nonstop from New York to San Francisco if the airline simply discarded the entire passenger aircraft after each flight. There would be very few "frequent flyers," and only the most essential defense-related, scientific, or commercial trips would receive approval and funding. Today, excessively high transportation costs impose very similar constraints on the expanded use of space.

The earliest design concepts for a reusable, delta-winged aerospace vehicle appeared in the American space program of the late 1960s and represented the long-sought solution to the cost-inefficient throwaway-vehicle problem. With presidential approval, NASA officially began its space shuttle program in 1972. However, the early optimism and unrealistically low projected payload-hauling costs (some early estimates suggested less than $100 per kilogram) soon gave way to engineering setbacks and political realities. The end product was a partially reusable aerospace vehicle that reflected fiscally pressured design compromises and the overall political vagueness characteristic of a culturally paralyzed NASA in the post-Apollo era. It was very difficult for NASA personnel to enthusiastically pursue the construction of a reusable aerospace vehicle while trying to respond to the question "Where do we go after we've been to the Moon?" Within the fading glory of the spectacular Moon-landing missions, trying to rally national attention and international space-technology prestige on "a taxi ride into low Earth orbit" hardly seemed possible. In fact, when NASA's space shuttle emerged as the world's first aerospace vehicle in 1981, it did not even have a regular destination in low Earth orbit since a program to construct a permanent space-station was still almost two decades away.

There is little debate within today's aerospace community that the space shuttle vehicle is an incredibly complex machine that has expanded the uses of space technology in a number of important areas. It is also generally recognized that the space shuttle, with its more gentle ascent-acceleration environment (about 3 g maximum), opened spaceflight to a larger population of human beings, male and female, who now represent many technical professions, ethnic groups, and nationalities. However, despite these important accomplishments, the "reusable" space shuttle system failed to significantly reduce the cost of sending objects into space. The transportation fee for a typical shuttle-carried payload is about $20,000 per kilogram for delivery into low Earth orbit.

What went wrong? Why did the aerospace engineers and project managers miss this important economic target so badly? Engineering difficulties, material limits, and an overall loss of political vision following the spectacular Apollo Project sit high on the list of factors that helped keep the dream of inexpensive, reusable space transportation vehicles from becoming a reality. Numerous design compromises, forced by unrealistic budget cuts during the vehicle's development phase, produced a space shuttle that was only partially reusable and one that required very time-consuming and expensive overhauls between each spaceflight.

As the history of the American space program amply reveals, a trip into space, using either an expendable launch vehicle or the space shuttle, requires complex, high-performance rocket engines and large quantities of potentially explosive chemical propellants. Customized manufacturing of rocket hardware, precision assembly, and careful preflight testing and inspection make each flight into space a very expensive and labor-intensive activity. Yet despite all the special handling and precautionary steps, a launch attempt will sometimes end in dramatic failure as the rocket vehicle explodes, destroying its expensive payload in the process.

The physics solution for this issue is quite straightforward. Since the beginning of the space age (1957), the well-demonstrated way of getting into space has involved the use of a powerful rocket vehicle capable of lifting itself, large amounts of propellant, and any intended payload to velocities of about 8 kilometers per second (km/s). Rocket scientists use the term *mass fraction* as a performance indicator to describe how efficient a particular rocket design is in sending a payload into orbit. They define the mass fraction as the mass of propellant the vehicle needs to acquire a specific orbit divided by the gross liftoff mass of the vehicle (including its propellant load, structure, and payload). The smaller the mass fraction, the better the performance of the rocket vehicle. As an example, the U.S. Air Force's mighty Titan IV expendable multistage launch vehicle has a mass fraction of about 0.87. This means that approximately 87 percent of the vehicle's total mass is propellant.

To make access to space more affordable, aerospace engineers must obey the laws of physics while still performing modern feats of magic as they bend metal and shape new materials into future launch vehicles that can provide a significantly less expensive and more reliable ride into orbit. One concept under consideration is the fully reusable launch vehicle (RLV). The term *fully reusable* means just that. Between space missions, the proposed vehicle would simply require inspection, refueling, and payload processing, much like the handling of a modern jet airliner at a typical airport terminal. Unlike the partially reusable space shuttle, the conceptual single-stage-to-orbit (SSTO) RLV would not have any major components requiring extensive refurbishment after each flight. However, since a reusable vehicle has no throwaway (expendable) components, it must carry more propellant than if it discarded empty propellant tanks during launch ascent.

Within the U.S. government, both NASA and the Department of Defense (U.S. Air Force) recognize that dramatic improvements in launch-vehicle technology are necessary if space transportation is to become more affordable and safer. The best available expendable launch vehicles have a typical reliability somewhere between 90 and 95 percent. This means that a space-mission planner must anticipate (on a statistical basis) that between 5 and 10 rocket vehicles will fail out of every 100 launch attempts. Accessing space is still a very high-risk business. Engineering improvements during the first decade of the twenty-first century will improve these ELV reliability figures somewhat and reduce the overall cost of delivering payloads into LEO by perhaps 10 percent (say, from about $10,000 per kilogram to $9,000 per kilogram), but there is only so much improvement that the aerospace engineers can infuse into the existing family of vehicles.

Within the American aerospace industry, a debate goes on involving another important issue embedded within that of overall access to space. This derivative issue centers around this question: How far should the aging space shuttle fleet be pushed in extended service over the first two decades of the twenty-first century? An intense search is on for a replacement aerospace vehicle that can more safely and reliably carry people and cargo into low Earth orbit at a significantly reduced cost, but aerospace-industry experts also realize that even under the most optimistic development scenarios, an advanced launch vehicle capable of replacing the functions of the space shuttle is at least a decade away from providing reliable service. A dangerous space transportation gap could develop within the next decade if the space shuttle fleet becomes unserviceable and a suitable replacement vehicle is not yet available to take human beings and cargo into low Earth orbit. Under such unfavorable circumstances, routine access to the permanently crewed *International Space Station* would then require extensive use of foreign (primarily Russian) human-rated expendable launch vehicles and foreign launch sites.

Today, the United States is pursuing demonstration efforts that will pioneer the advanced technologies needed to move space transportation closer to the airline style of operations with horizontal takeoffs and landings, quick turnaround times, small ground support crews, and greatly reduced costs. Some of these contemporary projects have ended in expensive failure, while others continue in the hopes of achieving their important objectives.

The X-33 vehicle program attempted to simultaneously demonstrate several of the advances in space technology needed to increase launch-vehicle safety and reliability and to lower the cost of placing a kilogram of mass into space from about $20,000 to about $2,000. However, after several years of disappointing setbacks and serious cost overruns, anticipated project milestones were not being achieved, so NASA canceled the effort in March 2001. The X-33 vehicle was a half-scale prototype of a proposed single-stage-to-orbit (SSTO), reusable launch vehicle (RLV) the Lockheed-Martin company called the *VentureStar*. The goal of that commercial program was to flight-demonstrate all the critical technologies needed for private industry to build and operate a successful RLV early in this century. That particular dream and several billion dollars are now lost. NASA's abrupt cancellation of the X-33 program made a very definitive statement: inexpensive and reliable access to space still remains the most important, difficult, and challenging space-technology issue facing the modern aerospace industry.

NASA's X-34 vehicle served as a suborbital flying laboratory for several technologies and operational procedures applicable to the development of future, low-cost, reusable launch vehicles. The program involved a partnership between NASA and the Orbital Sciences Corporation of Dulles, Virginia. Despite progress in X-34 vehicle development and testing, NASA decided to cancel the program in early 2001 due to funding constraints and other factors.

NASA's X-37 vehicle is an advanced technology demonstrator that will help define the future of reusable space transportation systems. It is a cooperative program involving participation by NASA, the U.S. Air Force, and the Boeing Company. The X-37 is a reusable launch vehicle designed to operate in both the orbital and reentry phases of spaceflight. The robotic space plane can be ferried into orbit by the space shuttle or by an expendable launch vehicle. It will demonstrate many new airframe, avionics, and operational technologies that support future spacecraft and launch vehicle designs. The X-37's shape is a 120 percent scale derivative of the U.S. Air Force's X-40A space vehicle, also designed and built by Boeing.

To permanently resolve the space transportation dilemma, however, aerospace engineers now look beyond the aging space shuttle and the current "X" rocket-plane programs to a third generation of reusable launch vehicles. Envisioned as information-age vehicles, they will incorporate a wide variety of cutting-edge technologies, including smart materials and intelligent vehicle health-management systems that allow the launch vehicle to determine its own operational status without human intervention or inspection. Smart sensors embedded throughout the vehicle will send signals to determine if any damage or threatening wear and tear has occurred during a particular flight. Upon landing, the vehicle's onboard computer will download a detailed vehicle health-status report to a ground controller's computer, recommend specific maintenance tasks, and even let the launch site know when it is ready for the next journey into space.

Rocket scientists are also exploring the use of air-breathing rocket engines and magnetic levitation (maglev) to push the cost of accessing space to well under $100 per kilogram by the year 2025. An air-breathing engine (or rocket-based, combined-cycle engine) would obtain its initial takeoff thrust from specially designed rockets, called air-augmentation rockets, that boost performance about 15 percent over conventional rocket engines. When this vehicle's velocity reached twice the speed of sound (a condition called Mach 2), the air-augmentation rockets would be turned off, and then (like the engines on a modern jet aircraft) the combined-cycle engines would totally rely on oxygen extracted from Earth's atmosphere to combust the fuel (most likely liquid-hydrogen). Finally, once the vehicle's speed increased to about 10 times the speed of sound (Mach 10), the combined-cycle engines would convert back to conventional rocket-engine systems that combust both fuel and oxidizer obtained from tanks carried by the vehicle. The combined-cycle engines would then continue to function in this rocket-engine mode until the vehicle achieved orbital velocity.

Magnetic levitation (maglev) technologies may help send payloads into orbit by using electromagnetic forces to initially accelerate the launch vehicle along a track. Just as high-strength electromagnets lift and propel high-speed trains and amusement-park roller coasters above a guideway, a maglev launch-assist system would use electromagnetic forces to drive an advanced space launch vehicle along a special, upward-sloping track. When the magnetically levitated vehicle reached a speed of about 1,000 kilometers per hour, its rocket engines would ignite and send it into orbit.

The development of a reusable launch vehicle in the twenty-first century also introduces some interesting derivative issues related to space transportation. Because of its anticipated operational simplicity, the SSTO RLV can (in theory) fly into space from just about any place on Earth, and its routine operation will require only a minimal launch-complex infrastructure. Consequently, nations that lie along or near Earth's equator (like Brazil) might suddenly become very competitive spacefaring countries because of the favorable west-to-east spin of the planet that gives a natural boost to any vehicle launched eastward from an equatorial (low-latitude) location.

Future commercial RLV initiatives within a global free-market economy might make today's government-sponsored space agencies and their sprawling billion-dollar launch complexes obsolete. Rogue nations or well-financed terrorist groups might even acquire instant access to space and then try to perform hostile acts against the military or commercial space systems of more powerful countries. Finally, as more and more RLV flights ascend into orbit from launch sites scattered all over the planet, the first space traffic jam could occur. Prevention of such orbital gridlock and of catastrophic collisions between RLVs, operational space platforms, derelict spacecraft, and chunks of space debris could require the services of an international space traffic agency empowered to monitor and regulate space-vehicle activities in low Earth orbit.

Hazards of Human Spaceflight

Another critical issue centers around this probing question: "Why send people on high-risk missions when machines can do the job, often at a far lower cost?" This debate extends across the entire spectrum of space missions from planetary exploration to in-orbit assembly, construction, and repair. It is a very difficult question to ignore and an even more difficult question to respond to. On one side are technical experts who strongly advocate the exploration and use of space, but want to minimize the cost and the risk of human life by using smart machines (i.e., an evolutionary family of space robots) to the greatest extent possible. On the other side are the human spaceflight advocates who point out that people are far more flexible and intelligent than machines. They also suggest that it is an inherent characteristic of human beings to explore the unknown. Between these opposing positions, there is a middle-ground group of space advocates who see the successful expansion of space activities in the twenty-first century as a symbiotic partnership between human beings and their very smart machines.

Human spaceflight remains a hazardous, high-risk endeavor. To survive, each space traveler must practice BYOB (bring your own biosphere). There are several significant health and safety-related issues that space-mission planners must address to accommodate an expanding extraterrestrial population. These issues include (1) preventing launch-abort, spaceflight, and space-construction accidents; (2) preventing failures of life-support systems; (3) protecting human beings from chronic exposure to excessive quantities of the space (ionizing) radiation environment; (4) protecting space vehicles and habitats from lethal collisions with space debris and meteoroids; (5) minimizing or avoiding the long-term physiological consequences of continuous exposure to microgravity (weightlessness); and (6) designing spacecraft crew compartments

This unusual photograph, taken during the *Apollo 12* lunar landing mission, shows an astronaut visiting the *Surveyor 3* robot spacecraft. NASA used several Surveyor spacecraft between 1967 and 1968 to carefully examine the lunar surface prior to sending human beings to the Moon. *Apollo 12* astronauts Charles (Pete) Conrad and Alan Bean visited this robot precursor on November 20, 1969, during their own lunar-surface activities. (The astronauts' lunar excursion module [LEM] appears on the horizon in the background of this picture.) The image heralds the synergistic partnership between human explorers and smart robot exploring machines necessary for the successful exploration and settlement of the solar system in the twenty-first century and beyond. Photograph courtesy of NASA.

and permanent habitats with good-quality living and working conditions that minimize psychological stress.

In the twentieth century, several human-crewed space missions conducted by either the United States or the former Soviet Union ended in tragedy. For example, on January 27, 1967, Apollo astronauts Virgil (Gus) Grissom, Edward White II, and Roger Chaffee died during a training accident (flash fire) in their *Apollo 1* spacecraft at Complex 34 in Cape Canaveral, Florida. On April 23, 1967, the Soviet Union lost Cosmonaut Vladimir Komarov during a reentry accident. A second Soviet space tragedy occurred in June 1971 at the end of the *Soyuz 11* mission. This fatal accident claimed the lives of cosmonauts Georgi Dobrovolsky, Viktor Patsayev, and Vladislav Volkov and occurred during their automatic reentry operation. Finally, on January 28, 1986, the space shuttle *Challenger* lifted off from the Kennedy Space Center, and just under 74 seconds into the flight, an explosion destroyed the vehicle and claimed the lives of its seven crew members.

The microgravity environment of an orbiting spacecraft can and does affect its human occupants. The condition of continuous free fall, while considered a pleasant, even euphoric, experience, also plays tricks on the body. Many space travelers become disoriented. Without the tug of gravity, there is no up or down sensation to help the body's sensors navigate the new environment. The inner ear will often send confusing data back to the brain, and the astronaut's eyes can experience a variety of optical illusions. As a result of this sensory mix-up, about half of astronauts and cosmonauts experience a form of motion sickness called *space adaptation syndrome*.

The onset of space sickness normally occurs within a few minutes or possibly hours after launch, when the space traveler initially encounters the continuous microgravity environment of orbital flight. Space-sickness symptoms include nausea, vomiting, and general malaise. Fortunately, this uncomfortable condition is only temporary, usually lasting no more than 24 hours or so. Despite years of investigation by space-medicine specialists, medications can still only treat the symptoms, but cannot prevent the occurrence of space sickness. An individual's spaceflight experience offers no immunity, since this common space-travel malady afflicts veteran and rookie astronauts without distinction.

The most critical biological hazard associated with prolonged exposure to microgravity is now considered to be bone loss due to calcium depletion. Healthy astronauts have strong bones until they fly in space and experience continuous exposure to microgravity, which triggers bone loss. In microgravity, the density in load-bearing (weight-bearing) bones appears to decline at a rate of about 1 or 2 percent a month. Strategies are now being developed to overcome this adverse effect of microgravity. Astronauts and cosmonauts can participate in a rigorous, though boring, daily exercise program while in orbit. Medication and nutritional supplements can also lessen bone depletion.

One way around this problem, especially for very long-term missions, is to provide the spacecraft or platform with acceptable levels of artificial gravity by slowly rotating portions of the crewed vehicle or facility. Very large space settlements will most likely offer the inhabitants a wide variety of gravity levels, ranging from microgravity up to and including normal Earth gravity (one-g) levels. This multiple-gravity-level option should make space-settlement lifestyles more interesting and diverse than those found on Earth (or the Moon or Mars). The orbiting settlement's multiple-gravity-level condition will also help prepare planetary pioneers for life on their new worlds or help condition other space explorers who are returning to Earth.

Maintaining the physical and psychological health of Mars-expedition astronauts during an anticipated three-year interplanetary journey will be a major space-medicine challenge in the twenty-first century. For example, after about 500 days in microgravity, one typical 45-year-old astronaut could experience a sufficient amount of bone deterioration so that his/her leg bones would resemble the bones of a person severely afflicted with osteoporosis. Extending this spaceflight hazard scenario a bit further, such extensive bone deterioration might transform the astronaut's historic first steps onto the soil of the Red Planet into an incapacitating leg-bone-snapping event that would all but destroy the surface-exploration portion of the long-awaited expedition. Therefore, some Mars-mission advocates suggest avoiding this problem entirely by outfitting the expedition spacecraft with artificial gravity (i.e., slowly rotating crew compartments) for the expedition. With an advanced-design, simulated-gravity-level space vehicle, Mars-expedition astronauts might use some of the time on the outbound journey to fully condition themselves to living and working in a new surface-gravity environment. Then, on the long return journey back to Earth, the crew could gradually raise the artificial gravity level to approximately one g in preparation for their homecoming celebrations and debriefings.

The ionizing-radiation environment found in outer space is extremely hazardous and, without proper protection, can ultimately limit human activities there. Scientists characterize the space radiation environment as follows: geomagnetically trapped particles (primarily electrons and protons), solar-flare events (primarily energetic protons and alpha particles), and galactic cosmic rays. In low Earth orbit, electrons and protons are trapped by Earth's natural magnetic field, forming the Van Allen belts. The amount of ionizing radiation in LEO varies with altitude and solar activity. The trapped radiation belts are normally of concern only when astronauts attempt to travel from low Earth orbit (above about 500 kilometers altitude) to geostationary Earth orbit (GEO) or to the Moon. Outside Earth's protective magnetosphere, however, at geostationary altitude and beyond, unpredictable solar-particle events (SPEs) represent a major threat to space travelers. On journeys through interplanetary space, astronauts will also experience constant bombardment by galactic cosmic rays—an all-pervading collection of very energetic atomic particles, such as protons, helium nuclei, and very energetic heavy nuclei (called HZE particles). Space vehicles with extra-shielding "safe" areas, solar-flare warning systems, and highly reliable radiation-detection equipment should help prevent a twenty-first-century space traveler from experiencing an acute, lethal dose of ionizing radiation.

In addition to physical dangers, spaceflight also causes mental stress. Astronauts on long-duration missions and the first space settlers on a planet can develop a number of debilitating psychological disorders, including the solipsism syndrome and the shimanagashi syndrome. Psychologists describe the solipsism syndrome as the state of mind in which a person begins to feel that everything around him or her is a dream and is not part of reality. This disorder could easily arise among people serving on extended missions in a confined, artificial (human-made) environment, such as that of an interplanetary spaceship or an orbiting space platform. The shimanagashi syndrome (a term derived from an exile punishment in feudal Japan) is a feeling of isolation in which individuals begin to feel left out of the mainstream of life. This can easily occur among active, intelligent people who now find themselves confined to physical environments that, while perhaps comfortable, are nonetheless very remote. Careful design of future space living quarters and readily available, high-quality communications links with Earth should help prevent the onset of such psychological disorders or at least relieve some of a person's emotional distress until more extensive psychological treatment is available.

Sending humans into space and keeping them there is a very expensive undertaking. Despite the very strong pro-human-spaceflight culture that is deeply embedded within the American civilian (but not military) space program, there is also an active prorobot culture. These space officials (along with their academic and industrial partners) are advocates of space exploration but oppose excessive emphasis on human spaceflight, which, they feel, takes valuable resources away from other important projects. They endorse the relatively inexpensive and low-risk alternative of sending smart robot spacecraft on long-duration, hazardous journeys throughout the solar system. They further suggest that with more balanced funding and programmatic (political) support, contemporary developments in computer-science-related disciplines like artificial intelligence will soon pave the way for a new generation of more intelligent and versatile machine explorers. From their perspective, intelligent space robots could make many exciting discoveries on distant worlds decades before human explorers can ever reach these destinations.

Consequently, the prorobot space-exploration faction boldly challenges the human spaceflight bias within NASA with this probing question:

"Why delay exciting and important discoveries and why spend huge sums of money just to send a 'man' to do a 'machine's' job?" They point to the human-centered *International Space Station*, its $90-billion spiraling cost, and its apparently very low-percentage return of scientific discovery per dollar expended as strong justifications for more emphasis on robot exploration, on teleoperation, and on the well-planned synthesis of human space operations and efficient, automated operations by smart machine companions.

Space Debris

Space debris, or derelict human-made objects in orbit around Earth, represents a hazard to astronauts, spacecraft, and large space platforms, like the *International Space Station* (*ISS*). Aerospace engineers also use the term *orbital debris* to refer to this collection of material that now orbits around Earth as a result of human space initiatives, but is no longer serving any useful function.

At the beginning of the space age in 1957, aerospace engineers considered the potential danger posed by meteoroids to a spacecraft or space traveler. In a general sense, meteoroids represent a natural debris hazard in interplanetary space. Ranging from large chunks of rock to specks of dust, they sweep through Earth's orbital space at an average speed of about 20 kilometers per second. Space-science data indicate that at any moment a total of approximately 200 kilograms of meteoroid mass lies within some 2,000 kilometers of Earth's surface. Aerospace-industry personnel call this region of space low Earth orbit (LEO), and it represents the portion of outer space most frequently visited and used. The majority of the meteoroid mass consists of relatively tiny objects that are about 0.01 centimeter in diameter or less. However, a diminutive portion of this total mass involves a distributed population of larger objects in which the largest objects are the smallest in number. It is common for space scientists to refer to meteoroids that are less than one gram in mass as *micrometeoroids*.

Fortunately for space travelers, small meteoroids large enough to significantly damage or destroy an orbiting spacecraft are somewhat rare. For example, the estimated time between collisions of a one-gram-mass meteoroid and a space shuttle orbiter operating at an altitude of 300 kilometers is approximately 25,000 years. In comparison, space-operations analysts estimate that a shuttle orbiter at this altitude has a chance of colliding with a 0.01-gram-mass meteoroid about once every 180 years.

Because of the relatively high encounter velocities, collisions with even a tiny object can prove to be a shattering experience. Have you ever been in an automobile traveling at interstate-highway speeds when suddenly a small pebble hits the windshield? Astronauts and cosmonauts have reported experiencing similar sudden "thuds" or "bangs" while on board Earth-orbiting space vehicles. Mission analysts attribute most of these random "impact noises" to noncatastrophic collisions between tiny particles or specks of dust (of natural or human origin) and the space vehicle.

Where did the natural bulletlike "cosmic rocks" come from? From analyzing the chemical composition of *meteorites* (the name given to an Earth-impacting meteoroid), scientists believe that most meteoroids originate from asteroids and comets that have perihelia (portions of their orbits nearest the Sun) near or inside Earth's orbit around the Sun. Scientists further hypothesize that the parent celestial objects probably disintegrated into the current population of small natural debris bodies (i.e., meteoroids) as a result of shattering collisions that occurred millennia ago. More recently formed (on a geologic time scale) meteoroids appear concentrated in a debris cloud along the orbital path of their parent celestial body. As a consequence, the natural meteoroid flux varies in intensity as Earth travels around the Sun and encounters these swarms of small space rocks (more formally called stream meteoroids) that produce the well-known meteor showers seen at certain dates from Earth, such as the Perseids (August) and the Geminids (December).

Human-caused space debris is equally hazardous, but differs from the natural meteoroid threat in several important aspects. First, space debris remains in Earth orbit during its entire lifetime and is not a transient phenomenon like the stream meteoroid showers that occur as Earth travels through interplanetary space around the Sun. Second, at altitudes less than 2,000 kilometers above Earth's surface, the space-debris population dominates the natural meteoroid population for objects of 0.001 meter diameter (one millimeter) and larger. As a point of reference, at the close of the twentieth century, the total estimated mass of human-made objects orbiting within this region (i.e., within about 2,000 kilometers of Earth's surface) was three million kilograms. This quantity of human-launched mass is approximately 15,000 times greater than the average total mass associated with the natural meteoroid population within the same region of space at any instant. Therefore, near Earth, human-generated space junk represents a greater collision threat to operational spacecraft and space travelers than the natural meteoroid population. Third, the space-debris population is quite dynamic and will grow or decline as a consequence of future space-technology activities.

For example, unexpected spacecraft breakups, spent upper-stage rocket explosions, and unavoided collisions between existing pieces of space debris and operational spacecraft could create enormous increases in the space-debris population at various orbital locations. The use of advanced weapon systems in space that attack and destroy enemy spacecraft would also create large, lingering clusters of orbiting debris particles and fragments. (The overall issue of weapons in space is discussed later in this chapter.) However, aerospace engineers and spacecraft operators can also implement "litter-free" design changes and operational procedures that greatly minimize or totally prevent new sources of space debris from contaminating near-Earth space and endangering the growth of expanded activities in Earth orbit.

There are many space-technology-related sources of human-made space debris. One major source is intentionally discarded space hardware. Upper-stage rockets from modern launch vehicles often go into orbit after they have expended all or most of their propellant supply. These spent upper stages then tumble around Earth along an orbital trajectory governed by their particular burnout velocity and the laws of celestial mechanics. Similarly, mission controllers frequently abandon satellites in orbit when they reach the end of their operational life and are no longer functional. Once abandoned, these spacecraft become uncontrolled and untended orbiting derelicts. Derelict spacecraft now make up a significant portion of the objects tracked by military and civilian space surveillance networks.

Another major source of space debris is the "operational litter" that accompanies launch and orbital activities. This category of debris includes expendable interstage hardware, separation devices (such as clamps, springs, and explosive bolts), lens caps, payload fairings, adapter shrouds, auxiliary motors, and even an occasional piece of lost equipment (such as a camera or a space-suit glove). The space-debris population also includes the family of tiny particles attributed to a wide variety of space-activity sources. This miniparticle debris family includes particulates such as paint flakes and bits of multilayer insulation produced by the material degradation of spacecraft components due to atomic oxygen interactions, solar radiation (especially ultraviolet), and solar-heating phenomena.

The solid-rocket motors (sometimes called apogee kick motors) used to boost payloads from low Earth orbit to their final operational orbits also generate an assortment of space-debris items. These include spent motor casings, motor-liner residuals, unburned solid-fuel fragments, aluminum oxide exhaust particles, and small fragments of the nozzle or exhaust cone's surface that eroded or ablated during propellant burn.

Space-object breakup represents another major contributor to the space-debris population. Since the start of the space age, space surveillance experts have identified more than 130 breakups. A space object experiences breakup when it explodes or suffers an orbital collision with another high-speed object. According to debris experts, explosions have caused the majority of space-object breakups observed to date. An explosive breakup can take place for a variety of reasons. The spacecraft's batteries might become overpressurized and explode, fragmenting the spacecraft into thousands of smaller objects that then orbit around Earth. Similarly, residual liquid-rocket propellant in an orbiting spent upper stage can become explosively overheated due to solar radiation. On other occasions, a leak in the propellant tank allows the residual onboard propellant supply to inadvertently mix with any remaining oxidizer. When this occurs, the resultant detonation suddenly shatters the derelict upper-stage vehicle into thousands of tiny fragments. Finally, spacecraft operators sometimes deliberately detonate satellites. During the Cold War, for example, both Soviet and American military space authorities tested antisatellite (ASAT) systems against target satellites. Although these ASAT tests did not violate existing treaties or international space law, they did aggravate the space-debris problem and heighten political tensions.

Space-object breakup (explosions) can also occur when the object suffers a high-speed orbital collision with a piece of space debris or a random meteoroid. Since 1957, space surveillance analysts believe that there have been at least three confirmed collisions. However, the initiating cause for many (more than 20 percent) of the observed space-object breakups still remains a mystery.

An unusual collection of space debris consists of the radioactive cores from spent (used) space nuclear-reactor systems. The former Soviet Union "parked" several such spent nuclear-reactor cores from its series of low-altitude space nuclear-powered ocean surveillance spacecraft called RORSATs (radar ocean reconnaissance satellites). Following normal operations, the reactor core would automatically separate from the main portion of the spacecraft, and then the Soviet controllers boosted the spent reactor core to a higher parking orbit at an altitude of about 900 kilometers above Earth. Parking the reactor core in a higher, longer lifetime orbit provided time (millennia) for the radioactive fission products in the core to decay to harmless levels. However, using contemporary space surveillance observations, NASA space experts now estimate that about 70,000 objects (approximately 2 centimeters in diameter) form a debris band between 850 and 1,000 kilometers in altitude. Observers think that these bits of space debris are frozen chunks of reactor coolant that have leaked from a number of the Soviet space reactor cores sent into parking orbits about 900 kilometers in altitude.

In January 1978, one of these Soviet nuclear-reactor spacecraft, a satellite called *Cosmos 954*, focused worldwide attention on the issue of nuclear space debris plunging back to Earth. At the end of its low-altitude surveillance mission, this particular Soviet spacecraft malfunctioned and did not separate its reactor core properly. Instead, the entire spacecraft with the spent nuclear-reactor core still attached disobeyed all commands from its Soviet ground controllers and entered an unstable, declining-altitude orbit. For days, this highly radioactive derelict spacecraft wobbled in its unstable polar orbit around Earth. Then, on January 24, 1978, the global anxiety on exactly where this dangerous spacecraft would impact ended when *Cosmos 954* finally entered the denser regions of Earth's atmosphere over Canada's Northwest Territories. Much of the derelict spacecraft burned up in the atmosphere, but enough of it survived to leave a trail of radioactive debris scattered across the frigid regions of northern Canada, primarily near a remote area known as the Great Slave Lake. We discuss the use of space nuclear power and the general concern about possible radioactive contamination of portions of Earth's biosphere as a result of a launch accident or spacecraft reentry in the section of this chapter on volatile issues.

Contrary to popular belief, many orbiting objects do not remain in space indefinitely but gradually return to Earth as they lose energy through frictional encounters with the residual upper limits of Earth's atmosphere. Our planet's atmosphere does not end abruptly but extends upwards for several hundred kilometers in progressively thinner amounts. Over time, this atmospheric resistance causes a derelict object to fall through a series of progressively lower-altitude orbits. Eventually the object encounters a sufficiently dense region of the upper atmosphere and starts its final, fiery plunge toward Earth's surface. Once such an object enters the sensible atmosphere, greatly enhanced atmospheric drag slows it more rapidly and causes it either to burn up completely or fall through the atmosphere and impact on Earth's surface or in its oceans. Space surveillance specialists report that between 100 and 200 tracked space objects reenter Earth's atmosphere each year. Despite the severe frictional heating encountered during reentry, some unprotected spacecraft components can and do survive the fiery plunge and impact on Earth.

One of the most celebrated reentries of a large human-made object happened on July 11, 1979, when the then-decommissioned and abandoned first American space station, *Skylab*, plunged back to Earth over Australia and the Indian Ocean. Fortunately, this spectacular space-debris reentry event took place without harm to property or human life. Space-mission planners use the term *decayed satellite* (or piece of orbital debris) to describe a human-made object that reenters Earth's atmosphere under the influence of natural phenomena such as atmospheric drag. However, when mission controllers intentionally remove an object from orbit around Earth, the term *deorbit* is used to describe this human-directed process.

One contemporary approach to cleaning up low Earth orbit while avoiding the hazard of falling space objects on the ground is to perform a carefully planned deorbit maneuver. If a satellite or spent rocket stage has any residual propulsive capability, mission controllers can command the object to perform one burn or a series of engine burns that sufficiently lower the orbital altitude at a predetermined location, causing reentry to occur over a specific isolated or uninhabited area. Unfortunately, in the past, aerospace engineers did not usually provide upper rocket stages or satellites with a surplus end-of-life propulsive capability so the object could assist in its own disposal. However, that is one of the contemporary space-vehicle design and/or operational approaches being chosen within the international aerospace community to lessen the overall space-debris problem.

Very large, massive space objects in low Earth orbit definitely represent a significant falling-debris hazard to people and property on Earth's surface. Therefore, if possible, spacecraft operators prefer to use a controlled deorbit maneuver at the end of the mission. This operational procedure not only removes the large, nonfunctioning object from orbit, but also assures that any pieces of the object surviving reentry will impact harmlessly in a remote ocean area. For example, after the failure of one of its three gyroscopes in December 1999, NASA personnel decided to follow this approach with the massive (14,000-kilogram) Compton Gamma Ray Observatory (CGRO). Calculations suggested that about 35 percent of the large spacecraft would survive reentry and impact somewhere on Earth. To avoid risk to life or property, NASA personnel executed four carefully planned thruster burns, and the CGRO safely deorbited in a controlled manner. Following reentry on June 4, 2000, the surviving pieces of this massive spacecraft splashed down harmlessly in a remote area of the Pacific Ocean. Russian flight controllers successfully performed a similar maneuver on the very large, but abandoned, *Mir* space station in March 2001. Surviving components of *Mir* also fell harmlessly into a remote area of the Pacific Ocean.

Aerospace engineers find it helpful to characterize space debris in three size ranges: debris objects less than 0.01 centimeter in diameter that produce surface erosion; debris ranging from 0.01 centimeter to 1.0 centimeter in diameter that might cause potentially serious impact damage; and debris objects greater than 1 centimeter in diameter that can easily produce catastrophic damage to a satellite, space vehicle, or platform. Impact with a debris object larger than approximately 0.1 centimeter in diameter can cause considerable structural damage. For example, a small 0.3-centimeter-diameter sphere of aluminum that is traveling at 10 kilometers per second contains about the same amount of kinetic energy as a large bowling ball traveling at 100 kilometers per hour. Impact with such an object would do severe damage to a spacecraft or space vehicle.

The primary mechanism now removing low-altitude space debris is the natural decay of orbiting objects due to interaction with the very thin upper layers of Earth's atmosphere. Solar activity greatly affects this process. High levels of solar activity heat Earth's upper atmosphere, causing the thin atmosphere to extend farther into space. This atmospheric expansion helps reduce the orbital lifetimes of space objects found below about 600 kilometers in altitude. However, above 600 kilometers in altitude, the atmospheric density is so low that solar-induced atmospheric expansion produces no noticeable effects on debris-population lifetimes. Unfortunately, this solar-cycle-based natural cleansing process for space debris in low Earth orbit is also extremely slow and by itself cannot offset the current rate of space-activity debris generation. Spacefaring nations must take active steps, like controlled deorbit maneuvers at the end of mission life, to reduce this growing hazard.

One important step is to require aerospace engineers of all nations to pay close attention to the growing space-debris problem when they design a new spacecraft, launch vehicle, or permanent space platform. They should make the launch and operation of the spacecraft as litter free as possible. For example, aerospace personnel can develop spacecraft deployment procedures that avoid or minimize the ejection of objects. Bolt catchers for explosive bolt debris and tethered lens caps for sensors are simple examples of debris-minimizing design approaches. Aerospace engineers can also integrate "passivation" techniques at the end of an object's useful life to avoid explosions and breakups. Mission managers can either vent or burn to depletion any residual propellant supply in an upper-stage vehicle or spacecraft. They can also include simple design features, like special external fittings that accommodate retrieval and removal at the end of the mission by an advanced telerobotic space-debris collection system.

Mission managers could send spacecraft operating at an altitude of 2,000 kilometers or greater to an internationally agreed-upon long-lifetime "graveyard" orbit. This debris parking orbit would only be an interim solution, however, until advances in space technology permit efficient collection and removal. An informal precedent for this approach already exists. Today, many spacecraft controllers simply boost a geosynchronous-orbit satellite to a slightly higher-altitude parking orbit upon mission completion. While this action does not remove the decommissioned spacecraft from space, it does prevent undesirable debris accumulation in this very busy and very valuable location, an orbit for which there is no significant natural debris-cleansing mechanism.

VOLATILE ISSUES

Space Nuclear Power

One of the most controversial space-technology issues involves the use of nuclear power sources. Space nuclear power systems use the thermal energy (heat) released by either of two general nuclear processes: the predictable decay of a radioisotope or the controlled fission (splitting) of certain heavy atomic nuclei such as uranium 235. A direct conversion or thermodynamic cycle appropriate for the space mission then converts this nuclear heat into electric power. Advocates point out that such nuclear-generated electric power enables deep-space exploration far away from the Sun, allows a robot lander to operate continuously on a hostile planetary surface, and represents a compact source of large quantities of reliable electric power for future planetary bases on the Moon and on Mars. Opponents point to past operational incidents as indicative of future threats to Earth's environment—technical threats that in their opinion represent an unnecessary level of involuntary risk to Earth's population.

The use of nuclear power systems is permitted under international space law. However, national governments and sponsoring space agencies are expected to use technical designs and operational procedures that protect Earth's environment from radiological contamination and that minimize any potential ionizing-radiation risk to human life. Aerospace professionals, working with the United Nations Committee on the Peaceful Uses of Outer Space (COPUOS), recognize that nuclear power supplies, such as radioisotope thermoelectric generators (RTGs), have a role on certain interplanetary missions. Nuclear power systems may also be used in Earth orbit if they are stored in a high orbit after conclusion of the operational part of their mission—a procedure that eliminates any long-term radioactive-debris threat to our planet. The international space community (through COPUOS) also recommends the use of RTG designs that contain the radioisotope fuel under all normal launch and operational conditions, as well as any credible accident conditions.

Both the United States and the former Soviet Union have developed nuclear generators (RTGs) and nuclear reactors to meet some of the energy requirements of their spacecraft. Since the early 1960s, for example, agencies of the U.S. government have used RTGs to perform successful missions to the surface of Mars (*Viking 1* and *2* landers), the polar regions of the Sun (*Ulysses*), and the giant outer planets (*Pioneer 10* and *11*, *Voyager 1* and *2*, *Galileo*, and *Cassini*). It is the success of these important space-exploration missions—made possible by compact nuclear power generators—that encourages space nuclear power advocates to endorse further use of the controversial technology on appropriate missions in the twenty-first century. By the close of the twentieth century, the United States had launched a total of 23 nuclear-powered spacecraft (22 with RTGs and 1 with a reactor), while the former Soviet Union is estimated to have launched approximately 30 Cosmos series spacecraft with nuclear power supplies (some with RTGs and many others with nuclear reactors). People who hold the space nuclear power advocacy position regard the scientific benefits derived from nuclear-power-enabled space missions as far outweighing any possible environmental risk—often calculated to be essentially negligible because of inherent aerospace nuclear safety design features.

Within the risk-assessment discipline, a "negligible" risk is often regarded as equivalent to an event called an "act of God"—a very low-probability event for which no further precaution or preventative action is warranted. Typically, an act-of-God level of risk could produce about one fatality per million persons in the at-risk population per year. This level of risk is sometimes quantified as a risk of 10^{-6} serious injuries or fatalities per person-year.

In the twentieth century, several launch failures, failures to achieve orbit, and accidental reentries through Earth's atmosphere did occur involving spacecraft with nuclear power supplies. While none of these aerospace nuclear accidents caused measurable health effects to the human population, some radiological contamination of the terrestrial environment did take place. Therefore, it is the persistent concern about the occurrence of similar, perhaps even more serious accidents that drives opponents of space nuclear power to vigorously oppose any further use of the technology in future space missions. People who embrace this opposition position generally consider any scientific benefits from such missions as insufficient to balance the possible environmental risks. Debate often centers around the interpretation of a person's freedom to accept voluntary risks (like smoking or skydiving) but to reject any society-imposed involuntary risks that are perceived as unnecessary. In a democracy, the imposition of such involuntary risks by the society on its citizens often involves a formal decision-making process that seeks compromise between individual freedom and the anticipated common good or benefit to the society at large. Frequently, a detailed environmental impact statement (EIS) is part of the decision-making process.

Since its inception in 1960, the U.S. nuclear power program has experienced three major incidents involving radioisotope power sources. On April 21, 1964, the *Transit-5BN-3* spacecraft failed to achieve orbit due to a launch-vehicle failure. Upon reentry, the plutonium-238-fueled SNAP-9A RTG completely burned up (as designed) and dispersed its radioisotope inventory in the upper atmosphere (between 45 and 60 kilometers) over the West Indian Ocean near Madagascar. The U.S. government performed air- and soil-sampling operations to assess the extent of the radiological release in the atmosphere. Although there was no measurable harm to the human population and the nuclear generator's safety features performed as intended, the United States abandoned the technique of atmospheric dispersal of the radioisotope fuel as a safety design option after the SNAP 9-A incident. Consistent with the recommendations of the international aerospace community, the objective of current American RTG design philosophy is for full fuel containment—that is, in the event of an abort during launch or the in-orbit phase of a mission, the RTGs are designed to retain their nuclear material.

The second U.S. aerospace nuclear incident occurred on May 18, 1968, and involved a SNAP-19 RTG on board the *Nimbus B-1* meteorological spacecraft. In this case, erratic behavior of the launch vehicle forced its intentional destruction by the air force's range-safety officer when the vehicle and its payload were at an altitude of 30 kilometers and traveling downrange from the launch site at Vandenberg Air Force Base, California. Following command destruct, the resultant debris impacted in the Santa Barbara Channel about five kilometers north of San Miguel Island off the California coast. Aerospace safety engineers had designed the SNAP-19 RTG for intact reentry, and a recovery team found the uncompromised nuclear generator on the floor of the Pacific Ocean five months later. Aerospace safety officials stated that this incident verified the RTG's safety features—that is, the generator could remain in a marine environment for long periods of time following a launch/mission abort without concern for nuclear fuel release. Postincident examination of the plutonium-238 fuel capsules revealed that the fuel suffered no detrimental effects following the destruction of the launch vehicle, impact into the ocean, and nearly five months' residency on the ocean bottom.

The third American aerospace nuclear incident involved the famous *Apollo 13* lunar mission abort in April 1970. While the world waited for and then cheered the survival of the three *Apollo 13* astronauts (James Lovell, John Swigert, and Fred Haise), the mission's SNAP-27 RTG reentered Earth's atmosphere on board the astronauts' lifesaving *Aquarius* lunar module (LM). This particular radioisotope generator was intended for use on the Moon's surface as the long-term power supply for an astronaut-deployed experiment station. The *Apollo 13* SNAP-27 fuel capsule was housed in a graphite cask attached to the lunar module. Consequently, both reentered Earth's atmosphere at approximately 122 kilometers above the South Pacific. Atmospheric monitoring at several high- and low-altitude locations indicated that none of the plutonium-238 fuel had leaked out of the capsule during the fiery reentry plunge. Aerospace safety officials concluded that the SNAP-27 generator impacted intact (as designed) in the Pacific Ocean south of the Fiji Islands and now lies on the bottom somewhere near the Tonga Trench beneath between 6 and 9 kilometers of water.

In contrast, the former Soviet Union flew many space nuclear reactors to provide power to its family of low-altitude ocean surveillance spacecraft, called RORSATs. There have been at least three major aerospace nuclear safety incidents involving these reactor systems. The most significant of these incidents was mentioned earlier in this chapter (see the space-debris issue), and additional details relevant to this section are provided here. At the end of one RORSAT mission, called *Cosmos 954*, the reactor core failed to properly separate from the spacecraft. Instead of being boosted up into a higher, parking orbit so the radioactive fission-product inventory in the core could safely decay, the misbehaving spacecraft circled Earth in an unstable polar orbit that kept losing altitude. People around the globe watched anxiously as the derelict, highly radioactive Soviet spacecraft passed overhead. Where would it land? The answer came on January 24, 1978, when *Cosmos 954* entered Earth's atmosphere over Canada's Northwest Territories.

Soviet officials reported that the reactor was equipped with a backup safety system that would disperse the contents of the core in the upper atmosphere during any unplanned reentry following powered operation. The Canadian Atomic Energy Control Board mounted a massive airborne and ground search and recovery program. This large emergency response effort included teams of nuclear-accident experts from the United States, provided to Canada under a special project called Operation Morning Light. Except for small radioactive particles that were scattered over about 100,000 square kilometers of frigid, snow-covered territory, most of the surviving debris from *Cosmos 954* fell on the then-frozen Great Slave Lake or on other unpopulated regions to the northeast. Despite its relatively minor environmental impact, the spectacular reentry of *Cosmos 954* focused world attention on the use of nuclear power sources in space. This adverse public opinion survives to the present day and even impacts future space nuclear power programs totally unrelated (in technology or safety philosophy) to the highly controversial Soviet practice of operating nuclear reactors on low-altitude-orbit military ocean reconnaissance spacecraft.

Since the United States first used nuclear power sources in space, the companion field of aerospace nuclear safety has placed great emphasis on protecting people and the environment. Consistent with federal regulations and law, no nuclear power source has been or can be launched into space by any government agency without presidential approval. The president grants that approval only after detailed safety reviews and analyses by many panels of experts. Like any complex decision-making effort within a large bureaucracy, however, the extensive safety review process is not perfect. According to federal officials, this detailed process represents a formal technical effort by the sponsoring agencies (like NASA and the Department of Energy) to use these controversial power sources in a safe and responsible manner and only to enable certain beneficial space missions for which there are currently no other practical power alternatives. However, not every American citizen agrees with the federal government's risk-assessment approach (as is their right in a democracy). These people suggest that government-sponsored risk-assessment studies are inherently flawed and biased in favor of a protechnology outcome by virtue of their federal agency sponsorship.

To appreciate the dilemma, you should recognize that any risk-assessment study can only provide a quantitative estimate of some particular risk. The *acceptability* of that particular risk (no matter how numerically large or insignificant it may be when judged on various relative scales of comparison) involves a very subjective judgment made by each individual. Rigorous technical and mathematical arguments alone cannot and do not force a person to accept an involuntary risk he or she feels is unacceptable. Furthermore, no aerospace nuclear safety specialist (or any other technical expert, for that matter) can prove that something is safe. All they can hope to do is objectively identify the probability that something can or cannot happen (such as a launch abort) and then quantify (within available data and accident-model limits) the probable consequences of that adverse event or sequence of undesirable events.

Detailed treatment of the principles and psychology of risk acceptance and risk aversion goes far beyond the scope of this book. Risk assessment (or hazard guesstimating) is a complex, controversial, yet fascinating discipline. The following brief example may help clarify the general approach. Remember that the acceptability of any potential risk involves personal judgment and choice. The acceptance of whether something is "safe" cannot be forced by mathematical reasoning or legislation.

Risk-assessment studies in support of the launch of NASA's *Cassini* mission (October 1997) defined the risk of using the spacecraft's plutonium-fueled RTGs as the probability (per unit of radiation dose) of producing, in an individual or affected population, a radiation-induced detrimental health effect, such as cancer. The government-sponsored prelaunch risk analyses concluded that a launch accident early in the flight that caused a release of plutonium dioxide (the current RTG fuel form) had a probability of occurrence of 1 in 1,400 and would cause 0.1 fatality per million people. A launch accident later in the flight or during spacecraft reentry (during the Earth gravity-assist flyby) had a probability of 1 in 476 and might cause 0.04 fatality per million people. However, mathematical arguments such as these failed to satisfy certain portions of the scientific community and members of the general public. Despite the continuing protests, President Bill Clinton, after final consultation with his senior advisors, released the mission for launch in October 1997. The controversy ebbed as the nuclear-powered spacecraft traveled through interplanetary space for its rendezvous with Saturn in 2004. While the *Cassini* nuclear generator debate is over, the overall debate about space nuclear power remains.

NASA is currently planning missions to Pluto and to Europa—missions that require the use of nuclear generators to achieve their scientific objectives. Options to further minimize launch risks include improving launch-vehicle reliability and making further safety-related improvements (wherever possible) in current nuclear generator design. Another option, of course, is simply to abandon undertaking these missions for several decades and wait for some advanced alternative power supply to emerge.

Sovereignty, Personal Privacy, and Commercial High-Resolution Satellite Imagery

As mentioned in chapter 5, the arrival of the military photoreconnaissance satellites in the early 1960s gave both the United States and the former Soviet Union unhindered "open-sky" access to previously denied territories. From the very beginning, both superpowers accepted the right of spacecraft to travel freely through outer space and to observe any territory on Earth. For decades, the skilled evaluation of the high-resolution spy-satellite images provided clandestine agencies within the U.S. government an exceptional information edge. Civilian Earth-observing spacecraft, like the Landsat series, were intentionally limited to relatively low spatial resolution (typically about 10 meters or larger in each dimension). Consequently, while remaining scientifically useful, the openly distributed low-resolution imagery from such civilian systems could identify only large-

A seven-year journey to the ringed planet Saturn began with the successful liftoff of a Titan IVB/Centaur expendable launch vehicle from Launch Complex 40 at Cape Canaveral Air Force Station, Florida, on October 15, 1997. The mighty rocket carried the controversial nuclear-powered *Cassini* orbiter spacecraft and its attached *Huygens* probe. The *Cassini* mission is an international scientific effort involving NASA, the European Space Agency (ESA), and the Italian Space Agency (Agenzia Spaziale Italiana [ASI]). The two-story-tall *Cassini* spacecraft is scheduled to reach Saturn in July 2004. Photograph courtesy of NASA.

scale natural and human features and could not be used as a source of intelligence information by potentially unfriendly nations.

However, this policy began to erode in the early 1990s when space-technology competitors from France, Russia, and India began to commercially distribute overhead imagery that broke through the nominal 10-meter-resolution barrier. Responding to this economic competition from foreign commercial space entities, the U.S. government suddenly changed its rules and started licensing American companies to collect, sell, and widely distribute high-resolution satellite imagery as a commercial enterprise. For example, Space Imaging, a Colorado-based American corporation, successfully launched its *IKONOS* commercial Earth-observing spacecraft in September 1999. The spacecraft's one-meter-resolution panchromatic imagery created the civilian equivalent of the transparent globe previously enjoyed by superpower intelligence agencies.

While high-resolution satellite imagery stimulated a broad range of exciting new civil and commercial applications, this advancement in commercial space technology also aroused a deep sense of uneasiness among certain government officials and national security experts. The tragic losses from the September 11, 2001, terrorist attacks on New York City and the Pentagon further elevated this apprehension. Security experts in and out of government voiced renewed concern that such good-quality overhead images might unintentionally become a valuable source of intelligence information for unfriendly nations, rogue states, and technically sophisticated terrorist groups. In addition to such national security concerns, prickly issues of sovereignty and personal privacy versus freedom of information also zoomed in from outer space along with the new high-resolution commercial imagery.

On one side of the volatile issue were those in various government agencies, professional security experts, diplomats, and personal-privacy legal groups who wanted to limit "public spying" from space. They emphasized that high-resolution commercial satellite imagery could disseminate sensitive information concerning ongoing U.S. or friendly-nation military operations, might cause a loss of political control during a crisis, and could be used as a valuable intelligence source by rogue nations and terrorist groups. There was also fear that such imagery might even stimulate political retaliation (for example, the closing of an important military base) from a previously friendly foreign government that felt embarrassed or threatened by information revealed in the commercially released overhead images.

Those opposed to high-resolution public spying from space also voiced another major concern. With abundant, high-quality overhead imagery, members of the news media could now quickly and independently see a story unfolding in a denied area. Then, without any censorship or control, reporters might misinterpret the imagery in a rush to deliver a breaking news story. An erroneous story involving serious issues such as suspected troop movements or arms-control violations might then cause significant disruption in international affairs in politically unstable regions of the world. Traditional antagonists might use these independent, but incorrect reports as a pretext to trigger a deadly regional conflict. Finally, opponents of high-resolution public spy satellites point out that there is simply no telling how much more harm rogue states and terrorist groups can cause once they are given easy visual access to previously denied political, military, and industrial areas around the world.

The potential assault on the right of personal privacy—a vaguely defined but long-cherished American legal principle—also concerns those who oppose the widespread distribution of commercial high-resolution satellite imagery. For example, law-enforcement officials might use such imagery to conduct warrantless searches of a person's property, carefully taking note of any suspicious activity that is now in "plain view," even though the private property may be totally surrounded by a very high privacy wall. Paparazzi and entertainment-industry reporters might learn how to spy from space on unsuspecting celebrity personalities while they frolic in their most private outdoor hideaways. Within the American legal system, several constitutional challenges could arise concerning the First Amendment right of the news media to collect and publish information (including commercially available information revealed by satellites) versus an individual's right to privacy.

In strong contrast, advocates who support and encourage the collection and distribution of high-resolution commercial images from space point out that such information supports the spread of freedom throughout the world, stimulates technical progress, encourages economic growth, and promotes openness in government. Ruthless, despotic regimes, for example, that deny news media access to their nation are suddenly under the close scrutiny of global news agencies who can now watch and report suspicious activities without official permission or support. There are an amazing number of modern information-services opportunities in the commercial, technical, or civil sector that are being stimulated by the availability of high-resolution imagery. During the next decade, these opportunities will continue to grow as more and more information-age entrepreneurs learn how to extract the data content of high-resolution satellite images in support of specialized needs.

Commercial high-resolution satellite images are now part of the space-technology equation—there is no turning back. However, several derivative issues need swift resolution. Should news agencies be totally uncontrolled in their use of these data? Freedom of the press is a long-cherished, constitutionally protected American principle, but what happens when a zealous news agency decides to embellish a breaking news story and releases a satellite image that shows American troops secretly deployed in a hostile area? If release of this information endangers American troops and gives combat assistance to the enemy, should members of that news agency be prosecuted for treason?

To avoid such compromising situations during a national security crisis, the U.S. government could use the terms of any licensing agreement to exercise some form of "shutter control" on companies doing business in the United States, but what about foreign companies that decide to release similar imagery into the global marketplace? Should the U.S. government pressure the foreign satellite-imagery company through diplomatic or economic channels? If such overtures fail, do national security interests encourage the use of space weapons to temporarily blind or totally negate the offending foreign "eye-in-the-sky"?

Weapons in Space

The final volatile issue concerns the placement and use of weapons in space. Many people mistakenly associate the term *weaponization of space* with the term *militarization of space*. This is an inaccurate comparison, because since the earliest days of the space age, both the United States and the former Soviet Union have deployed military satellites. These early systems were generally highly classified surveillance and reconnaissance satellites that had specific information-gathering, defense-related tasks. Consequently, outer space was "militarized" from the very beginning, and to suggest that space should not "become" militarized is an incorrect, after-the-fact argument.

However, advances in space technology now offer a wide spectrum of functions for modern military space systems, ranging from traditional, passive information-gathering roles to active, offensive systems that can destroy enemy spacecraft. The deployment and use of weapons in space poses a critical international issue in the twenty-first century. Semantics also plays a key role in properly understanding this issue. The vagueness of such basic terms as *weaponization of space* and *space weapon* often leads to misinterpretation and confusion during any debate on this controversial issue. Within the language of the international aerospace community, a space (orbiting) system that interferes with or harms another space system is called a *space weapon*. Unfortunately, there is a good deal of ambiguity with this widely used definition. For example, a military rocket fired from Earth's surface or from a high-altitude aircraft that directly ascends into space and destroys an enemy satellite is clearly an antisatellite system (ASAT), but the rocket-propelled, direct-ascent ASAT is not normally regarded as a "space weapon" because it does not need to be in orbit around Earth prior to performing its lethal attack.

While the international community has acknowledged and accepted the right of a sovereign nation to use military space systems in support of its national defense needs, there is considerable disagreement about whether this "right" should now include the placement of weapon systems in space. Space-weapon advocates compare outer space to Earth's oceans and extend the analogy by mentioning the fact that the cherished international doctrine of "freedom of the seas" is periodically protected and reinforced through the projection of naval power, offensive and defensive. Those who oppose the deployment of offensive weapons in space point out that spacefaring nations have peacefully deployed civilian and military systems in outer space for more than four decades and that any introduction of weapons in space would serve to destroy this long-established, cooperative, peaceful environment. They further suggest that all spacefaring nations must strive to keep outer space a "weapons-free zone," similar to the way the family of nations (by international treaty) now maintains the continent of Antarctica as a scientific, peaceful sanctuary as part of the common heritage of humankind.

None of the early American or Soviet military satellites were offensive weapon systems, that is, designed to destroy other satellites or to attack ballistic missiles as they rose up through Earth's atmosphere on their way to their strategic targets. In the late 1950s, the space technology was simply not mature enough to support the development of such offensive space weapons. Furthermore, both superpowers found it politically more expedient to support the freedom of outer space to accommodate use of the emerging reconnaissance and surveillance satellites. Both nations publicized only the civilian side of their space programs and elected to keep the contributions of any emerging military space systems well hidden from public view.

President Eisenhower also believed that space without offensive weapon systems was in the best interests of the United States. He pursued an open, civilian space program (through NASA) that championed the freedom of passage through space and the peaceful exploration and use of space "for the good of all humankind." The American public quickly embraced this "space for peace" policy. Lost in the publicity glare of many spectacular space-exploration accomplishments of the 1960s and 1970s was the fact that maturing military space systems had become an integral component in national defense. This dependence on military space systems continued to grow in the late 1970s and early 1980s and culminated with the effective use of space systems to enhance allied ground, sea, and air forces in Desert Storm (February 1991). The Gulf War brought the operational power and role of military space systems clearly into public view for the first time. Today, the United States follows a military doctrine of *space control* that ensures access to and operation in space by friendly forces, but denies such access and use to enemy forces. Implied in this twenty-first-century doctrine is the assumption that should an enemy nation threaten U.S. war-fighting or peacekeeping interests, that enemy nation will be prevented from interfering with American space assets by diplomacy, economic sanctions, or appropriate military action. Furthermore, any foreign space system that threatens American or friendly-nation civilian or military spacecraft will be similarly dissuaded or else negated. One way of negating a threatening enemy spacecraft is to deploy and use a space weapon against it.

However, the mere suggestion of deploying a space weapon to maintain space control evokes much controversy. Those in favor of the space-control doctrine point out the need to defend and protect the many valuable military and civilian space systems that have now become attractive targets in any twenty-first-century conflict. The opponents of weapons in space prefer to maintain the traditional image of space as a peaceful environment and suggest that any orbiting weapons would violate international law.

Part of this current controversy stems from mistaken or divergent interpretations of the terms of the international agreements that now govern the use of outer space by all nations. The widely accepted and signed Outer Space Treaty of 1967 places activities in space under international law and encourages the exploration and use of outer space for the benefit and in the interest of all humankind. According to this treaty, the use of space for peaceful purposes and the passage through space and across celestial bodies must remain free from interference. In addition, both the regions of outer space and the natural bodies it contains cannot be subjected to the sovereignty of any state (nation).

However, the treaty also recognizes that human-made objects placed in space are the property of the nation (or organization) that paid for them and are the responsibility of the country that registered them (with the United Nations) or whose government authorized their launch by commercial entities. The treaty further elaborates that space is open to exploration and peaceful exploitation by all countries. Warlike activities are forbidden in space or on celestial bodies except in self-defense or the defense of allied nations. However, military personnel and military spacecraft are not considered warlike in and of themselves, and the operation of military satellites is clearly legal under the treaty. Signatories to this treaty (including the United States and Russia) also agreed not to place nuclear weapons or other weapons of mass destruction in orbit around Earth. Unfortunately, during debate, people often misconstrue or misinterpret this particular portion of the treaty to mean "all weapons," but that is simply not what the actual text of the treaty states.

As a result, an orbiting antisatellite (ASAT) system that does not use a nuclear weapon or other type of weapon of mass destruction (WMD) to attack and destroy an enemy satellite clearly represents the use of a space weapon, but is not prohibited by the terms of the existing space treaty. In fact, between October 1968 and June 1982, the former Soviet Union on several occasions successfully launched and tested an operational, co-orbiting ASAT system. This space weapon would hunt down its target (a Russian test satellite), perform a closing maneuver, and then once within range detonate a conventional (high-explosive) warhead to destroy the target satellite with a blast of high-speed pellets. The United States also developed an antisatellite weapon. The F-15-aircraft-launched, rocket-boosted American ASAT homed in on the target satellite and performed a kinetic-energy (proximity) kill. Having demonstrated such space weapons (which did not violate the Outer Space Treaty of 1967), both superpowers then mutually agreed to suspend their ASAT activities in the mid-1980s.

Current U.S. government interest in developing a limited (theater-level) ballistic missile defense system raises a thorny derivative issue. Some aerospace weapon experts now believe that a space-based laser (SBL) weapon would substantially contribute to the doctrine of space control and also enhance any ground-based antimissile defense system (see Figure 6.1). They also suggest the development and use of another space-weapon system, called the defensive antisatellite system (D-ASAT). This proposed military satellite would function like an "armed bodyguard," orbiting inconspicuously near a capital (major) military or civilian space asset. The D-ASAT would attack and destroy any unknown or threatening space object that enters declared "keep-out zones" in outer space. These keep-out zones would surround certain high-value military or civilian satellites. Opponents state that such proposed weapons are well beyond existing space-technology levels. They levy similar arguments against ongoing plans to develop and deploy a number of high-speed, ground-based interceptor missiles that would shoot down any incoming ballistic missile warhead launched by a rogue state or terrorist group. Critics point out that what amounts to hitting a high-velocity bullet with another high-velocity bullet is a very difficult, almost impossible, technical challenge, and certainly well beyond what can be accomplished in the first or second decade of the twenty-first century. Advocates for such a ballistic missile defense system, including space-based weapon components, argue that the proliferation of ballistic missile technology into the hands of rogue states and terrorist groups imperils civilization and must be thwarted.

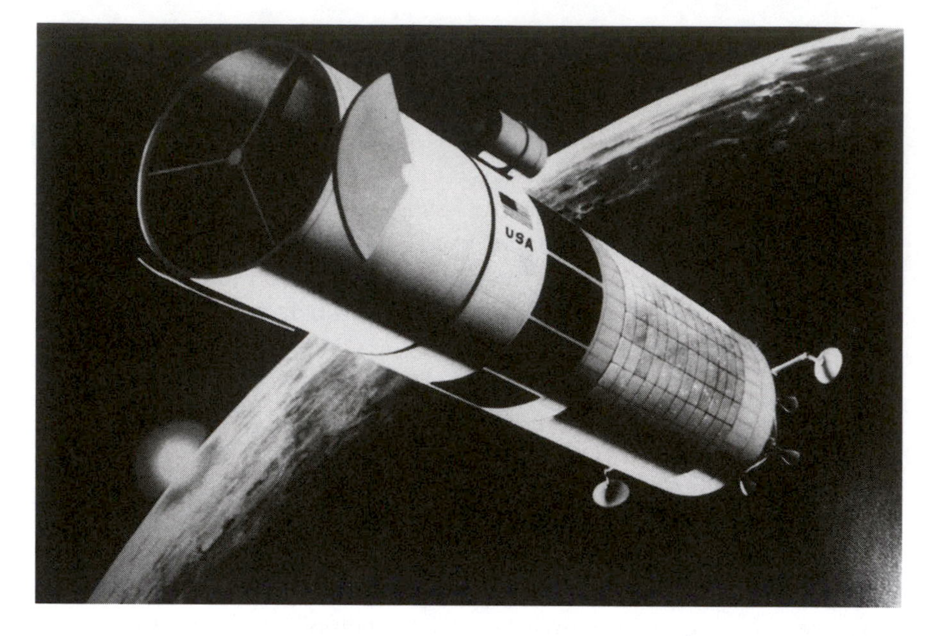

Figure 6.1 Artist's rendering of a space-based laser (SBL) (circa 2015). Advocates for this type of "surgical" space weapon system champion its role in space control as a major defensive-antisatellite weapon system (D-ASAT) that can protect important military and civilian space assets from attack by enemy ASATs and space mines. Opponents suggest that this type of orbiting weapon system will have a destabilizing impact on the peaceful applications of outer space and encourage warfare beyond the boundaries of Earth. Artist rendering courtesy of U.S. Department of Defense.

HORIZON (TWILIGHT) ISSUES

Extraterrestrial Contamination

Concern about extraterrestrial contamination and planetary protection is a horizon (or twilight) issue that straddles the worlds of science fact and science fiction. In H.G. Wells's classic science-fiction novel *The War of the Worlds*, the invading Martians are eventually defeated and Earth saved by terrestrial microorganisms that prove deadly to the hostile aliens. With the arrival of the space age came the very real scientific concern about protecting Earth's biosphere from a similar invasion of (microscopic) alien life forms should a robot sample-return mission or human-crewed expedition to Mars encounter any. There is also the important scientific and ethical companion issue about protecting potential life-bearing alien worlds (such as Mars, Europa, and Titan) from biological invasion by "hitchhiking" terrestrial microorganisms that could arrive on board robot exploration spacecraft (especially landers).

Scientists define extraterrestrial contamination as the biological contamination of one world by life forms (especially microorganisms) from another world. Using Earth's biosphere as the reference, they call this process *forward contamination* if terrestrial microorganisms contaminate the alien world or a returned sample of rock or soil from that world. Conversely, they call the process *back contamination* if alien organisms come in contact with and contaminate the terrestrial biosphere.

The issue of forward contamination is not new. For years, space scientists have responded to international law (e.g., the Outer Space Treaty of 1967) and planetary-protection protocols (established through the United Nations Committee on Space Research [COSPAR]). Mission planners continue to take the precautionary steps necessary to prevent the invasion of Mars or other potential life-bearing bodies in the solar system by hitchhiking microorganisms from Earth. Strict cleanliness and mission design criteria are imposed on lander robot spacecraft and to a lesser extent on flyby and orbiter spacecraft that might accidentally crash-land on the target planetary body.

However, the issue of back contamination is a bit more challenging and its resolution a little less straightforward. No one knows whether life now exists on Mars or on some other solar-system body. As a result, the presence of potentially pathogenic alien life forms remains an open scientific question. Just how much precaution, expense, and effort should mission managers and engineers invest when they design robot space missions that return material samples from these worlds for detailed study on Earth? There is a wide range of technical and operational responses to this question within the scientific community. Concerned members of the public worry about the possibility that a sample container from Mars might crash while returning to Earth and release unknown, potentially dangerous microorganisms directly into the terrestrial biosphere. Aerospace mission designers call such missions "restricted Earth return" missions because they involve extensive design and operational constraints to guarantee the safety of Earth's biosphere.

Stimulated in part by NASA's Moon-landing project, scientists began to seriously address the issue of extraterrestrial contamination in the 1960s. With limited initial knowledge, space scientists established early planetary-protection procedures (or quarantine protocols) to prevent the forward contamination of alien worlds by outbound, robot spacecraft. There was also concern about the possibility of back contamination of the terrestrial biosphere when the Apollo astronauts returned with their lunar samples, so NASA officials exercised a limited crew-quarantine protocol as part of the early landing missions. Depending on what robot missions discover, returning Mars-mission astronauts (in about 2025) could undergo a more rigorous quarantine protocol.

The currently recommended planetary-protection guidelines require that all space-exploring nations design and configure outbound robot planetary missions to minimize the probability of alien-world contamination by terrestrial microorganisms. Aerospace engineers must use special design techniques, decontamination procedures, and physical isolation (i.e., a prelaunch quarantine) to meet the objectives and intent of this internationally endorsed protocol. Because implementation of these contamination-control procedures is expensive and time consuming, aerospace program managers are sometimes tempted to challenge the validity of such planetary-protection requirements, especially when faced with serious cost overruns and other more clearly understood technical difficulties. But planetary scientists and exobiologists who seek alien life forms oppose such budget-saving measures and counterpunch with reasonable speculations about the possible presence of extremophiles in trapped water (or ice) below the Martian surface.

The recommended quarantine policy for samples returned to Earth remains focused toward containing all potentially hazardous material from Mars or any other world. Proposed sample-return missions continue to generate lingering concerns about the possible existence of a difficult-to-control pathogen capable of directly infecting human hosts or a life form capable of upsetting the Earth's biosphere. According to the technical experts, there are three fundamental approaches for handling extraterrestrial samples to prevent back contamination. First, the sample-return spacecraft could automatically sterilize the entire sample of material while it is en route to Earth from its native world. However, this approach would most likely destroy the sample's exobiological value. Second, mission managers could deliver the returned sample in a sealed reentry capsule directly to the vicinity of a remotely located, maximum-confinement laboratory facility on Earth. There scientists would examine the sample under tight biological security. Third, the sample-return mission managers might choose to perform all preliminary hazard analyses (called protocol testing) on the alien-world material in an Earth-orbiting quarantine facility. Such challenge testing would have to demonstrate that any detected alien life forms presented no biological threat before the mission managers gave permission to bring the material into the terrestrial biosphere for more detailed study. Each approach has several distinct advantages, as well as several inherent disadvantages. Mission planners also recognize that any final decision on the disposition of inbound alien-world materials includes political and economic as well as scientific dimensions.

A space-based quarantine facility offers several unique advantages, even though it is the most costly approach. First, use of an orbiting quarantine facility eliminates the possibility that a sample-return spacecraft would crash within the biosphere and accidentally release potentially hazardous alien microorganisms. Second, the physical and biological isolation of an orbiting facility also guarantees that should any alien microorganisms escape confinement during protocol testing or prove extremely hazardous to handle, these microorganisms would not endanger the terrestrial biosphere. Finally, all quarantine workers on the orbiting complex would remain in total physical and biological isolation from other human beings during protocol testing. Therefore, an exobiologist could not accidentally take home a (hypothetical) case of the "Martian plague," thereby causing an uncontrollable, rapidly spreading epidemic that would endanger life on Earth.

Who Speaks for Earth?

The next twilight issue involves the very interesting, but highly speculative, concern as to whether the people of Earth should respond to a recognizable "welcome" signal sent into our portion of the galaxy by an obviously intelligent and powerful extraterrestrial civilization. Who speaks for Earth? Do we answer at all, or should we just quietly listen? About one century ago, the mere attempt by any scientist to publicly raise this speculative question would trigger instant ridicule in the popular press and might incur professional ostracism. Today, when scientists openly discuss the search for extraterrestrial intelligence (SETI), their deliberations are tolerated, but still quietly cataloged as a highly speculative component of humanity's overall space-technology effort. Tomorrow, a successful SETI event could yield humanity's first invitation to join a much larger family of intelligent beings who reflect different aspects of the evolving levels of consciousness within our large, mysterious, and beautiful universe.

The thought of a universe bristling with intelligent life excites some people, but threatens others. Space technology is our physical bridge into that universe. Advanced space-exploration activities over the next few decades might easily reveal that we as a species are definitely not alone. This monumental scientific discovery would trigger an entire suite of companion issues—issues generating deep philosophical and religious discussions regarding who human beings really are in the context of an intelligent universe. One example of such a nonlinear event would be the discovery of a derelict alien spacecraft found drifting in the asteroid belt. This space-technology-enabled discovery would trigger all manner of interesting discussions.

However, such a nonlinear event might not be the discovery of an alien artifact, but rather the reception of a signal. If a coherent signal from an intelligent alien civilization is ever received, we must then ask ourselves: "Who speaks for Earth?" An advanced intelligent civilization somewhere in the Milky Way galaxy might search for other intelligent civilizations by sending out signals across the interstellar void in hopes that other creatures would intercept, decode, and respond to their messages. SETI scientists here on Earth now suggest that an intelligent alien civilization might use certain portions of the electromagnetic spectrum (such as microwave or optical radiation) to transmit their messages.

Since the dawn of the space age, a new wave of science-fiction literature and speculative science papers have addressed the intriguing possibilities and consequences of "making contact" with an intelligent alien civilization. Of course, the overall impact of this contact will depend to a great extent on the circumstances surrounding the initial discovery. The reception and decoding of a signal from an alien civilization could offer the promise of practical and philosophical benefits for humanity. Consequently, some people recommend a swift response to the signal in an effort to establish a valuable (but time-delayed) dialogue that could link our planetary civilization to a galaxywide information exchange, involving many other intelligent species. Other individuals take a more conservative approach and point out that response to such a signal could easily involve a significant risk to our planetary civilization.

There is, therefore, a significant amount of speculative discussion within the technical and legal portions of the aerospace community concerning this issue. Some advocate that only a global organization, like the United Nations, should respond and speak on behalf of all the people of the planet. Others suggest that if a special government agency, like the U.S. National Security Agency (NSA), intercepts and decodes an alien message, it is the responsibility of that agency and government to protect the contents of the signal until national leaders, perhaps in consultation with other leaders from around the world, make a decision on whether to respond or not. A third group, including some of the independent scientists and private research organizations who are now conducting various SETI programs, feel that it is their privilege as discoverers to announce the findings to the world and perhaps even to send an initial response acknowledging the reception of the signal.

Historians and anthropologists, who study past contacts between two terrestrial cultures, point out that in general (but not always), the stronger, more technically advanced culture ends up dominating or overwhelming the weaker one. To avoid this circumstance, they suggest that if a non-governmental team should happen to intercept a valid alien signal, that team should report the event to proper government authorities or to a reputable portion of the global news media, or both. As a planet, the people of Earth should decide to respond or choose not to respond. The ensuing public and private debates would most likely touch every aspect of terrestrial culture, philosophy, religion, politics, and military (defense) thinking. A rogue state or group of political extremists might even attempt to send their own response message in the hopes of gaining a decisive "information-flow" advantage. However, if the majority of Earth's people feel suspicious about the possible motives of the (hypothetical) alien civilization that sent the message, we are under no obligation to respond. In fact, there would be no practical way for "them" to realize that the people of planet Earth intercepted, decoded, and understood their signal.

Simply listening for such signals does not pose any great danger to our planetary society—save the possible cultural shock of finding out that we are indeed not alone in the universe. The real hazard concerning this issue occurs if we decide to respond to such a signal. Determining who speaks for Earth and what we initially say to an unknown alien civilization could lead to a golden age of enlightenment or subversion, submission, and even invasion. Many of the scientists and space lawyers who address this speculative issue suggest that for now, the people of Earth should follow a conservative "listen-only" strategy. However, a few more adventuresome individuals point out that our planetary civilization has been broadcasting its presence to the galaxy with over five decades of ultra-high-frequency (UHF) television signals. Radio astronomers also intentionally beamed one very powerful radio signal, called the Arecibo Interstellar Message, to the fringes of the galaxy on November 16, 1974. Since we have already announced our presence to the galaxy, should we be surprised or hesitant if someone or something eventually responds?

Who Owns Outer Space?

The final horizon issue centers around this intriguing question: "Who owns outer space?" Within contemporary space law, there are two possible answers: no one or everyone. Under the terms of the widely accepted 1967 Outer Space Treaty, all signatory nations (including the United States and Russia) have renounced any claims of national sovereignty over regions of outer space and over other celestial bodies in the solar system, including the Moon, Mars, asteroids, and comets. Therefore, no nation (and by extrapolation, no citizen of that nation) can own a celestial object or a region in outer space.

However, the United Nations Committee on the Peaceful Uses of Outer Space (COPUOS) also encourages each nation to explore outer space and to use space for the benefit of humankind. The phrase the "common heritage of mankind" or "CHM" appears quite often in contemporary space-law literature, clearly suggesting that space resources that may be found on the Moon, on an asteroid, or on Mars actually belong to everyone.

This creates a bit of a futuristic dilemma, especially for terrestrial companies that may want to mine lunar oxygen or harvest lunar-water ice as part of an expanding free-market activity in the commercial space economic sector. Space resource harvesting for profit would clearly exert ownership on the material being extracted. However, before any terrestrial company will be willing to invest and risk the billions of dollars in necessary advanced space-technology equipment, the current vagueness of "ownership" of space resources must be resolved by international agreement.

For example, will a hypothetical Moon-mining company have to give every person on Earth a portion of its profits, since space belongs to everyone? Some advocates of the CHM principle vigorously say yes. Others who support a free-market approach to space commercial development suggest that some type of international licensing agreement is more appropriate. In this way, space resources are harvested for profit by privately owned companies, but under rules that share the profits in a reasonable way. There is also a group of space-exploration advocates and scientists who oppose any commercial exploitation of solar-system bodies. This group wants these objects to remain pristine scientific reserves, much as Antarctica is protected for scientific research by treaty and international cooperation.

Chapter 7

The Future of Space Technology

> It's hard to make predictions, especially about the future.
>
> Yogi Berra

Despite such a sage warning from the famous baseball philosopher, this chapter provides a comprehensive look at tomorrow and where the applications of advances in space technology might take the human race. For intellectual convenience, we will divide tomorrow into three parts: the near term (the next 10 or so years), the midterm (between 10 and 50 years into the future), and the far term (from the mid-twenty-first century out to 2101 and beyond). With this somewhat arbitrary division of the future, we can comfortably engage in the "reasonable" extrapolation of current technology into the near term, exercise a slight stretch of the imagination to generate the midterm projections, and then marshal our deepest creative talents to entertain several far-term possibilities.

In the process of technical forecasting, however, there will be some unavoidable overlaps. For example, a major space-technology effort, like the *International Space Station (ISS)*, resides partially in the near term, but also extends well into the midterm future because of its planned two-decade-plus operational lifetime. Similarly, a permanent lunar base constructed in the midterm future could evolve into a cluster of important space facilities that would exert special influence on far-term space-technology activities.

This overlap must be expected and tolerated because the future is a smooth continuum sprinkled with interesting breakthroughs and interrelated activities rather than a rigorous, time-ordered collection of discrete, unrelated events. Anticipating surprises, both pleasant and unpleasant, is an essential part of technology forecasting, but is also extremely difficult to accomplish with any degree of accuracy. Without question, some of the most conservative near-term space-technology projections presented here will encounter delays and difficulties, while some of the currently "incredible" midterm and far-term projections will be accomplished with ease. Furthermore, many of these midterm and far-term "fantastic" speculations will be viewed several decades from now as rather commonplace technologies.

This pattern has occurred throughout the evolution of space technology. At the start of the twentieth century, the three founding fathers of astronautics—Robert Goddard, Hermann Oberth, and Konstantin Tsiolkovsky—independently made many very bold space-technology conjectures and projections. They were professionally ignored and sometimes publicly ridiculed for their "fantastic" ideas, yet many of these incredible concepts, like powerful rockets going into space, people walking on the Moon, and smart machines exploring the solar system, became technical realities in only a few decades of focused social commitment and properly funded engineering effort.

In the first part of this chapter, the near-term future, the dominant space-technology activities are the *International Space Station* (*ISS*) and sophisticated robotic spacecraft, like NASA's *Cassini* mission to Saturn and a variety of Martian missions. These projects incorporate lessons learned from the past to perform more detailed scientific investigation of some of the most interesting objects in the solar system.

The second part of the chapter introduces two major midterm projections: a return to the Moon and a human expedition to Mars. To become reality, either or both of these future space activities will require major political and social commitments, even though they are technically achievable (in principle) within our current understanding of science and engineering. Of course, unanticipated breakthroughs in space technology, such as the rapid development of a practical reusable launch vehicle, would greatly facilitate the accomplishment of these midterm activities and set the stage for many far-term opportunities.

The last part of the chapter describes several interesting far-term space-technology projections, including the satellite power system, the large space settlement, and a planetary defense system capable of protecting all life on the home planet from the threat of impact annihilation by a marauding asteroid or comet. The first two far-term concepts would also function as major engines of economic development and promote social transitions for the human race—much like what occurred with the arrival of electricity in the late nineteenth century and the information-technology revolution in the late twentieth century. The third far-term concept uses space technology to create a planetary insurance policy, protecting Earth's biosphere from cosmic disruption or destruction.

NEAR-TERM SPACE-TECHNOLOGY PROJECTIONS

International Space Station

At the beginning of the twentieth century, the Russian space visionary Konstantin Tsiolkovsky became the first person to describe all the essential technical components needed to maintain a permanent human presence in outer space. Tsiolkovsky's far-reaching recommendations included the use of solar energy, rotation of the station to provide artificial gravity, and creation of a closed ecological system that included a "space greenhouse."

The dream of human beings living in platforms beyond Earth's atmosphere continued to evolve with the publication of Hermann Oberth's *The Rocket into Planetary Space* (1923). In this book, the German space-travel visionary described several possible applications, including use of the space station as an astronomical observatory, an Earth-monitoring facility, and a scientific research platform.

After World War II, the German-American rocketeer Wernher von Braun, assisted by the famous space artist Chesley Bonestell and the entertainment visionary Walt Disney, helped popularize space exploration and the concept of a wheel-shaped space station in the United States. In a groundbreaking 1952 article in *Collier's* magazine, for example, Braun combined fantasy with physics to create a technical vision of how he would use contemporary (but pre-space-age) technology to place a permanent space station into orbit around Earth. He proposed an inflated wheel-shaped station made of reinforced nylon. Rotation of this 75-meter-diameter station would provide the crew with the comforts of artificial gravity. This station would function as a navigation aid, meteorological station, military platform, and transportation node from which to explore space beyond Earth orbit.

A half century later, the *International Space Station* (*ISS*) is being assembled in low Earth orbit. Although the *ISS* looks considerably different from the previous visionary concepts of Tsiolkovsky, Oberth, and Braun, it represents a major space-technology pathway into the future—perhaps

Figure 7.1 Backdropped against the blackness of outer space and Earth's horizon, the *International Space Station* was photographed by the crew of the space shuttle *Endeavour* as the aerospace vehicle separated from the station during the STS-108 shuttle mission (December 15, 2001). Illustration courtesy of NASA.

even the sine qua non in the development of a future human civilization that extends throughout the solar system (see Figure 7.1).

The arrival of the Expedition 1 crew at the *ISS* on November 2, 2000, marked the start of a continuous and permanent human presence in outer space in the twenty-first century. The first resident crew consisted of the expedition commander, American astronaut William Shepherd, the Soyuz commander, Russian cosmonaut Yuri Gidzenko, and the flight engineer, Russian cosmonaut Sergei Krikalev. Such international cooperation in space at the start of the new millennium represents a refreshing difference from the Cold War superpower confrontation that prevailed at the dawn of the space age (1957).

The *ISS* represents a global partnership of 16 nations. The governments of the United States, Canada, Europe, Japan, Russia, and Brazil are collaborating with their commercial, academic, and international affiliates in the design, operation, and use of the *ISS*. European states are represented through membership in the European Space Agency (ESA). The current ESA members cooperating on the ISS project are Belgium, Denmark,

France, Germany, Italy, the Netherlands, Norway, Spain, Sweden, Switzerland, and the United Kingdom.

The $95-billion *ISS* project is considered by its sponsors to be the engineering, scientific, and technological marvel that will usher in a new era of human space exploration over the first few decades of the twenty-first century. When fully assembled (circa 2005), the approximately 454,000-kilogram-mass space station will include six main laboratories, providing more opportunity for research in a prolonged microgravity environment than any spacecraft ever built. The internal (pressurized) volume in the six laboratories of the space station is approximately 1,220 cubic meters—roughly equivalent to the passenger-cabin volume of a Boeing 747 jumbo-jet aircraft.

With a projected operational lifetime of more than two decades, the *ISS* will perform three concurrent roles: first, it will serve as a test bed for advanced technology and human performance in space; second, it will function as an out-of-this-world premier research facility; and third, it will operate as a commercial platform that accommodates profit-oriented, applied research and development. Consequently, for scientists, engineers, and entrepreneurs, the *ISS* represents an unprecedented platform on which to perform complex, long-duration experiments in the unique environment of an Earth-orbiting spacecraft. As currently envisioned, this station will operate for many years in a way that maximizes its major assets—prolonged exposure to microgravity and the physical presence of human beings in orbit to participate in the research process.

While an international crew lives and works in space, an expanding space-station user community on Earth will also take advantage of "telescience" to control and manipulate certain groups of experiments located on the station. Dramatic advances in communications and information technologies will enable Earth-bound investigators to enjoy a "virtual presence" on the station. In this way, the *ISS* can serve as a unique orbital laboratory for long-term research, where one of the fundamental forces of nature (gravity) is greatly reduced. The station's complement of modern laboratory tools will support important research in biology, chemistry, physics, ecology, and medicine—all performed in a challenging space-frontier environment.

Some early research results may be anticipated by extrapolation of the work performed during previous, but relatively brief, orbital missions, yet the most interesting research results cannot even be imagined at this time. We can confidently forecast, however, that such revolutionary breakthroughs will occur in a serendipitous manner, as have many great discoveries in the past when inquisitive human beings probed the boundaries of the physical universe with new tools. Used wisely, the *ISS* represents a key discovery tool for many different scientific disciplines.

For example, the medical benefits of conducting in situ life-science research on the space station can lead to new pharmaceuticals and a better understanding of the building blocks of life. Researchers also anticipate developing a more thorough understanding of the comprehensive effects of long-term (chronic) exposure to microgravity on human beings. Several important questions need answers. For example, on extended human missions, can rigorous in-flight exercise programs and new medications be combined to eliminate the highly undesirable consequences of severe muscle degradation and bone loss? Unfortunately, these deleterious conditions now occur when astronauts or cosmonauts remain continuously exposed to microgravity for many months. If the consequences of prolonged microgravity exposure cannot be mitigated or avoided, will extended future human missions, such as an expedition to Mars, require the presence of artificial gravity in the crew habitat? Only extended research in a well-equipped microgravity research laboratory can provide valid answers to such important questions.

Throughout history, breakthroughs in materials science have often led the way to major improvements in the overall quality of life for human civilization. With the space station, many different commercial, academic, and government researchers can routinely use the microgravity conditions of an orbiting spacecraft to understand and possibly control gravity's influence during the production and processing of such important materials as metals, semiconductors, polymers, and glasses. In particular, the space station's pressurized laboratory modules will allow researchers to carefully examine such interesting phenomena as solidification, crystal growth, fluid flow, and combustion with unprecedented clarity under a continuous microgravity environment.

The *International Space Station* is only the first of many possible permanent stations in space. In a visionary sense, it serves as the pioneer that leads to ever more sophisticated human-inhabited space platforms in the mid- to far-term portions of the twenty-first century. For example, one future space station might be designed for and dedicated to the servicing and repair of operational satellites and spacecraft; another to serve as an operational logistics facility for propellants, supplies, spare parts, and special equipment. Yet another permanently crewed space platform might become the training and transportation base for future human missions to Mars and beyond. An isolated portion of this Mars-expedition space platform might even function as an orbiting quarantine facility that handles returning crews and alien-world samples. Finally, other far-term space stations might serve as initial assembly bases for the in-orbit construction of very large space structures, such as satellite power systems. These exciting far-term projections appear a little later in the chapter.

Cassini Mission

The *Cassini* mission to Saturn is the most ambitious effort in planetary exploration ever undertaken by a robotic spacecraft. It is a joint endeavor of NASA, the European Space Agency (ESA), and the Italian Space Agency (Agenzia Spaziale Italiana [ASI]) and involves a very sophisticated spacecraft that will go into orbit around Saturn in July 2004 and begin studying this majestic ringed planet and its complement of 30 currently known moons in detail for a period of four or more years. The *Cassini* spacecraft also carries a deployable scientific probe named *Huygens*—a special spacecraft that is scheduled for release from the main (orbiter) spacecraft in late December 2004. Upon release, *Huygens* will parachute down through Titan's opaque atmosphere, collecting in situ atmospheric data as it descends to the surface of Saturn's largest and most interesting moon.

Launched from Cape Canaveral by a mighty U.S. Air Force Titan IVB/Centaur rocket combination on October 15, 1997, the *Cassini* spacecraft will travel for seven years through interplanetary space before reaching its planetary destination in July 2004. Along the way, it will have flown past Venus, Earth, and Jupiter in a series of carefully planned gravity-assist maneuvers that increase its speed.

Saturn is the second-largest planet in the solar system and, like the other gaseous giant outer planets (Jupiter, Uranus, and Neptune), has an atmosphere made up mostly of hydrogen and helium. Saturn's distinctive, bright rings consist of ice and rock particles that range in size from grains of sand to railroad boxcars. The butterscotch-colored planet has a family of natural satellites that range in diameter from less than 20 kilometers up to 5,150 kilometers for Titan, which is larger than the planet Mercury and is the only moon in the solar system possessing a sensible atmosphere.

Discovered by Christiaan Huygens (1629–1695) in 1655, Titan lies beneath an opaque atmosphere that is more than 50 percent denser than Earth's atmosphere. Like that of Earth, Titan's atmosphere has two major components, nitrogen and oxygen, but is filled with a brownish orange haze made of complex organic molecules. This condition fascinates planetary scientists, who postulate that complex organic molecules might literally fall like rain from the cold Titanian sky, forming lakes of methane or ethane on the large moon's surface.

When the *Cassini* spacecraft arrives at Saturn in July 2004, it will go into orbit around the planet, and the 12 scientific instruments on the orbiter spacecraft will begin performing in-depth studies of the planet and its moons, rings, and magnetic environment. The spacecraft's complement of instruments includes an imaging-science subsystem to take pictures in the visible, near-ultraviolet, and near-infrared portions of the electromagnetic spectrum and an imaging radar system to pierce the veil of haze surrounding Titan and map its surface.

In late December 2004, the *Cassini* mother ship will release the *Huygens* probe and send it on a one-way journey down through Titan's atmosphere. The six instruments on the deployable probe will provide the first direct (in situ) sampling of Titan's atmosphere and investigate the physical properties of its hidden surface. If the probe survives landing on Titan's surface, it might continue to provide additional data for several minutes (depending on the limits of battery life). However, the probe could also splash into a lake of liquid methane or ethane and immediately cease functioning upon submersion in the icy liquid.

The *Cassini* mission is an example of the very sophisticated robotic spacecraft that will explore the solar system in the twenty-first century. The complete spacecraft includes the orbiting mother ship and the *Huygens* probe. It is one of the largest and the most complex interplanetary spacecraft ever constructed. The orbiter spacecraft alone has a mass of 2,125 kilograms and the *Huygens* probe a mass of 320 kilograms. Because of Saturn's distance from the Sun, power is supplied to the spacecraft by a set of radioisotope thermoelectric generators (RTGs) that use plutonium 238 as their radioisotope fuel.

The *Cassini* orbiter has advanced and extended the American space-technology base with several important innovations in engineering and information systems. Previous interplanetary spacecraft used onboard tape recorders to store mission data, but *Cassini* is pioneering a new solid-state data recorder with no moving parts. Depending on the successful outcome of this mission, the new solid-state recorder will eventually replace tape recorders on all future NASA deep-space-exploration missions in the twenty-first century. Similarly, the main onboard computer that directs the operations of the *Cassini* orbiter spacecraft and its 12 scientific instruments uses a novel design that draws upon new families of electronic chips. Among these new chips are very high-speed, space-mission-qualified integrated circuits.

The near-term space-technology advances pioneered by the *Cassini* mission should significantly influence future (midterm) robotic missions to the Jovian moon Europa, detailed science missions to Uranus and Neptune, and pathfinder missions to the outer fringes of the solar system in which advanced robotic spacecraft initially explore Pluto and its moon Charon and then travel beyond to investigate several icy bodies in a region beyond the orbit of Neptune, called the Kuiper belt.

With respect to Europa, NASA is already planning to send a new spacecraft, called the *Europa Orbiter*, to study this intriguing moon of Jupiter in 2008. If currently projected schedules are maintained, the *Europa Orbiter* will be launched in 2008 from Cape Canaveral, arrive in the Jovian system in 2010, and end its primary scientific mission in 2012. This robot explorer has three primary scientific objectives: first, to determine the presence or absence of a subsurface liquid-water ocean; second, to characterize the three-dimensional distribution of any subsurface liquid water and its overlying ice layers; and third, to understand the formation of various surface features to help select candidate landing sites for future exploration missions by more advanced robotic spacecraft.

Robotic Explorers to Mars

Prior to any human expedition to Mars in the twenty-first century (discussed a little later in this chapter), an armada of robot spacecraft, including orbiters, landers, surface rovers, and sample-return vehicles, will pave the way and attempt to resolve some of the planet's most intriguing issues. Previous robotic missions to Mars have uncovered a host of tantalizing clues that perhaps a billion years ago the Red Planet hosted great rivers, lakes, and possibly even an ocean, but the fate of any ancient water on Mars is now a hotly debated scientific topic.

Based on what previous NASA spacecraft have observed, Mars appears to be a frozen desert that is too cold for liquid water to exist on its surface and too cold for rain to occur. Furthermore, the planet's atmosphere is too thin to permit any significant amount of snowfall. Even if some internal heat source warmed a portion of the planet up enough for ice to melt, the process would still not yield liquid water. Because the Martian atmosphere is so thin, if the temperature of an ice-bearing location somehow rose above the freezing point, the Martian ice would sublime or change directly into water vapor in a process much like how a solid block of carbon dioxide sublimes directly into gaseous CO_2 when it is heated.

Why all the intense scientific interest about the fate of Martian water? The answer is simple. Without water there can be no life as we know and understand it. If it has been more than three billion years since liquid water was present on Mars, the chance of finding life there is now probably quite remote. But if liquid water is now present on Mars—perhaps in some hidden, well-protected subsurface niche—then life (at the microscopic level, at least) may still be tenaciously clinging to survival there.

Responding to these questions about possible life on Mars (extinct or existent), NASA, with invited international participation, intends to carry out a long-term exploration program that uses progressively more advanced robotic spacecraft. This plan incorporates the lessons learned from past and ongoing Mars-mission successes and setbacks. Within this plan, several new robotic-spacecraft missions will visit the Red Planet during the first decade of the twenty-first century and pave the way for multiple sample-return missions in the second decade of the century. This era of intense Red Planet reconnaissance will attempt to uncover profound new insights about past environments on Mars, the history of its rocks and interior, the many roles and abundances of water, and, quite possibly, evidence of past and present life.

NASA's *Mars Global Surveyor* mission is now orbiting the Red Planet and collecting more valuable information than any previous Mars mission. Since late March 1999, it has been providing quality images from which scientists have extracted tantalizing hints of recent liquid water at the Martian surface. Imaging operations by this orbiter spacecraft will continue until 2002 under the currently approved extended mission schedule.

The 2001 *Mars Odyssey* orbiter spacecraft was launched from Cape Canaveral on April 7, 2001, and successfully went into orbit around Mars on October 24, 2001. This spacecraft's scientific mission is to help provide another vital piece of information toward solving the "Martian water puzzle." It will do so by mapping the basic elements and minerals that are present in the uppermost layer (the first few centimeters) of the planet's surface. The 2001 *Mars Odyssey* will be the first spacecraft to make direct observations of the element hydrogen near and within the surface of Mars. Hydrogen mapping can provide the strongest evidence for the presence of water on or just below the Martian surface, since it is one of the two key elements within the water molecule (H_2O) (see Figure 7.2).

NASA plans to send identical twin rovers to Mars in 2003. The Mars Exploration Rovers will land at two separate sites and conduct close-up inquiries about the climate history of the planet and the role of water in locations where conditions may once have been favorable for life. Through teleoperation, human beings on the rover science team on Earth will participate in exploring Mars. Within the limits imposed by minutes of radio-frequency-signal-transmission time delays, these people will select surface targets of interest, such as certain rocks and soils. Each robot rover will then evaluate the composition, texture, and morphology of various rocks and soils, using its sophisticated sets of instruments and access tools. Data

Figure 7.2 This artist's rendering shows NASA's *2001 Mars Odyssey* spacecraft starting its science mission around Mars in January 2002. Illustration courtesy of NASA (Jet Propulsion Laboratory).

collected by the *Mars Global Surveyor* and the 2001 *Mars Odyssey* orbiter spacecraft will help NASA scientists select a suitable landing site for each rover.

Mars Reconnaissance Orbiter (MRO) is now planned for launch in 2005. The MRO will be able to view interesting locations within the Martian landscape at a spatial resolution of between 20 and 30 centimeters—that is, good enough to observe rocks the size of beach balls. The spacecraft also will carry a specialized, high-resolution sounding radar that will probe hundreds of meters into the Martian subsurface in search of clues of frozen pockets of water or other unique geophysical layers.

NASA plans to develop and launch (in about 2007) a new generation of mobile surface laboratory, called the Smart Lander, that can wander about 10 kilometers from its original landing site and operate for about one year on the Martian surface. The space-technology advances demonstrated by this roving robot laboratory will pave the way for a successful wave of sample-return missions in the subsequent decade.

Also starting in 2007, NASA intends to deploy a new family of small airborne and lander robotic vehicles, collectively referred to as Smart Scout missions, to explore or scout interesting areas of Mars worthy of intense future investigation. This family of robot scouts will maintain an active scientific presence on Mars for at least a decade and will engage in concentrated investigations of the most scientifically promising and intriguing places on the planet.

In the second decade of the twenty-first century, NASA plans to send its first robotic mission to return samples of promising Martian materials (rocks and soil) back to Earth. As presently envisioned, NASA will launch the first sample-return mission in about 2014 and a second in 2016. Of course, discoveries from any of the previous missions could modify and accelerate this schedule of exploration, perhaps encouraging the first sample-return mission as early as 2011.

This planned robust, two-decade campaign of exploration by advanced robotic spacecraft centers on "following the water on Mars." The results will help scientists try to resolve two age-old questions: "Did life ever arise on the Red Planet?" and "Does life exist on Mars now?" Beyond 2020, an expanded response to these questions could form the scientific rationale and social catalyst for sponsoring the first human expedition through interplanetary space to the surface of the Red Planet. This exciting human-crewed mission is discussed among the midterm projections of this chapter.

MIDTERM SPACE-TECHNOLOGY PROJECTIONS

Returning to the Moon

When human beings return to the Moon in the twenty-first century, it will not be for a brief moment of scientific inquiry and political demonstration, as occurred during NASA's Apollo Project. Rather, these pioneering men and women will go there as the first inhabitants of a new world. They will build lunar-surface bases from which to complete exploration of Earth's nearest planetary neighbor, establish innovative science and technology centers that take advantage of the interesting properties of the lunar environment, and efficiently harvest the Moon's native resources (especially minerals and suspected deposits of lunar ice in the permanently shadowed polar regions). All of these activities represent a feasible but challenging extrapolation of space technology about two to three decades from now. Clearly, the Apollo Project demonstrated that human beings could go to the Moon. What lies ahead, however, is to develop the social will and advanced-technology infrastructure to establish a permanent, eventually self-sustaining presence on Earth's nearby "planetary" companion. From the perspective of anticipated future history, the Moon should serve as humanity's gateway to the solar system.

During the first stage of a general lunar-development scenario, human beings along with their smart machines will go back to the Moon to perform detailed resource evaluations and site characterizations. Intelligent robots, teleoperated from Earth, will serve as the advanced scouts for this effort. Corollary developments in space robotics and planetary-surface teleoperations should be enormous. These highly focused efforts, driven by both scientific objectives and economic interests, will pave the way for the first permanent lunar-surface base. The physical return of human beings to the Moon's surface will quickly follow this series of successful robot missions.

The establishment of the first permanent base camp is a critical event in the formation of a self-sufficient lunar civilization in the far-term future. Operating out of this base, a select cadre of humans and their smart-machine companions would undertake the task of completely exploring the Moon from pole to pole, both nearside and farside.

Inhabitants of the first permanent lunar base will take advantage of the Moon as a large, natural platform from which to conduct high-quality science in space. Early on, they will also perform those fundamental technical and engineering studies that will more accurately define the Moon's special role as human beings decide to expand out into the solar system. The confirmed presence of water ice on the Moon should greatly influence the design and operation of this base and its ability to grow and evolve.

In March 1998, a team of NASA scientists announced that data from the *Lunar Prospector* spacecraft indicated the presence of water-ice deposits in the perpetually frozen recesses of the Moon's polar regions. If confirmed, this discovery will significantly change all future strategic planning for human migration into space. The availability of lunar ice in sufficient quantity (perhaps thousands of tons) greatly simplifies surface-base logistics. In situ water resources could accelerate base expansion and promote the early formation in the far term of a self-sufficient lunar civilization. This is an extremely important point. Should future missions confirm that there are ample quantities of lunar ice available for harvesting, then, because of its strategic position near Earth and its inherently low gravity level, the Moon could easily become the main logistics center, supplying human missions in cislunar space, to Mars, and beyond.

Space visionaries have suggested many interesting applications for future lunar bases and settlements. The first of these suggestions involves the construction and operation of a large scientific laboratory complex that takes full advantage of the Moon as an alien world with unique properties and characteristics. Some of the Moon's special characteristics include a low gravity (one-sixth that of Earth), seismic stability, low temperatures (especially in the permanently shadowed portions of the polar regions), and very high-vacuum conditions. Another feature of particular importance to astronomers is the fact that the lunar farside enjoys a very low radio-frequency (RF) noise environment. Since the Moon's synchronous orbit keeps one face (the nearside) constantly turned toward Earth, the bulk of its mass shields the farside from all terrestrial RF signals.

A farside scientific facility would capitalize on the Moon's unique environment to accommodate high-precision platforms for astronomical, solar, and space-science (e.g., solar wind) observations. These very large and powerful observatories could be set up and maintained by a small cadre of workers from the nearside lunar base. Astronomers back on Earth would use high-capacity telecommunications links to remotely operate these high-precision instruments and enjoy direct views of the universe simultaneously gathered across the entire electromagnetic spectrum.

A well-equipped lunar research complex would also provide life scientists with the unique opportunity of simultaneously studying biological processes in a reduced gravitational environment (1/6 g) and in the absence of geomagnetism. As an additional scientific bonus, these scientists would perform their experiments in completely regulated, artificially created environments that remain physically isolated from Earth's biosphere.

Genetically engineered "lunar plants," developed on the Moon in special greenhouse facilities, might become the settlement's major food supply. Greenhouse lunar plants would also assist in the regeneration of the breathable atmosphere within all the habitats. Finally, exobiologists would have the opportunity to experiment with microorganisms and plants under a variety of simulated alien-world conditions. The researcher might be a member of the lunar-base team or simply a "virtual" guest scientist from Earth who uses telepresence and teleoperation techniques to perform his or her experiments on the Moon. Like the *International Space Station* in the near term, a midterm lunar base will stimulate many exciting discoveries and promote advances in space technology.

Figure 7.3 In the NASA-funded artwork titled *The Deal*, artist Pat Rawlings depicts surface operations at a permanent lunar base (circa 2025). Artist rendering courtesy of NASA.

Space visionaries now believe that the first large lunar settlement will emerge from an initial lunar base and grow (or spin off) into a dynamic complex that successfully demonstrates industrial-scale applications of native Moon materials. A fledgling lunar economy could arise as pilot factories start supplying selected raw materials and finished products to customers both on the Moon and in a variety of orbital locations within cislunar space. With a thriving lunar spaceport as part of the permanent Moon settlement, access to all points in cislunar space could actually become easier and less energy intensive than from the surface of Earth. Consequently, this future lunar spaceport might become the busiest launch complex in the Earth-Moon system, far exceeding any of its counterparts on Earth's surface (see Figure 7.3).

From the surface of the Moon, traditional chemical-rocket propulsion techniques might be supplemented by more exotic launch techniques involving electromagnetic-mass drivers, mechanical catapults, and compressed-gas systems. These high-impulse launch techniques are especially suitable for shock-resistant, "dumb-mass" payloads that are quite literally thrown into space at very high velocities from the surface of an airless

Moon. At the close of the twenty-first century, it might cost just a few dollars per kilogram to provide bulk lunar materials to orbital destinations throughout cislunar space. Such a favorable economic condition would clearly influence all far-term space-technology developments and applications.

With the rise of highly automated lunar agriculture (performed by robots in special greenhouses), the Moon may even become an extraterrestrial breadbasket, satisfying the food needs of all human beings living beyond the boundaries of Earth. When the combined population of several large lunar bases and settlements reaches about 10,000, a demographic critical mass is attained that could encourage self-sufficiency from Earth. This moment of self-sufficiency for the lunar civilization would also represent a very historic moment in the history of the human race, for from that moment on, people would live in two distinct and separate "planetary niches." We would be terran and nonterran (or extraterrestrial).

As mentioned earlier in this book, the potential role of the Moon in human destiny was beautifully summed up as follows: "The Creator of our Universe wanted human beings to become space travelers. We were given a Moon that was just far enough away to require the development of sophisticated space technologies, yet close enough to allow us to be successful on our first concentrated attempt" (Krafft Ehricke, Washington, D.C., 1984).

Mars Expeditions and Surface Bases

Outside of the Earth-Moon system, Mars is the most hospitable planet for humans and is the only practical candidate for human exploration and settlement in the midterm portion of the twenty-first century. Unfortunately, although Mars may once have been warm and wet, today it is a frozen wasteland. The mean annual temperature on the Red Planet is −55° C. For comparison, that corresponds to the temperature at Earth's south pole during winter. Because of this, many scientists believe that it is highly unlikely that any living creatures—even the hardiest microorganisms—could survive for long on its surface. Therefore, when humans first travel there to explore the planet's surface up close and to establish surface bases, they will have to grow their food in airtight, heated greenhouses. The Martian atmosphere appears far too cold and dry for edible plants to grow in the open, unprotected Martian environment.

Planning the logistics for the first crewed mission to Mars is a very complex process. Mission organizers must consider many different factors before a team of human explorers sets out for the Red Planet with an acceptable level of risk and a reasonable expectation of returning safely to Earth. The expedition could depart from low Earth orbit (LEO) or perhaps from the vicinity of the Moon if there is a permanent lunar base that can conduct crew training and provide supplies such as air, water, and chemical propellants.

In either case, the first Mars explorers would embark on a round-trip interplanetary voyage that would take somewhere between 600 and 1,000 days, depending on the propulsion system and orbital trajectories chosen. Mission planners must strike a balance among such competing factors as the overall objectives of the expedition, the available transit vehicles and desired flight trajectories, the amount of time the explorers should spend on the surface of Mars, the primary surface site to be visited, an optimum logistics strategy to keep costs at an acceptable level, and the maintenance of crew safety and health throughout the extended journey.

With respect to logistics for this challenging interplanetary expedition, one suggested strategy takes advantage of lunar resources to provide most of the mission's outbound supply needs and then harvests similar supplies from in situ Martian resources for the return journey to Earth. In another approach, the Mars expedition is assembled in LEO near a space station with all the necessary equipment and supplies delivered to LEO by a fleet of heavy-lift launch vehicles flown from Earth's surface.

Whatever logistics strategy Mars-expedition planners ultimately select, one important factor will dominate—self-sufficiency in the face of uncertainty. Once the crew departs from the Earth-Moon system and heads for Mars, they must be totally self-sufficient and yet flexible enough (perhaps with contingency equipment and supplies) to adapt to unplanned situations. Because of the general nature of interplanetary travel along minimum-energy trajectories, there is no turning back, there is no quick return to Earth, and there is not even a reasonable possibility that timely help could be sent from Earth should misfortune occur during transit or while the crew is visiting Mars. However, to alleviate "point-of-no-return" stress on the crew, another logistics strategy recommends that surface-operations equipment and Earth-return supplies be safely prepositioned on Mars before the human explorers depart the Earth-Moon system on their outbound journey. That way, a distressed outbound crew would know that if they could just make it to Mars, resupply and additional equipment would await them there.

Exactly what happens after the first human expedition to Mars is a question subject to a great deal of speculation, but any confirmed discovery of life on Mars, extinct or perhaps even existent, could have as powerful a so-

Figure 7.4 Artist Pat Rawlings's rendering, titled *20/20 Vision*, shows an anticipated event of great philosophical and scientific significance as an astronaut discovers evidence of ancient life on Mars circa 2025. Artist rendering courtesy of NASA.

cial impact as the Copernican hypothesis did some five centuries earlier (see Figure 7.4).

People on Earth might recognize that this accomplishment is the precursor to permanent human migration into heliocentric space. Then Mars could become the central object of greatly expanded human space activities—perhaps surface-base-development activities complementing the rise of a self-sufficient lunar civilization.

Frontier settlements of about 1,000 persons each would eventually dot the surface of the Red Planet at critical locations, including sites related to extinct or existent life on Mars. By the end of the twenty-first century, native Martian resources could continue to fuel economic growth on the planet and simultaneously support a new wave of human expansion into the mineral-rich main asteroid belt and beyond. Martian-grown food, processed metals, manufactured products (including robots and spacecraft), and bulk quantities of air, water, and propellants for nuclear (hydrogen) and chemical (hydrogen and oxygen) rockets would dominate space-technology activities and interplanetary trade beyond cislunar space.

Scientists have already suggested an interesting variety of mechanical and biological terraforming (planetary-engineering) approaches, including the introduction of biologically engineered microorganisms and hardy plants (lichens) into the Martian environment. Of course, social and ethical issues will also arise concerning the impact of such terraforming activities on any (as yet undiscovered) native Martian life forms. Human-initiated greening of the Red Planet would represent a major far-term task for the native Martians, and life for those "pioneering" human beings would certainly assume many interesting technical, social, and philosophical dimensions.

This is just a brief glimpse of how advances in space technology throughout the twenty-first century will certainly provide many interesting new challenges and opportunities by which human beings can grow and mature as an emerging intelligent species. Taking a thousand-year future perspective, Mars—thoroughly explored, scientifically studied, and economically developed—opens up the remainder of the outer solar system for human exploration and use. As they set about the task of making this currently hostile planet more suitable for life, the Martian settlers will form a close partnership with many types of advanced, intelligent space robots. As the descendants of the original Martian settlers spread out into the more distant regions of the solar system, they will develop and use even more advanced space technologies. The development and application of very smart space robots, possibly even the initial versions of John von Neumann's long-postulated self-replicating machine, will become a major Martian industry.

By coincidence, these are the very same space technologies and skills needed to undertake the first wave of interstellar exploration via both robot and eventually human-crewed spacecraft. It is, therefore, not too much of an extrapolation to suggest here that the successful settlement of Mars could also represent a key milestone on the pathway to the stars for the human race.

FAR-TERM SPACE-TECHNOLOGY PROJECTIONS

Satellite Power Systems

The proposed satellite power system (SPS) uses the unlimited supply of sunlight found in outer space to make energy available in the form of electricity on Earth. An illustrative calculation for a reference design 5-kilometer-by-5-kilometer SPS platform suggests that some 34 gigawatts of raw sunlight are available for collection by a satellite of this size. Space-technology planners and visionaries have suggested that a constellation of between 50 and 100 such giant platforms be built in the twenty-first century and placed all around Earth's equator at an altitude of approximately 35,900 kilometers. In this geostationary orbit, each SPS would experience sunlight more than 99 percent of the time. (There is a brief period when relative orbital motions take each individual SPS into Earth's shadow and incoming sunlight is temporarily blocked.) These large orbiting space structures could then send the intercepted solar energy back to Earth in any of three basic ways: mirror transmission, microwave transmission, or laser transmission.

In the mirror-transmission technique, very large (kilometers-across) orbiting mirrors are used to beam concentrated raw sunlight directly down to suitable solar-energy-conversion facilities on Earth's surface. Because sunlight is being sent continuously from the large space platform, the solar-energy-conversion plant on the ground can operate continuously. One candidate mirror system would employ about 900 orbiting mirrors, each perhaps 50 square kilometers in surface area, to support a planetary power grid capable of providing 800 gigawatts of electricity. In this particular SPS concept, the space component of the overall system is quite simple. However, this concept also requires the use of a very large contiguous land area at the ground site. In addition, the large orbiting mirrors would illuminate the night sky within at least a 150-kilometer radius of the geometric center of the ground site.

In the microwave-transmission concept, the SPS collects sunlight and converts it into electricity. This electric energy is transformed immediately into radio-frequency (RF), or microwave, energy and then beamed to an appropriate ground site on Earth. The giant satellite accomplishes the sunlight-to-electricity conversion either directly by using photovoltaic (solar) cells or indirectly by using a solar-thermal heat engine to drive appropriate turbogenerator equipment. The heat engine is a machine that operates in a thermodynamic cycle and converts a portion of the heat (thermal-energy) input into mechanical work. Microwave-generating tubes called klystrons then convert the solar-generated electricity into microwave energy that is beamed to a special receiving station on Earth. In the final step of this wireless power-transmission process, a receiving antenna called a rectenna collects the incoming beam of microwave energy and transforms it into direct current (DC) electricity. Other power-conditioning equipment at the ground site inverts the DC electricity into high-voltage alternating current (AC) that can flow into a regional power grid.

For this particular SPS concept to send five gigawatts of electricity into a terrestrial power grid, the rectenna site requires a ground area of about 100 square kilometers. To protect human and animal life from continuous exposure to low-level microwave radiation, the receiving site should have an adjacent exclusion area of about 70 square kilometers. For example, a reference SPS study performed by NASA and the U.S. Department of Energy (DOE) showed that the intensity of the incident an RF beam at 2.45-gigahertz frequency would range from approximately 20 milliwatts per square centimeter (mW/cm^2) at the center of the ground site to 0.1 mW/cm^2 at the outer edge of the exclusion area.

The third SPS concept is called the laser-transmission technique. Here, incoming solar energy is converted into infrared-wavelength laser radiation that the satellite beams down to a special ground facility. Like the microwave-transmission approach, the incoming laser beam is collected at the site and its energy content used for the production of electricity. The use of laser radiation represents an alternative approach to wireless power transmission. Compared with microwave radiation, shorter-wavelength laser radiation is more precise and involves a much smaller beam diameter for the same amount of energy transmitted.

Since the SPS concept first appeared in 1968, a variety of studies have explored whether these giant orbiting systems can actually help relieve global energy needs in an environmentally friendly manner. Investigations sponsored by NASA and DOE in the 1970s provided a particularly sweeping vision of companion advanced space-technology developments that included the construction of satellite power systems from lunar materials. Within these far-term future space-technology scenarios, all the necessary SPS manufacturing and construction activities might be accomplished by thousands of space workers who would live with their families in large space settlements at L_4 and L_5, as discussed later.

More recent (mid-1990s) NASA studies suggest the use of smaller, modular-design SPS units that can be assembled in orbit by robot systems with only minimal use of human workers. Either SPS development strategy requires significant advances in space technology and offers exciting outer-space energy options to sustain an environmentally responsible, planetary civilization.

Large Space Settlements

Conceptually, a future space settlement would be a large, orbiting habitat in which from 10,000 to more than 50,000 people might live, work, and play. Citizens of this planetoid-sized, space city-state would engage in a variety of extraterrestrial operational, scientific, and commercial activities. The term *city-state* comes down through history from ancient Greece. It means a sovereign state that consists of an independent city and its surrounding area—making the term very appropriate for describing large, politically autonomous space settlements that contain significant numbers of people.

The inhabitants of such space city-states would experience life in a variety of human-engineered, miniworld environments. Some space settlements would focus on producing a major line of products or providing selected services. The construction of giant satellite power systems or the operation of a large space manufacturing complex that produces everything from rocket ships to the latest-model space construction robot are examples of the specialized-mission space settlement. Other settlements would function primarily as deep-space, frontier city-states and provide humans scattered throughout some distant region of the solar system with a central habitable location for scientific, economic, social, medical, and administrative services.

One frequently suggested far-term scenario involves the use of lunar materials as the feedstock for a giant space manufacturing complex located at Lagrangian libration point 4 or 5 (popularly known as L_4 or L_5). The French mathematician Joseph-Louis Lagrange (1736–1813) correctly postulated the existence of five different points in outer space where a small object can have a stable orbit in spite of the gravitational attractions exerted by two much more massive celestial objects when the objects orbit about a common center of mass.

Lagrangian points L_1, L_2, and L_3 represent points of unstable equilibrium, while L_4 and L_5 are points of stable equilibrium. Considering the Earth-Moon system, the Lagrangian points L_4 and L_5 lie on the Moon's orbit round Earth, about 60 degrees ahead of (leading) and 60 degrees behind (trailing) the Moon, respectively. The condition of inherently stable mechanical equilibrium for objects placed at these two points (imagine a bowl-shaped "gravity valley") makes the L_4 and L_5 locations especially attractive for very large space settlements and manufacturing complexes.

One possible industrial activity for a large space settlement at the L_4 or L_5 orbital locations would be the manufacture and assembly of satellite power system (SPS) components (discussed in the previous section).

An interesting candidate design for a large space settlement is a torus-shaped configuration that can accommodate about 10,000 people. In this particular concept, the inhabitants are members of a space manufacturing complex located at Lagrangian libration point 5 (L_5). After work, they would return to their homes located on the inner surface of the torus, which is nearly 1.6 kilometers in circumference. The settlement would rotate to provide the inhabitants with an artificial gravity level similar to that experienced on the surface of Earth. A nonrotating shell of material, possibly the accumulated waste material from processed lunar ores, would shield the habitat from cosmic rays and solar-flare radiation. Outside the shielded area, agricultural crops would be grown by taking advantage of the intense continuous stream of sunlight available in cislunar space. Docking areas and microgravity industrial zones would be located at each end of the settlement; so would be the large flat surfaces needed to reject waste thermal energy (heat) away from the complex into outer space.

By the late twenty-first century, large space settlements might start serving as important "frontier cities" in the main asteroid belt and beyond. However, for deep-space locations that lie beyond the orbit of Mars, the greatly diminished intensity of sunlight will make the harvesting of solar energy impractical for satisfying the energy demands of a major space settlement. Consequently, when such settlements begin to appear at strategic locations in the Jovian, Saturnian, Uranian, and Neptunian systems, they will most likely be nuclear-powered facilities that derive all their prime energy needs from advanced devices that imitate (on a much smaller scale) the thermonuclear energy processes taking place within the Sun. In the far term, harnessing the power of the stars by means of controlled nuclear fusion opens up the rest of the solar system to human occupancy.

Possibly at the start of the twenty-second century, the final link in a growing chain of outpost space cities will appear in the Neptunian system some four light-hours' distance from the Sun. To the inhabitants ("Neptunian settlers") of this distant frontier space city, the Sun will be only a bright star, and their main space-technology activity may involve the assembly, fueling (possibly using hydrogen or helium extracted from Neptune's frigid atmosphere) and launching of humankind's first robotic interstellar probe. Just as the initial lunar base served as our gateway to the riches and wonders of the solar system, this remote Neptunian space settlement and its hardy inhabitants could function as our race's center of future interstellar space-technology excellence (see Figure 7.5).

Figure 7.5 Near the end of the twenty-first century, the human race might apply advances in space technology to send the first robotic interstellar probe on a historic journey of scientific investigation. Artist rendering courtesy of NASA.

Planetary Defense System

Far-term space technology also has a very special role to play in protecting all forms of life on Earth. Our planet resides in a swarm of comets and asteroids that can and do impact its surface. The entire solar system contains a long-lived population of asteroids and comets, some fraction of which are perturbed into orbits that then cross the orbits of Earth and other planets. Although the annual probability of a large (about one-kilometer-diameter or greater) asteroid or comet is extremely small, the environmental consequences of just one such cosmic collision would be catastrophic on a global scale.

How real is this threat? From archaeological and geological records, scientists suspect that a tremendous catastrophe occurred on Earth about 65 million years ago. As a result of that ancient cataclysm, more than 75 percent of all life on our planet, including the dinosaurs, disappeared within a very short period. One popular explanation for this mass extinction is that a large asteroid (about 10 kilometers in diameter) slammed into Earth. This "killer" asteroid partially vaporized while passing through the atmosphere and then impacted with great force in the shallow sea near the Yucatán region of Mexico. The collision created a huge tidal wave and left a

180-kilometer-diameter impact-ring structure on the bottom of the Gulf of Mexico. Geologists call this impact structure Chicxulub.

About 50,000 years ago, a 50- to 100-meter-diameter iron mass, traveling at more than 11 kilometers per second, impacted on Earth and formed the well-preserved, bowl-shaped crater near Flagstaff, Arizona. Meteor Crater, as the impact site is popularly known, is approximately 1 kilometer across and 200 meters deep. In July 1994, a cluster of fragments, some about 20 kilometers in diameter, from the disintegrating comet Shoemaker-Levy 9 smashed into Jupiter's southern hemisphere. Scientists around the world observed this cosmic collision and then estimated that the comet's fragments collectively deposited the explosive energy equivalent of about 40 million megatons of trinitrotoluene (TNT). The planet's colorfully striped atmosphere heaved for months after this violent sequence of collisions that also sent dark material showering 1,000 kilometers into space above the frigid cloud tops.

The proposed planetary defense system is a future space-technology system that functions as a planetary life- and health-insurance policy against large asteroids or comets whose trajectories take them on a collision course with Earth or another important planetary body. The system would consist of two major components: a surveillance function and a mitigation function. The surveillance function uses a variety of strategically located optical and radar tracking systems to continuously monitor all regions of space for threatening near-Earth objects (NEOs). An NEO is an Earth-approaching asteroid or comet whose projected trajectory takes it on a collision course with Earth at some point in the future. Scientists consider an object with a diameter greater than one kilometer to be a "large" impactor with the potential for causing severe regional damage.

Should a large impactor be detected by the surveillance function, then the available warning time and existing level of space technology would determine appropriate defensive operations. Early warning is key to success. The surveillance portion of the planetary defense system must detect and identify all possible impactors in sufficient time so that the system can attempt one or more mitigation techniques. To deal effectively with all possible cosmic collision threats (but especially the very short-notice arrival of long-period comets), the interception/mitigation portion of the defense system should be deployed in space and ready to function on demand.

Mitigation techniques depend significantly on the warning time available and fall into two broad areas: techniques that deflect the threatening object and techniques that destroy or shatter the threatening object. Within the technology horizon of the first half of the twenty-first century, nuclear explosives appear to be the logical technical tool for deflecting or disrupting a large impactor. Later in the century, with sufficient warning time and reaction time, more advanced space technologies might be used to nudge the threatening object into a harmless (or even useful) orbit. Scientists have suggested a variety of interesting "nudging options," including carefully emplaced and focused nuclear detonations, very high-thrust nuclear-thermal rockets, low-thrust (but continuous) nuclear-electric propulsion systems, and even innovative mass-driver systems that use chunks of the threatening impactor as reaction mass).

Should deflection prove unsuccessful or impractical due to the warning time available, then the defense system must attempt the physical destruction or fragmentation of the approaching impactor. Multiple-megaton-yield nuclear detonations, including explosive devices precisely emplaced within the object at critical fracture points, might be used to shatter it into smaller, less dangerous projectiles that miss Earth. However, the human and robot "planet defenders" must perform this demolition process with great care; otherwise, the approaching cosmic cannonball would simply become an equally deadly cluster bomb. With enough reaction time, the defense system might even maneuver a smaller asteroid (about 100 meters in diameter) directly into the path of the threatening object, causing an incredible collision that should shatter both objects at a safe distance away from the Earth-Moon system.

Stimulated by the natural violence and incredible energy releases displayed during Jupiter's encounter with the remnants of the comet Shoemaker-Levy 9, members of the U.S. Congress requested in the mid-1990s that various agencies of the U.S. government, including the Department of Defense, the Department of Energy, and NASA, conduct appropriate studies concerning the planetary risk posed by large impactors and on ways to guard our planet from any cosmic collision catastrophe in the future. The requested studies identified the near-Earth-object (NEO) risk as real but not immediate and suggested the development of a planetary defense system that used surveillance and mitigation techniques, centered around the application of advanced space technology and nuclear explosive technology. International involvement in the planetary defense system was also suggested.

The use of space technology in this unique application would provide future generations of humans control over the destiny of the entire planet with respect to real natural hazards from outer space. In the past, random comet or asteroid impacts have dramatically altered Earth as well as all the other planetary bodies in the solar system. It is certain that sometime in the future, Earth will again be threatened by collision with a large object from space. This time, however, future human beings will have the benefit of advanced space technology to defend and protect their world or possibly any other threatened planetary body in the solar system.

CONCLUDING REMARKS

Our planetary civilization is at an important crossroads. Without space technology, without a focused and determined off-planet expansion of future human activities, we are in danger of rejecting our potential role as conscious, intelligent creatures in the cosmic scheme of things. As a planetary society, we can decide that lunar bases and Mars settlements simply should not play an integral part in any future global civilization. Why waste so much money going to a distant and hostile world when we have so many problems (like international terrorism) and needs (like third-world poverty) right here on Earth? That shortsighted question has been raised about space exploration in the twentieth century and will probably continue to be raised throughout this century and beyond. Perhaps future terrestrial technologies will even allow us to achieve some level of sustainable global society in the twenty-first century. But without a focused decision to support off-planet expansion, our species elects to isolate itself on a single world and denies future generations the rich technical and cultural opportunities of a multiworld, solar-system-level civilization.

The consequences of not embracing the full potential of space technology involve more than the loss of technical options. The introvert philosophy of a closed-world civilization contrasts sharply with the extrovert philosophy of an open-world civilization empowered by space technology. Over the next few decades, human beings will make important decisions concerning the expanded use of space technology—decisions that could ultimately influence the spread of life and intelligence throughout the galaxy. We are intelligent creatures in whom the consciousness of the universe resides. Through space technology, we can discover whether we are alone in this vast universe or else share it with many other intelligent species.

Centuries of scientific activity and technical progress have now led our species into the age of space. Was it part of some grand plan or a mere coincidence (stimulated by political conflict) that human beings left footprints on another world? Properly understood and applied, space technology can provide human beings with an expanded sense of evolutionary purpose. With space technology, our descendants can dream of traveling to and inheriting the stars. By consciously pursuing the beneficial applications of space technology in the twenty-first century, we will also provide many exciting new options for continued development of the human species. Space technology allows us to give the following response to the ancient questions of "Who are we?" and "Where are we going?": We are made of star dust, and to the stars we are returning.

Chapter 8

Glossary of Terms Used in Space Technology

abort To cut short, break off, or cancel an action, operation, or procedure with an aircraft, spacecraft, or aerospace vehicle, especially because of equipment failure. For example, NASA's space shuttle system has two types of abort modes during the ascent phase of a flight: the intact abort and the contingency abort. An intact abort is designed to achieve a safe return of the astronaut crew and orbiter vehicle to a planned landing site. A contingency abort involves a ditching operation in which the crew is saved, but the orbiter vehicle is damaged or destroyed.

acceleration (symbol: *a*) In physics and engineering mechanics, the time rate of change of velocity of an object. Acceleration typically has the unit of meters per second squared (m/s^2). At the surface of Earth, the acceleration due to gravity (symbol: g) of a free-falling object has been given the standard value of 9.80665 m/s^2 by international agreement.

accretion The gradual accumulation of small particles of gas and dust into larger material bodies, mostly due to the influence of gravity. For example, in the early stages of stellar formation, matter begins to collect or accrete into a nebula (a giant interstellar cloud of gas and dust), and eventually stars are born in this nebula. When a particular star forms, any small quantities of residual matter collect into one or more planets orbiting the new star.

acronym A word formed from the first letters of a name, such as STS, which means Space Transportation System, or a word formed by combining the initial parts of a series of words, such as lidar, which means light detection and ranging. Acronyms are frequently used in space technology.

active satellite A satellite that is functioning; a satellite that transmits a signal, in contrast to a passive satellite.

acute radiation dose The total dose of ionizing radiation received by a person at one time or over so short a period that biological recovery cannot occur prior to the onset of acute radiation syndrome (i.e., radiation sickness).

adapter skirt A flange or extension on a launch-vehicle stage or spacecraft section that provides a means of fitting on another stage or section.

adiabatic In general, happening without gain or loss of thermal energy (heat); in the context of thermodynamics, a process in which thermal energy (heat) is neither added to nor removed from the system involved.

aeroassist The use of a planet's atmosphere to provide the lift or drag needed to maneuver a spacecraft. Aeroassist maneuvers allow a spacecraft to change direction or slow down while flying through the thin, upper regions of a planet's atmosphere without expending control-rocket propellant. When an orbiter spacecraft arrives at a planet, for example, it needs to slow down to be captured by the planet's gravity field and to achieve a desirable orbit around the planet. Aeroassist provides a major propellant-savings advantage versus the all-propellant "braking" equivalent to achieve such a planetary capture and orbital insertion.

aerobrake A spacecraft structure designed to deflect rarefied (very low-density) airflow around a spacecraft and thereby support aeroassist maneuvers. The aerobrake helps reduce or even eliminates a spacecraft's need to perform the large propulsive burns traditionally used to make orbital changes around a planet. *See also* aeroassist.

aerodynamic vehicle A craft that has lifting and control surfaces to provide stability, control, and maneuverability while flying through a planet's atmosphere. For example, a glider or an airplane is an aerodynamic vehicle that is capable of flight only within a sensible atmosphere. Such heavier-than-air vehicles rely on aerodynamic forces to maintain flight through the atmosphere.

aeronautics The science of flight within the atmosphere.

aerospace A term, derived from aeronautics and space, meaning of or pertaining to Earth's atmospheric envelope and outer space beyond it. These two separate physical regimes are considered as a single realm for space-technology activities involving the launch, guidance, control, and recovery of vehicles and systems that can travel in both regions. For example, NASA's space shuttle orbiter vehicle is defined as an aerospace vehicle because it operates both in the atmosphere and in outer space.

aerospace ground equipment (AGE) All equipment needed on Earth's surface to make an aerospace system function properly in its intended environment.

aerospace vehicle A vehicle that is capable of operating both within Earth's sensible (measurable) atmosphere and in outer space. The space shuttle orbiter vehicle is an example.

aerospike nozzle A rocket nozzle design that allows combustion to occur around the periphery of a spike (or center plug). The thrust-producing, hot exhaust flow is then shaped and adjusted by the ambient (atmospheric) pressure.

afterbody 1. A companion body that trails a satellite or spacecraft. 2. A portion or section of a launch vehicle, rocket, or missile that enters the atmosphere unprotected behind the nose cone or other component that is protected against the aerodynamic heating during atmospheric entry. 3. The aft section of a vehicle.

airlock Generally, a small chamber with "airtight" doors that is capable of being pressurized and depressurized. The airlock serves as a passageway for crew members and equipment between places at different pressure levels—for example, between a spacecraft's pressurized crew cabin and outer space.

albedo The ratio of the amount of electromagnetic energy reflected by a surface to the total amount of electromagnetic energy incident upon the surface. The albedo is usually expressed as a percentage; for example, the albedo of Earth is about 30 percent. This means that approximately 30 percent of the total solar radiation incident upon Earth is reflected back to space. A perfect reflector has an albedo of 100 percent (1.0), while a totally black surface (perfect absorber of all incident radiation) has an albedo of 0 percent (0.0).

angstrom (symbol: Å) A unit of length commonly used by scientists to measure wavelengths of electromagnetic radiation in the visible, near-infrared, and near-ultraviolet portions of the spectrum. This unit is named after Anders Jonas Ångström (1814–1874), a Swedish physicist, who provided quantitative descriptions of the solar spectrum in 1868. One angstrom (Å) = 10^{-10} meter = 0.1 nanometer.

aphelion In orbital mechanics, the point in an object's orbit around the Sun that is most distant from the Sun. (The orbital point nearest the Sun is called the perihelion.)

apogee 1. In orbital mechanics, the point at which a missile's trajectory or a satellite's orbit is farthest from the center of the gravitational field of the controlling body or bodies. 2. Specifically, the point that is farthest from Earth in a geocentric orbit. The term applies to both the orbit of the Moon around Earth and to the orbits of artificial satellites around Earth. At apogee, the orbital velocity of a satellite around Earth is at a minimum. To enlarge or "circularize" the orbit, aerospace engineers fire a spacecraft's thruster (often called the apogee kick motor) at apogee. This gives the spacecraft the necessary increase in velocity. Aerospace engineers often call this velocity increment a "delta v" or symbolize it as Δv. Apogee is the opposite of perigee. 3. The highest altitude above Earth's surface reached by a sounding rocket.

artificial gravity Simulated gravity conditions established within a spacecraft, space station, or space settlement. Rotating the crew cabin or habitable modules about an axis of the space vehicle or station creates this condition, since the centrifugal force generated by the rotation produces effects similar to the force of gravity within the vehicle. Certain spinning and looping rides at major amusement parks produce similar effects. Artificial gravity was first suggested by the Russian space visionary Konstantin Tsiolkovsky (1857–1935) at the start of the twentieth century.

asteroid A small, solid object (sometimes called a minor planet) that orbits the Sun but is independent of any planet. Most asteroids are found in the main asteroid belt, a region between Mars and Jupiter that extends approximately 2.2 to 3.3 astronomical units from the Sun. The largest asteroid is Ceres, which is about 1,000 kilometers in diameter. It was discovered in 1801 by the Italian astronomer Giuseppe Piazzi (1746–1826). Some asteroids, called Earth-crossers or near-Earth objects (NEOs), have orbits that take them near or across our planet's orbit around the Sun. These inner solar-system asteroids are divided into three general groups: Aten, Apollo, and Amor (each group bears the name of its most significant minor planet).

astro- A prefix that means "star" or (by extension) outer space or celestial; for example, astronaut, astronautics, or astrophysics.

astrobiology The search for and study of living organisms found on celestial objects beyond Earth. Also called exobiology.

astrodynamics The branch of engineering that deals with the trajectories of space vehicles; in particular, the practical application of orbital mechanics, propulsion theory, and related technical fields to the problem of planning and directing the movement of a space vehicle (e.g., a planetary flyby mission) or group of vehicles (e.g., orbital rendezvous and docking activities).

astronaut As used within the American space program, a person who travels in outer space; a person who flies in an aerospace vehicle to an altitude of more than 80 kilometers. The word comes from two ancient Greek words that literally mean "star (astro) sailor or traveler (naut)." *See also* cosmonaut.

astronautics 1. The branch of science dealing with space flight. 2. The professional skill, technical talent, or activity of operating spacecraft or aerospace vehicles.

astronomical unit (AU) In astronomy and space technology, a unit of distance defined as the semimajor axis of Earth's orbit around the Sun, that is, the mean distance between the center of Earth and the center of the Sun. One AU is equal to 149.6×10^6 kilometers or 499.01 light-seconds, approximately.

astrophysics The branch of science that investigates the nature of stars and star systems. Astrophysics provides the theoretical principles that enable scientists to understand astronomical observations. Through space technology, astrophysicists can place sensitive instruments above Earth's atmosphere and view the universe in all portions of the electromagnetic spectrum. High-energy astrophysics includes gamma-ray astronomy, cosmic-ray astronomy, and X-ray astronomy.

atmosphere 1. In general, gases and suspended solid and liquid materials that are gravitationally bound to the outer region (gaseous envelope) around a planet or satellite. 2. The breathable environment inside a space capsule, aerospace vehicle, spacecraft, or space station; the cabin atmosphere.

atmospheric drag The retarding force produced on a space vehicle as it passes through the upper regions of a planet's atmosphere. For a spacecraft orbiting Earth, this retarding force drops off exponentially with altitude. Consequently, atmospheric drag has only a small effect on spacecraft whose perigee (closest orbital point) is higher than a few hundred kilometers. However, for spacecraft with perigee values less than a few hundred kilometers, the cumulative effect of atmospheric drag eventually causes them to reenter the denser regions of Earth's atmosphere and be destroyed. Aerospace engineers prevent this from happening by giving such low-orbiting spacecraft onboard propulsion systems that provide periodic reboost.

atmospheric probe A special collection of scientific instruments (usually released by a spacecraft) for determining the pressure, composition, and temperature of a planet's atmosphere at different altitudes. For example, an atmospheric probe was released by NASA's *Galileo* spacecraft in December 1995. It plunged into the Jovian atmosphere and successfully transmitted scientific data to the *Galileo* spacecraft for about 58 minutes.

atmospheric window A wavelength interval within which a planetary atmosphere is transparent to (i.e., readily transmits) electromagnetic radiation.

attitude Generally, the position of an object as defined by the inclination of its axes with respect to a frame of reference. In space technology, the orientation of a vehicle (e.g., a spacecraft or aerospace vehicle) that is either in motion or at rest, as established by the relationship between the vehicle's axes and a reference line or plane (such as the horizon) or a fixed system of reference axes (such as the x-, y-, and z-axes in the Cartesian coordinate system). Attitude is often expressed in terms of pitch, roll, and yaw.

auxiliary power unit (APU) A power unit carried on a spacecraft, aerospace vehicle, or aircraft that supplements the main source(s) of power on the craft.

azimuth (common symbol: A) In astronomy, the direction to a celestial object measured in degrees clockwise from north around a terrestrial observer's horizon. On Earth, azimuth is 0° for an object that is due north, 90° for an object due east, 180° for an object due south, and 270° for an object due west.

backout In aerospace operations, to undo things or events that have already been completed during a launch-vehicle countdown; usually accomplished in reverse order.

ballistic missile Any missile propelled into space by rocket engines that does not rely upon aerodynamic surfaces to produce lift and therefore follows a ballistic trajectory when thrust is terminated; a missile that is propelled and guided only during the initial phase of its flight. During the nonpowered and nonguided portion of its flight, a ballistic missile assumes a trajectory similar to that of an artillery shell and behaves primarily in accordance with the laws of dynamics. Once thrust is terminated (usually at a predesignated time), a ballistic missile's reentry vehicles (RVs) are released, and these RVs (under the influence of gravity) then follow free-falling (ballistic) trajectories toward their targets. *See also* guided missile.

ballistic trajectory The path an object follows while being acted upon only by gravitational forces and the resistance of the medium through which it passes. A stone tossed into the air follows a ballistic trajectory. Similarly, after its propulsive unit stops operating, a rocket vehicle that does not have lifting surfaces describes a ballistic trajectory.

berthing The joining of two spacecraft using a manipulator or other mechanical device to move one object into contact (or very close proximity) with the other at an appropriate interface. For example, NASA astronauts use the space shuttle's remote manipulator system to carefully position a large free-flying spacecraft (like the Hubble Space Telescope) onto a special support fixture located in the orbiter's cargo bay during an in-orbit servicing and repair mission.

big-bang theory In cosmology, a theory concerning the origin of the universe. This theory suggests that a very large, ancient explosion, called by astrophysicists the *initial singularity*, started the space and time of the present universe, which has been expanding ever since. The big-bang event is currently thought to have occurred between 15 and 20 billion years ago. Astrophysical observations, especially discovery of the cosmic microwave background in 1964, tend to support this theory. *See also* cosmology.

biosphere The life zone of a planetary body; for example, that part of the Earth system inhabited by living organisms. On our planet, the biosphere includes portions of the atmosphere, the hydrosphere, the cryosphere, and surface regions of the solid Earth. *See also* ecosphere; global change.

bipropellant A rocket that uses two unmixed (uncombined) liquid chemicals as its fuel and oxidizer, respectively. The two chemical propellants flow separately into the rocket's combustion chamber, where they are combined and combusted to produce high-temperature, thrust-generating gases. The combustion gases exit the rocket system through a suitably designed nozzle.

black hole An incredibly compact, gravitationally collapsed mass from which nothing—light, matter, or any other kind of information—can escape. Astrophysicists believe that a black hole is the natural end product when a massive star dies and collapses beyond a certain critical dimension, called the *Schwarzschild radius*. Once the massive star shrinks to this critical radius, its gravitational escape velocity is equal to the speed of light, and nothing can escape from it. Inside this event horizon, an incredibly dense point mass, or singularity, is formed.

blueshift In astronomy, the apparent decrease in the wavelength of a light source caused by its approaching motion. The Doppler shift of the visible spectra of certain distant galaxies toward blue light (i.e., shorter wavelength) indicates that these galaxies are approaching Earth. *See also* Doppler shift; redshift.

boiloff The loss of a cryogenic propellant, such as liquid oxygen or liquid hydrogen, due to vaporization. This happens when the temperature of the cryogenic propellant rises slightly in the tank of a rocket being prepared for launch. The longer a fully fueled rocket vehicle sits on the launch pad, the more significant is the problem of boiloff.

booster rocket A rocket motor, using either solid or liquid propellant, that assists the main propulsive system (sustainer engine) of a launch vehicle during some part of its flight.

burn In aerospace operations, the firing of a rocket engine. For example, the "second burn" of the space shuttle's orbital maneuvering system (OMS) engines means the second time during a particular shuttle flight that the OMS engines have been fired. The *burn time* is the length of a thrusting period.

burnout The moment in time or point in a rocket's trajectory when combustion of fuels in the engine is terminated. This usually occurs when all the propellants are consumed.

canard The horizontal surface placed at the front of an aerodynamic vehicle (that is, ahead of the main lifting surface or wing) that helps control the vehicle during flight.

captive firing An aerospace-engineering operation in which a rocket propulsion system is test-fired at full or partial thrust while restrained in a test-stand facility; the propulsion system is completely instrumented, and data to verify design and demonstrate performance are obtained. Sometimes referred to as a *holddown test*.

Cartesian coordinate system Named after French philosopher and mathematician René Descartes (1596–1650), the Cartesian coordinate system is a system in which the locations of all points in space are expressed by reference to three mutually perpendicular planes that intersect in three straight lines called the x-, y-, and z-coordinate axes.

cavitation The formation of bubbles (vapor-filled cavities) in a flowing liquid. This condition occurs whenever the static pressure at any point in the moving fluid becomes less than the fluid's vapor pressure. The formation of these cavities (or vapor regions) alters the flow path of the liquid and can, therefore, adversely impact the performance of hydraulic machinery and devices, such as pumps. The collapse of these bubbles in downstream regions of high pressure creates local pressure forces that may result in pitting or deformation of any solid surface near the cavity at the time of collapse. In space technology, cavitation effects are most noticeable with high-speed hydraulic machinery, such as a liquid-propellant rocket engine's turbopumps.

celestial Of or pertaining to the heavens.

celestial mechanics The field of science that studies the dynamic relationships among bodies of the solar system and deals with the relative motions of celestial objects under the influence of gravitational fields.

chaser spacecraft In aerospace operations, the spacecraft or aerospace vehicle that actively performs the key maneuvers during orbital rendezvous and docking/berthing operations. The other space vehicle serves as the target and remains essentially passive during the encounter.

checkout 1. A sequence of actions (i.e., functional, operational, and calibration tests) performed to determine the readiness of a device or system to perform its intended task or mission. 2. The sequence of steps taken to familiarize a person with the operation of an airplane, aerospace vehicle, or other piece of aerospace equipment.

chemical fuel A fuel that depends upon an oxidizer for combustion or the development of thrust, such as liquid- or solid-rocket fuel or internal-combustion-engine fuel; as distinguished from nuclear fuel (such as the fissile radioisotope uranium 235).

chemical rocket A rocket that uses the combustion of a chemical fuel in either solid or liquid form to generate thrust. The chemical fuel requires an oxidizer to support combustion.

chugging A form of combustion instability in a rocket engine (especially a liquid-propellant rocket engine), characterized by a pulsing operation at a fairly low frequency. Aerospace engineers sometimes describe this phenomenon as occurring between particular frequency limits.

cislunar Of or pertaining to phenomena, projects, or activities happening in the region of outer space between Earth and the Moon. From the Latin word *cis*, meaning "on this side," and *lunar*, or, simply, "on this side of the Moon."

cold-flow test In aerospace operations, the thorough testing of a liquid-propellant rocket without actually firing (igniting) it. This type of test helps engineers verify the performance and efficiency of a propulsion system, since all aspects of propellant flow and conditioning except combustion are examined. For example, tank pressurization, propellant loading, and propellant flow into the combustion chamber (without ignition) are usually included in a cold-flow test.

Cold War A period from approximately 1946 to 1989 of intense ideological conflict and technological competition between the United States and the former Soviet Union. This era involved a missile and nuclear arms race, space-technology rivalry, deep mistrust, and armed hostility just short of head-to-head military conflict. The United States won the "space race" by successfully landing human beings on the Moon, starting on July 20, 1969. The symbolic end of the Cold War period was the demolition of the Berlin Wall (November 1989) and the reunification of Germany.

combustion chamber Any chamber for the combustion of fuel; specifically, that part of a rocket engine in which the combustion of chemical propellants takes place at high pressure. The combustion chamber and the diverging section of the nozzle comprise a rocket's *thrust chamber*.

comet A comet is a dirty ice ball made up of frozen gases and dust. As the comet approaches the Sun from the frigid regions of deep space, solar radiation causes these frozen materials to vaporize (sublime). While a comet's frozen nucleus is usually only a few tens of kilometers in diameter, the resultant vapors form an atmosphere (or coma) with a diameter that may reach 100,000 kilometers. When the comet travels through the inner solar system, the sublimed dust and ice also stream out in a spectacular, long, luminous tail. Space scientists believe that comets are remanent samples of the primordial material from which the planets were formed billions of years ago. The Dutch astronomer Jan Hendrik Oort (1900–1992) was the first to suggest that comets originate far from the Sun in a distant region (now called the Oort cloud) that extends out to the limits of the Sun's gravitational attraction. The Oort cloud forms a giant sphere with a radius of between 50,000 and 80,000 astronomical units (AUs) and contains billions of comets, whose total mass is thought to be roughly equal to the mass of Earth.

command destruct In aerospace operations, an intentional action that leads to the destruction of a rocket or missile in flight. The range-safety officer sends a command destruct signal whenever a malfunctioning vehicle's performance creates a safety hazard (on or off the rocket test range).

communications satellite A satellite that relays or reflects electromagnetic signals between two or more communications stations. An *active communications satellite* receives, regulates, and retransmits electromagnetic signals between stations, while a *passive communications satellite* simply reflects signals between stations. Active communications satellites, many of which operate in geostationary Earth orbit, have helped create today's global telecommunications infrastructure. The British space visionary and author Arthur C. Clarke first proposed the concept of the geostationary communications satellite in 1945.

companion body A nose cone, protective shroud, last-stage rocket, or similar object (e.g., payload-separation hardware) that orbits Earth along with an operational satellite or spacecraft. Companion bodies contribute significantly to a growing space-debris population in low Earth orbit, and aerospace engineers are now taking steps to reduce this problem through debris-minimization design approaches.

compressible flow In engineering (especially aerodynamics and fluid mechanics), flow at a speed sufficiently high that density changes in the fluid cannot be neglected.

console A desklike array of controls, indicators, and video display devices for the monitoring and control of a particular sequence of aerospace operations, as in the prelaunch checkout of a missile, a countdown action, or a launch procedure. A well-instrumented console allows an operator to efficiently monitor and control many different activating devices, data-recording instruments, and event sequences. During the critical phases of a space mission, the console becomes the central place from which to issue commands to or at which to display information concerning an aerospace vehicle, a deployed payload, an Earth-orbiting spacecraft, or an interplanetary probe. The mission control center generally contains clusters of consoles, each assigned to specific monitoring and control tasks. Depending on the nature and duration of a particular space mission, operators remain at a console continuously or only work there intermittently. Highly automated consoles often contain special alarms and flashing indicators to assist operations personnel in quickly identifying and correcting any performance anomalies that occur during the operation or mission.

continuously crewed spacecraft A spacecraft that has accommodations for continuous habitation during its mission. The *International Space Station* (*ISS*) is an example of a spacecraft designed for continuous human occupancy. Sometimes called a *continuously manned spacecraft*; but this term is not preferred.

continuously habitable zone (CHZ) The region around a star in which one or several planets can maintain conditions appropriate for the emergence and sustained existence of life. One important characteristic of a planet in the CHZ is that its environmental conditions support the retention of significant amounts of liquid water on the planetary surface.

control rocket A low-thrust rocket, such as a retrorocket or a vernier rocket, used to guide, to change the attitude of, or to make small corrections in the velocity of an aerospace vehicle, spacecraft, or expendable launch vehicle.

converging-diverging (C-D) nozzle A thrust-producing flow device for expanding and accelerating hot exhaust gases from a rocket engine. A properly designed nozzle efficiently converts the thermal energy of combustion into the kinetic energy of the combustion-product gases. In a supersonic converging-diverging nozzle, the hot gas upstream of the nozzle throat is at subsonic velocity (i.e., the Mach number $[M] < 1$), reaches sonic velocity at the throat of the nozzle ($M = 1$), and then expands to supersonic velocity ($M > 1$) downstream of the nozzle's throat region while flowing through the diverging section of the nozzle.

cosmic rays Extremely energetic subatomic particles (usually bare atomic nuclei) that travel through space and bombard Earth from all directions. While hydrogen nuclei (that is, protons) make up the highest proportion of the cosmic-ray population (approximately 85 percent), these particles range over the entire periodic table of elements, from hydrogen through uranium, and also include electrons and positrons. Galactic cosmic rays spiral along the weak lines of magnetic force found throughout the Milky Way galaxy. They bring astrophysicists direct evidence of important phenomena (like nucleosynthesis and particle acceleration) that occur as a result of explosive processes in stars throughout the galaxy. Solar cosmic rays consist of protons, alpha particles, and other energetic atomic particles ejected from the Sun during solar-flare events.

cosmology The branch of science that deals with the origin, evolution, and structure of the universe. Contemporary cosmology centers around the big-bang hypothesis, a theory stating that about 15 to 20 billion (10^9) years ago the universe began in a great explosion (sometimes called the "initial singularity") and has been expanding ever since. Astrophysical discoveries in recent years tend to support big-bang cosmology. For example, the 1964 discovery of cosmic microwave background radiation (at 2.7 degrees Kelvin) provided scientists with their initial physical evidence that there was a very hot phase early in the history of the universe.

cosmonaut The title given by Russia (formerly the Soviet Union) to its space travelers. *See also* astronaut.

countdown In aerospace operations, the step-by-step process that leads to the launch of a rocket or aerospace vehicle. A countdown takes place in accordance with a predesignated time schedule; inverse counting is generally performed during the process, with "0" being the "go" or activate time. For example, "T minus 60 minutes" indicates that there are 60 minutes to go until the launch event, except for holds and recycling activities. At the end of a countdown, T-time is given in seconds, namely, 4, 3, 2, 1, 0 (ignition and launch). In the launch of a rocket, *plus count* is the count (in seconds, such as plus 1, plus 2, plus 3, and so on) that immediately follows T-time. It is used to check on the sequence of events after the action of the countdown has ended.

crew-tended spacecraft A spacecraft that can only provide temporary accommodations for habitation during its mission. Sometimes referred to as a *man-tended spacecraft*. *See also* continuously crewed spacecraft.

cryogenic propellant A rocket fuel, oxidizer, or propulsion fluid that is liquid only at very low (cryogenic) temperatures. Generally, temperatures below the boiling point of liquid nitrogen (−195° C) are treated as cryogenic temperatures. Liquid hydrogen (LH_2) and liquid oxygen (LO_2) are examples of cryogenic propellants.

deadband An intentional feature that engineers design into a control system to prevent a flight-path error from being corrected until that error exceeds a specified size. This avoids a continuous or jittery response to minor errors.

deboost A retrograde (opposite-direction) burn or braking maneuver that lowers the altitude of an orbiting spacecraft.

deceleration Moving with decreasing speed; the opposite of acceleration.

degree of freedom (DOF) A mode of angular or linear motion with respect to a coordinate system that is independent of any other mode. For example, a body in motion has six possible degrees of freedom, three linear (sometimes called *x*-, *y*-, and *z*-motion with reference to linear [axial] movements in the Cartesian coordinate system) and three angular (sometimes called pitch, yaw, and roll with reference to angular movements with respect to the object's three axes).

delta V (symbol: Δv) In aerospace engineering, a velocity change; a useful numerical index of the maneuverability of a spacecraft or rocket. This term often represents the maximum change in velocity that a space vehicle's propulsion system can provide, for example, the delta-v capability of an upper-stage propulsion system used to move a satellite from a lower-altitude orbit to a higher-altitude orbit or to place an Earth-orbiting spacecraft on an interplanetary trajectory. Often described in terms of kilometers per second or meters per second.

destruct In aerospace operations, the deliberate destruction of a missile or rocket vehicle after it has been launched, but before it has completed its course. Destruct commands are executed by the range-safety officer whenever a missile or rocket veers off its intended (plotted) course or functions in such a way that it becomes a hazard to life or property.

destruct line A boundary line on a rocket test range that lies on each side of the downrange course. For safety reasons, a rocket or missile is not allowed to fly across this line. If the vehicle's flight path touches the destruct line, it is destroyed by the range-safety officer who enforces established command destruct procedures. The *impact line* is an imaginary line on the outside of the destruct line. It runs parallel to the destruct line and marks the outer limits of impact for a rocket or missile destroyed under command destruct procedures.

direct-broadcast satellite (DBS) A special type of communications satellite that receives broadcast signals (such as television programs) from points of origin on Earth and then amplifies and retransmits these signals to individual end users scattered throughout some wide area or specific region. The DBS usually operates in geostationary Earth orbit and has become an integral part of the information-technology revolution. For example, many American households now receive more than 100 channels of television programming directly from space by means of inconspicuous, small (less than 0.5 meter in diameter) rooftop antennas that are equipped to decode DBS transmissions.

direct conversion In engineering (thermodynamics), the conversion of thermal energy (heat) or other forms of energy directly into electrical energy without intermediate conversion into mechanical work. Direct conversion takes place without the use of the moving components usually found in a conventional electric generator system. The main approaches for converting heat directly into electricity include *thermoelectric* conversion, *thermionic* conversion, and *magnetohydrodynamic* conversion. Solar cells directly convert sunlight into electrical energy in a process called *photovoltaic* conversion. Finally, batteries and fuel cells directly convert chemical energy into electrical energy.

directed-energy weapon (DEW) A weapon system that uses a tightly focused intense beam of energy (either as electromagnetic radiation or elementary atomic particles) to kill its target. A very high-energy laser is an example of a DEW. Since the DEW device delivers the lethal amount of energy at or near the speed of light, it is sometimes called a *speed-of-light weapon*.

docking The act of coupling two or more orbiting objects. Often, two orbiting spacecraft are joined together by independently maneuvering one into contact with the other (called the target) at a designated interface. In human spaceflight, the process of tightly joining two crewed spacecraft together with latches and sealing rings so that two hatches can be opened between them without losing cabin pressure. This particular type of docking operation allows crew members to move from one spacecraft to another in "shirtsleeve" comfort. A special mechanical device, called a *docking mechanism*, often helps connect one spacecraft to another during an orbital docking operation. *See also* berthing; rendezvous.

doffing The act of removing wearing apparel or other apparatus, such as a space suit.

donning The act of putting on wearing apparel or other apparatus, such as a space suit.

Doppler shift The apparent change in the observed frequency and wavelength of a source due to the relative motion of the source and an observer. If the source is approaching the observer, the observed frequency is higher and the observed wavelength is shorter. This change to shorter wavelengths is often called the *blueshift*. If, however, the source is moving away from the observer, the observed frequency will be lower and the wavelength will be longer. This change to longer wavelengths is called the *redshift*. *See also* blueshift; redshift.

double-base propellant A solid propellant that uses two unstable chemical compounds such as nitrocellulose and nitroglycerin that are then chemically bonded together by a material that serves as an oxidizer.

drogue parachute A small parachute that pulls a larger parachute out of stowage; a small parachute used to slow down a descending space capsule, aerospace vehicle, or high-performance airplane.

dynamic pressure (common symbol: Q) In engineering, the pressure exerted on a body by virtue of its motion through a fluid; for example, the pressure exerted on a launch vehicle as it flies up through the atmosphere. The term *max-Q* refers to the condition of maximum dynamic pressure experienced by a rocket vehicle ascending through a planetary atmosphere.

early warning satellite A military spacecraft whose primary mission is the detection and notification of an enemy ballistic missile launch. This type of surveillance satellite uses sensitive infrared (IR) sensors to detect missile launches.

Earth-crossing asteroid (ECA) An inner solar-system asteroid whose orbital path takes it across Earth's orbital path around the Sun. *See also* asteroid.

Earthlike planet A planet around another star (i.e., an extrasolar planet) that orbits in a continuously habitable zone (CHZ) and maintains environmental conditions that resemble those of Earth. These conditions include a suitable atmosphere, a temperature range that permits the retention of large quantities of liquid water on the planet's surface, and a sufficient quantity of radiant energy striking the planet's surface from the parent star. Exobiologists hypothesize that with such environmental conditions, the chemical evolution and the development of carbon-based life as we know it on Earth could also occur on an Earthlike planet. In addition, the Earthlike planet should also have a mass somewhat greater than 0.4 Earth masses (to permit the production and retention of a breathable atmosphere), but less than about 2.4 Earth masses (to avoid excessive surface-gravity conditions). *See also* continuously habitable zone; ecosphere.

Earth-observing satellite (EOS) An Earth-orbiting spacecraft that has a specialized collection of sensors capable of monitoring important environmental variables. Data from such satellites help support Earth system science. Environmental-monitoring spacecraft generally use either of two orbits: *geostationary orbit* (which provides a simultaneous "big-picture" view of an entire hemisphere) or *polar orbit* (which provides a closer view of Earth, including very remote locations, such as the Arctic and Antarctic). Sometimes called an *environmental satellite* or a *green satellite*. *See also* Earth system science; global change; greenhouse effect.

Earth system science The modern study of Earth, facilitated by space-based observations, that treats the planet as an interactive, complex system. The four major components of the Earth system are the atmosphere, the hydrosphere (includes liquid water and ice), the biosphere (includes all living things), and the solid Earth (especially the planet's surface and soil).

ecosphere The habitable zone around a main-sequence star of a particular luminosity in which a planet could support environmental conditions favorable to the evolution and continued existence of life. For the chemical evolution of Earth-like carbon-based living organisms, global temperature and atmospheric-pressure conditions must allow the retention of a significant amount of liquid water on the planet's surface. Under favorable conditions, an effective ecosphere might lie between about 0.7 and 1.3 astronomical units (AU) from a star like the Sun. However, if all the surface water has completely evaporated (the *runaway greenhouse effect*) or if all the liquid water on the planet's surface has completely frozen (the *ice catastrophe*), then an Earthlike planet within an ecosphere cannot sustain life. *See also* continuously habitable zone.

electric propulsion An electric rocket engine converts electric power into reactive thrust by accelerating an ionized propellant (such as mercury, cesium, argon, or xenon) to a very high exhaust velocity. There are three general types of electric rocket engine: *electrothermal*, *electromagnetic*, and *electrostatic*. An electric propulsion system consists of three main components: (1) an electric rocket engine (thruster) that accelerates an ionized propellant; (2) a suitable propellant that can be easily ionized and accelerated; and (3) an electric power source.

electromagnetic radiation (EMR) Radiation composed of oscillating electric and magnetic fields and propagated with the speed of light. EMR includes (in order of increasing wavelength) gamma radiation, X-rays, ultraviolet, visible, and infrared (IR) radiation, and radar and radio waves.

encounter In aerospace operations, the close flyby or rendezvous of a spacecraft with a target body. The target of an encounter can be a natural celestial object (such as a planet, asteroid, or comet) or a human-made object (such as another spacecraft).

environment An external condition, or the sum of such conditions, in which a piece of equipment, a living organism, or a system operates; for example, vibration environment, temperature environment, radiation environment, or space environment. Environments can be natural or human engineered (artificial) and are often specified by a range of values.

escape velocity (common symbol: V_e) The minimum velocity that an object must acquire to overcome the gravitational attraction of a celestial object. The escape velocity for an object launched from the surface of Earth is approximately 11.2 kilometers per second (km/s), while the escape velocity from the surface of Mars is 5.0 km/s.

exoatmospheric Occurring outside Earth's atmosphere; in aerospace operations, events and actions that take place at altitudes above 100 kilometers.

expendable launch vehicle (ELV) A one-time-use-only rocket vehicle that can place a payload into Earth orbit or on an Earth-escape trajectory. None of an ELV's components, stages, engines, or propellant tanks are designed for recovery and reuse. Sometimes referred to in the aerospace industry as a *throwaway launch vehicle*. *See also* reusable launch vehicle.

explosive bolt A bolt that has an integral explosive charge that detonates on command, usually through a coded electrical signal. This action destroys the bolt and releases any pieces of aerospace equipment it was retaining. Aerospace engineers often use explosive bolts and spring-loaded mechanisms to quickly separate launch-vehicle stages or a payload from its expended propulsion system.

extraterrestrial Occurring, located, or originating beyond planet Earth and its atmosphere.

extraterrestrial catastrophe theory The popular modern hypothesis that a large asteroid (or possibly a huge comet) struck Earth some 65 million years ago. This collision created global environmental consequences that annihilated more than 90 percent of all animal species then living, including the dinosaurs.

extraterrestrial contamination The contamination of one world by life forms, especially microorganisms, from another world. If we make Earth's biosphere the reference, this planetary-contamination process is called *forward contamination* when an alien world (or returned soil sample) is contaminated by contact with terrestrial organisms and *back contamination* when alien organisms are released into Earth's biosphere.

extravehicular activity (EVA) Activities conducted by an astronaut or cosmonaut in outer space or on the surface of another world, outside of the protective environment of his/her aerospace vehicle, spacecraft, or planetary lander. Astronauts must don their space suits (which contain portable life-support systems) to perform EVA tasks. *See also* space suit.

extremophile A hardy terrestrial microorganism that exists under extreme environmental conditions, such as frigid polar regions or boiling hot springs. Exobiologists speculate that similar alien microorganisms could exist elsewhere in our solar system, perhaps within subsurface biological niches on Mars or in a shallow liquid-water ocean beneath the frozen surface of the Jovian moon Europa.

eyeballs in, eyeballs out Popular aerospace expression (derived from test pilots) describing the acceleration-related sensations experienced by an astronaut at liftoff or when retrorockets fire. The experience at liftoff is "eyeballs in" due to positive g forces on the body when the launch vehicle accelerates. The experience when the retrorockets fire is "eyeballs out" due to negative g forces on the body because of a spacecraft's or aerospace vehicle's deceleration.

fairing A low-mass structural component of a rocket intended to reduce air resistance during flight in the atmosphere by smoothing out various sections including the payload compartment. As such, it is similar to the windshield on an automobile.

fallaway section A section of a rocket vehicle that is cast off and separates from the vehicle during flight, especially a section that falls back to Earth.

farside The side of the Moon that never faces Earth.

ferret satellite A military spacecraft designed for the detection, location, recording, and analyzing of electromagnetic radiation (e.g., enemy radio-frequency [RF] transmissions).

film cooling In engineering, the process of cooling of an object or its surface by maintaining a thin fluid layer over the affected area; aerospace engineers often use film cooling to protect the inner surface of a liquid-propellant rocket's combustion chamber.

flame deflector Any engineered barrier that intercepts the hot exhaust gases of a rocket engine and deflects these gases away from the launch-pad structure, rocket-engine test cell, or the ground. A flame deflector may be a relatively small device fixed to the top surface of the launch pad, or it may be a heavily constructed piece of metal mounted at the side and bottom of a deep, cavelike launch-pad structure called a *flame bucket*. In the latter case, aerospace engineers also perforate the flame deflector with numerous holes. During thrust buildup and the beginning of the launch, a deluge of water pours through these holes onto the flame deflector to keep it from melting.

flyby An interplanetary or deep-space mission in which the flyby spacecraft passes close to its target body (e.g., a distant planet, moon, asteroid, or comet), but does not impact the target or go into orbit around it. Flyby spacecraft follow a continuous trajectory. Unlike a planetary orbiter, once a flyby spacecraft goes past its target, it cannot return to recover lost data. Therefore, mission controllers must carefully plan flyby operations months in advance of an encounter. NASA divides a typical flyby mission into four distinct phases: observatory phase, far-encounter phase, near-encounter phase, and postencounter phase.

free fall 1. In orbital mechanics, the free and unhampered motion of a body along a Keplerian trajectory. In this situation, the force of inertia counterbalances the force of gravity. For example, all people and objects inside an Earth-orbiting spacecraft experience a continuous state of free fall and appear "weightless." 2. In physics, the unimpeded fall of an object under the influence of a planet's gravitational field. For example, people inside an elevator car whose cable has snapped experience free fall until the car impacts the bottom of the elevator shaft.

fuel cell An engineered device that converts chemical energy directly into electrical energy by reacting continuously supplied chemicals. In the typical fuel cell, a catalyst promotes a noncombustible reaction between the fuel (such as hydrogen) and an oxidant (such as oxygen).

g In physics and engineering, the symbol used for the acceleration due to gravity. At sea level on Earth, g is approximately 9.8 meters per second squared (m/s^2). This value of acceleration represents "one g." In aerospace engineering, g values often characterize units of stress for accelerating or decelerating bodies. For example, when a rocket vehicle accelerates during launch, everything inside it experiences a force that can be as high as several g's.

gantry A frame that spans over something, such as an elevated platform that runs astride a work area, supported by wheels on each side. Often, the term is short for *gantry crane* or *gantry scaffold.*

geo A prefix meaning the planet Earth, as in geology and geophysics.

geostationary Earth orbit (GEO) A satellite orbiting Earth at an altitude of 35,900 kilometers above the equator revolves around the planet at the same rate as Earth spins on its axis. Communications, environmental, and surveillance satellites often use this important orbit. If the spacecraft's orbit is circular and lies in the equatorial plane, then, to an observer on Earth, the spacecraft appears stationary (or geostationary) over a given point on Earth's surface. If the satellite's orbit is inclined to the equatorial plane, then, when observed from Earth, the spacecraft traces out a figure-eight path every 24 hours and while still in synchronous orbit around Earth it does not appear to be stationary but appears to rise above and fall below the same spot on the equator. *See also* geosynchronous orbit; synchronous satellite.

geosynchronous orbit An orbit in which a satellite or space platform completes one revolution at the same rate as Earth spins, namely, 23 hours, 56 minutes, and 4.1 seconds. A satellite placed in such an orbit (approximately 35,900 kilometers above the equator) revolves around Earth once per day. *See also* geostationary Earth orbit; synchronous satellite.

gimbal A mechanism that allows an attached rocket motor to move in two mutually perpendicular directions.

global change The study of the combination of interactive linkages among our planet's major natural and human-made systems that appears to influence the planetary environment. Earth's environment is continuously changing. Many of these changes are natural and occur quite slowly, requiring thousands of years to achieve their full impact (for example, the building and erosion of mountains). However, other environmental changes (for example, the increase in atmospheric carbon dioxide) are the direct result of expanding human activities. Today, human-induced environmental change happens rapidly, in times as short as a few decades or less. *See also* Earth system science.

grain In rocketry, the integral piece of molded or extruded solid propellant that encompasses both fuel and oxidizer in a solid-rocket motor. Aerospace engineers design and shape the grain to produce (when burned) a specified thrust-versus-time relation.

gravitation In classical physics, the force of attraction between two masses. In accordance with Newton's universal law of gravitation, this attractive force operates along a line joining the centers of mass, and its magnitude is inversely proportional to the square of the distance between the two masses. From Einstein's general theory of relativity, gravitation is viewed as a distortion of the space-time continuum.

gravity assist In aerospace operations, a special maneuver by which a spacecraft changes its direction and speed using a carefully determined flyby trajectory through a planet's gravitational field. Some gravity-assist maneuvers require several planets, such as the Venus-Earth-Jupiter-gravity-assist (VEJGA) maneuver. Here, each planetary flyby contributes to the spacecraft's final trajectory.

greenhouse effect The general warming of the lower layers of a planet's atmosphere caused by the presence of "greenhouse gases," such as water vapor (H_2O), carbon dioxide (CO_2), and methane (CH_4). On Earth, the greenhouse effect occurs because our planet's atmosphere is relatively transparent to visible light from the Sun (which corresponds to the 0.3- to 0.7-micrometer-wavelength region of the electromagnetic spectrum), but is essentially opaque to the longer-wavelength thermal infrared radiation emitted by our planet's surface (typically near 10.6 micrometers in wavelength). Because of the presence of greenhouse gases in our atmosphere, such as carbon dioxide, water vapor, methane, nitrous oxide (N_2O), and human-made chlorofluorocarbons (CFCs), thermal radiation emitted by Earth's surface cannot escape into outer space. Instead, this outgoing thermal energy is absorbed in the lower atmosphere and causes a rise in temperature. As the atmospheric population of greenhouse gases increases, even more outgoing thermal radiation is trapped. This situation creates an overall global-warming trend.

green satellite A satellite in orbit around Earth that collects a variety of environmental data. *See also* Earth-observing satellite.

ground truth In surveillance and monitoring activities, measurements made on the ground to support, confirm, or calibrate remote sensing observations made from aerial or space platforms. Typical ground-truth data include local meteorology, soil conditions and types, vegetation-canopy content and condition, and surface temperatures. Scientists obtain the best results when they perform ground-truth measurements simultaneously with the airborne or spaceborne sensor measurements. *See also* Earth-observing satellite; remote sensing.

guidance In aerospace engineering, the process of directing the movements of an aerospace vehicle or spacecraft, with particular reference to the selection of a flight path. A variety of guidance options exist. These include preset, inertial, beam-rider, terrestrial-reference, celestial, and homing guidance.

guided missile (GM) A self-propelled (uncrewed) vehicle that moves above Earth's surface. After launch, a guided missile can control its trajectory during the flight. There are several general classes of guided missiles. An air-to-air guided missile (AAGM) is an air-launched missile for use against aerial targets. An air-to-surface guided missile (ASGM) is an air-launched missile for use against surface targets. A surface-to-air guided missile (SAGM) is a surface-launched guided missile for use against targets in the air. Finally, a surface-to-surface guided missile (SSGM) is a surface-launched missile for use against surface targets. Submarines launch both guided missiles (e.g., cruise missiles) and ballistic missiles. *See also* ballistic missile.

gyroscope A device that employs the angular momentum of a spinning mass (rotor) to sense angular motion of its base about one or two axes mutually perpendicular (orthogonal) to the spin axis. Also called a *gyro*.

hard landing A relatively high-velocity impact of a "lander" spacecraft or probe on a solid planetary surface. The impact usually destroys all equipment, except perhaps a very rugged instrument package or payload container.

hatch A tightly sealed access door in the pressure hull of an aerospace vehicle, spacecraft, or space station.

heliocentric With the Sun as a center, as in "heliocentric orbit" or "heliocentric space."

Hohmann transfer orbit In aerospace operations, the most efficient orbit-transfer path between two coplanar circular orbits. The maneuver consists of two impulsive high-thrust "burns" (or firings) of a spacecraft's propulsion system. Named after Walter Hohmann, a German engineer, who first suggested the technique in 1925.

hold In aerospace launch operations, to stop the sequence of events during a countdown until a sudden problem has been solved or an impediment has been removed, after which the launch countdown continues.

hot test In aerospace operations, a liquid-fuel rocket-system test during which the engines fire (usually for a short period of time) while holddown bolts secure the rocket vehicle/engine to the launch pad. *See also* cold-flow test.

housekeeping In aerospace operations, all the routine tasks mission controllers and/or astronauts must perform to keep an aerospace vehicle or spacecraft functioning properly.

human-factors engineering The branch of engineering dealing with the design, development, testing, and construction of devices, equipment, and artificial living environments to the anthropometric, physiological, and/or psychological requirements of the human beings who will use them. The human-factors portion of aerospace engineering involves such challenging tasks as the design of a reliable microgravity toilet suitable for both male and female crew persons.

hydrazine (N_2H_4) A toxic, colorless liquid often used as a rocket propellant because it reacts violently with many oxidizers. For example, hydrazine ignites spontaneously with concentrated hydrogen peroxide (H_2O_2) and nitric acid. When it decomposes through the action of a suitable catalyst, hydrazine also becomes a good monopropellant in simple small rocket engines, such as those used by aerospace engineers for spacecraft attitude control.

hypergolic fuel A rocket fuel that spontaneously ignites when brought into contact with an oxidizing agent (oxidizer); for example, aniline ($C_6H_5NH_2$) mixed with red fuming nitric acid (85 percent HNO_3 and 15 percent N_2O_4) produces spontaneous combustion. Also called hypergol.

HZE particles Very damaging cosmic rays, with high atomic number (Z) and high kinetic energy (E). Physicists define HZE particles as high-energy atomic nuclei with Z greater than 6 and E greater than 100 million electron volts (100 MeV). When an HZE particle passes through an astronaut or a piece of radiation-sensitive space hardware, it deposits a large amount of ionizing energy along its path. This action disrupts important molecular bonds in the bombarded substance.

ice catastrophe An extreme climate crisis in which all the liquid water on the surface of a life-bearing or potentially life-bearing planet has become frozen. *See also* greenhouse effect.

igniter In aerospace engineering, a device that begins combustion of a rocket engine; for example, a squib that ignites the fuel in a solid-propellant rocket.

inclination (symbol: *i*) One of the six Keplerian (orbital) elements, inclination describes the angle of an object's orbital plane with respect to the central body's equator. For Earth-orbiting objects, the orbital plane always goes through the center of Earth, but it can tilt at any angle relative to the equator. By general agreement, inclination is the angle between Earth's equatorial plane and the object's orbital plane measured counterclockwise at the ascending node.

infrared radiation (IR) The part of the electromagnetic (EM) spectrum between the optical (visible) and radio wavelengths. The IR region extends from about 0.7 micrometer to 1,000 micrometers wavelength.

injector In aerospace engineering, the device in a liquid-propellant rocket engine that drives (injects) fuel and/or oxidizer into the combustion chamber. The injector atomizes and mixes the propellants so they can burn more completely.

insertion In aerospace operations, the process of putting an artificial satellite, aerospace vehicle, or spacecraft into orbit.

integration In aerospace engineering and operations, the collection of activities and processes leading to the assembly of payload and launch-vehicle components, subsystems, and system elements into the desired final (flight) configuration. Verification of the compatibility of the assembled elements is a key part of the integration process.

intercontinental ballistic missile (ICBM) A ballistic missile with a range in excess of 5,500 kilometers.

intermediate-range ballistic missile (IRBM) A ballistic missile with a range capability from about 1,000 to 5,500 kilometers.

interplanetary Between the planets; within the solar system.

interstellar Between or among the stars.

inverse square law In physics, an important relation between physical quantities of the form x is proportional to $1/y^2$, where y is usually a distance and x terms are of two kinds, forces and fluxes. Newton's law of gravitation is an example of an inverse square law, since the force of gravitational attraction between two objects decreases as the inverse of their distance apart squared (i.e., as $1/r^2$). *See also* Newton's law of gravitation.

ion engine A reaction rocket system that achieves its thrust by expelling electrically charged particles (ions) at extremely high velocity. *See also* electric propulsion.

ionizing radiation Nuclear radiation that displaces electrons from atoms or molecules, thereby producing ions within the irradiated material. Examples include: alpha (α) radiation, beta (β) radiation, gamma (γ) radiation, protons, neutrons, and X-rays.

jansky (symbol: Jy) A unit describing the strength of an incoming electromagnetic wave signal. The jansky is commonly used in radio and infrared astronomy. It is named after Karl G. Jansky, an American radio engineer (1905–1950), who discovered extraterrestrial radio-wave sources in the 1930s. One jansky (Jy) = 10^{-26} watts per meter squared per hertz [W/(m^2-Hz)].

jettison To discard or toss away. For example, when the space shuttle's huge external tank no longer contains propellants for the orbiter's three main engines, astronauts jettison the tank, and it falls back to Earth, impacting in a remote ocean area.

joule (symbol: J) In physics and engineering, the basic unit of energy (or work) in the International System (SI). One joule is the work done by a force of one newton moving through a distance of one meter. This important unit honors the British scientist James Prescott Joule (1818–1889).

Jovian planet A large, Jupiter-like planet characterized by a great total mass, low average density, and an abundance of the lighter elements (especially hydrogen and helium). In our solar system, the Jovian planets are Jupiter, Saturn, Uranus, and Neptune.

Kepler's laws In physics and orbital mechanics, the three empirical laws describing the motion of the planets in their orbits around the Sun. The German astronomer Johannes Kepler (1571–1630) first formulated these important relationships using the detailed observations of the Danish astronomer Tycho Brahe (1546–1601). The laws are as follows: (1) the orbits of the planets are ellipses, with the Sun at a common focus; (2) as a planet moves in its orbit, the line joining the planet and the Sun sweeps over equal areas in equal intervals of time (sometimes called the law of equal areas); and (3) the squares of the periods of revolution of any two planets are proportional to the cubes of their mean distances from the Sun.

kilo- (symbol: k) A prefix in the SI unit system meaning multiplied by one thousand (1,000); for example, a kilogram (kg) is 1,000 grams, and a kilometer (km) is 1,000 meters.

kiloton (symbol: kT) A very large unit of energy, namely, 4.2×10^{12} joules (or 10^{12} calories). This unit often describes the energy released in a nuclear detonation or some other massive explosion. It is approximately the amount of energy released by exploding 1,000 metric tons (i.e., one kiloton) of the chemical high explosive trinitrotoluene (TNT).

kinetic energy (common symbols: KE or E_{KE}) The energy an object possesses as a result of its motion. In Newtonian (nonrelativistic) mechanics, kinetic energy is one-half the product of mass (m) and the square of its speed (v), that is, $E_{KE} = \frac{1}{2}mv^2$.

kinetic-energy weapon (KEW) A weapon that employs the energy of motion (kinetic energy) to disable or destroy a target. KEW projectiles strike their targets at very high velocities (typically more than 5 km/s) and cause lethal impact damage.

Kuiper belt A region in the outer solar system beyond the orbit of Neptune that contains millions of icy planetesimals (small solid objects). These icy objects range in size from tiny particles to Plutonian-sized planetary bodies. The Dutch-

American astronomer Gerard P. Kuiper (1905–1973) proposed the existence of this disk-shaped reservoir of icy objects in 1951. *See also* Oort cloud.

Lagrangian libration point In orbital mechanics, one of five points in space where a small object can have a stable orbit (i.e., remain in the orbital plane) of two much more massive celestial objects despite their gravitational attraction. Joseph-Louis Lagrange (1736–1813), a French mathematician, first postulated the existence of these libration points. Three of the points (called L_1, L_2, and L_3) lie on the line joining the two massive bodies and are actually unstable. As a result, any slight displacement in the position of a small object at these points results in its rapid departure. However, the fourth and fifth libration points (called L_4 and L_5) are stable. For example, members of the Trojan group of asteroids occupy such stable Lagrangian points 60 degrees ahead of (L_4) and 60 degrees behind (L_5) the planet Jupiter in its orbit around the Sun. Space visionaries have suggested placing large space settlements in the L_4 and L_5 libration points in cislunar space.

lander In aerospace engineering, a spacecraft designed to safely reach the surface of a planet and function there long enough to send useful scientific data back to Earth.

latch In engineering, a mechanical device that fastens one object or part to another. A latch is not a permanent connection, but is subject to ready release on demand so the objects or parts can be easily separated. For example, a latch (or several latches) can hold a small rocket on its launcher and then quickly release the rocket after engine ignition and sufficient thrust buildup.

launch pad The load-bearing base or platform from which aerospace engineers launch a rocket, missile, or aerospace vehicle. Often simply called "the pad."

launch site In aerospace operations, the specific, well-defined area used to launch rocket vehicles operationally or for test purposes, for example, the Cape Canaveral Air Force Station/Kennedy Space Center launch-site complex on the central east coast of Florida or the European Space Agency's launch site in Kourou, French Guiana, on the northeast coast of South America.

launch vehicle (LV) An expendable (ELV) or reusable (RLV) rocket-propelled vehicle that thrusts a payload into Earth orbit or sends a payload on an interplanetary trajectory to another celestial body or deep space. Sometimes called a *booster* or *space lift vehicle*.

launch window In aerospace operations, the specific time interval during which a launch can occur to satisfy some mission objective, usually a short period of time each day for a certain number of days.

life cycle In engineering, all the phases through which an item, component, or system passes from the time engineers envision and initially develop it until the object is either used (e.g., expended during a mission) or disposed of as excess to established requirements (e.g., a flight spare that is not needed because of the success of the original system). NASA engineers often use the following life-cycle phases: pre–phase A (conceptual study), phase A (preliminary analysis), phase B (definition), phase C/D (design and development), and the operations phase.

life-support system (LSS) A system that maintains life throughout the aerospace flight environment. Depending on the particular mission, the spaceflight environment can include travel in outer space, activities on the surface of another world (e.g., the Martian surface), and ascent and descent through Earth's atmosphere. The life-support system must reliably satisfy the human crew's daily needs for clean air, potable water, food, and effective waste removal.

liftoff The action of a rocket or aerospace vehicle as it separates from its launch pad in a vertical ascent. In aerospace operations, this term applies only to vertical ascent, while takeoff applies to ascent at any angle.

light-year (ly) In astronomy, a unit of distance based upon the distance light or other electromagnetic radiation travels in one year. One light-year is equal to a distance of approximately 9.46×10^{12} kilometers or 63,240 astronomical units.

line of sight (LOS) The straight line between a sensor or the eye of an observer and the object or point being observed. Sometimes called the *optical path*.

liquid hydrogen (LH_2) A liquid propellant used as the fuel in high-performance cryogenic rocket engines, with liquid oxygen commonly serving as the oxidizer. Hydrogen remains a liquid only at very low (cryogenic) temperatures, typically about 20 K (−253° C) or less.

liquid oxygen (LOX or LO_2) A cryogenic liquid propellant requiring storage at temperatures below 90 K (−183° C). Aerospace engineers use LOX as the oxidizer with liquid hydrogen (LH_2) as the fuel in contemporary high-performance cryogenic-propellant rocket engines.

liquid propellant Any combustible liquid fed into the combustion chamber of a liquid-fueled rocket engine.

low Earth orbit (LEO) An orbit, usually almost circular, just above Earth's appreciable atmosphere. In LEO operations, an altitude of 300 to 400 kilometers or more is usually sufficient to prevent the Earth-orbiting object from decaying rapidly because of atmospheric drag.

lunar Of or pertaining to Earth's natural satellite, the Moon.

lunar rover Human-crewed or automated (robot) vehicles that can travel across the Moon's surface and support scientific investigation or resource exploration.

Mach number (symbol: M) In physics and engineering, a dimensionless number expressing the ratio of the speed of an object with respect to the surrounding air (or other fluid) to the speed of sound in air (or other medium). Engineers often use the Mach number to describe compressible flow conditions. If $M < 1$, the flow is called *subsonic*, and local disturbances can propagate ahead of the flow. If $M > 1$, the flow is called *supersonic*, and disturbances cannot propagate ahead of the flow, with the result that shock waves form. Finally, if $M = 1$, the flow is *sonic*. This important dimensionless number honors the Austrian scientist Ernst Mach (1838–1916).

magnetosphere The region around a planet in which its own magnetic field significantly influences the behavior of charged atomic particles in concert with the

Sun's magnetic field, as projected by the solar wind. For example, Earth's magnetic field interacts strongly with atomic particles in the solar wind, producing a very dynamic and complicated magnetospheric region complete with trapped-particle radiation belts.

main stage In aerospace engineering, for a multistage rocket vehicle, the stage that develops the greatest amount of thrust, with or without booster engines; for a single-stage rocket vehicle powered by one or more engines, the period when "full thrust" (i.e., at or above 90 percent of the rated thrust) is attained.

manned vehicle An older (now obsolete) term in the aerospace literature, describing a rocket, aircraft, aerospace vehicle, or spacecraft that carries one or more human beings, male or female. This term helped distinguish that particular craft from a robot (pilotless) aircraft, a ballistic missile, or an automated (and uncrewed) satellite or planetary probe. Today, the preferred expression is either *crewed vehicle* or *personed vehicle*.

maria (singular: mare) Latin word for "seas." Originally used by Galileo to describe the large, dark, flat areas on the lunar surface, since he and other early astronomers thought that these darkened areas were bodies of water on the Moon's surface. Modern astronomers still preserve this nomenclature, although they recognize these dark formations as ancient lava flows triggered by large meteorite impacts.

mass (common symbol: m) The amount of material present in an object. Mass describes "how much" material makes up an object. The SI unit for mass is the kilogram (kg). The terms *mass* and *weight* are very often confused. However, these terms do not represent the same thing, since weight is a derived unit that describes the action of the local force of gravity on the mass of an object. It is important to recognize that an object with a mass of one kilogram on Earth will also have a mass of one kilogram on the surface of Mars or anywhere else in the universe. However, the weight of this one-kilogram-mass object will be quite different on the surface of each planet, since the local acceleration of gravity (g) is different in each location, namely, about 9.8 m/s^2 on Earth's surface versus 3.7 m/s^2 on Mars.

mass-energy equivalence The very significant postulation made by Albert Einstein (1875–1955) that "the mass of a body is a measure of its energy content," as an extension of his 1905 special theory of relativity. Careful experimental measurements of mass and energy in a variety of nuclear reactions quickly verified Einstein's profound hypothesis. The famous equation $E = mc^2$ summarizes this equivalence. For example, when the energy of an object changes by an amount E, its mass (m) will change by an amount equal to E/c^2. This equation is sometimes called the *Einstein equation*.

mass ratio In aerospace engineering, the ratio of the mass of the propellant charge of a rocket to the total mass of the rocket when it is charged with propellant. For example, if a sounding rocket has a total mass (including propellant) of 2,000 kilograms and a propellant charge of 1,800 kilograms, then the mass ratio of the rocket system is 0.90.

mate To join or fit together two major components of a system; for example, aerospace technicians mated the payload to the second-stage rocket vehicle.

mega- (symbol: M) A prefix in the SI system meaning multiplied by one million (10^6), for example, megahertz (MHz), meaning one million hertz.

meteoroid In space science and astronomy, an encompassing term that refers to natural (often rocky) solid objects found in space. These objects range in diameter from micrometers to kilometers and in mass from less than 10^{-12} gram to more than 10^{+16} grams, respectively. If the object has a mass of less than 1 gram, scientists call it a *micrometeoroid*. When objects with a mass of more than 10^{-6} gram reach Earth's atmosphere, they experience aerodynamic heating and glow, producing the visible effect commonly termed a *meteor*. If some of the original meteoroid survives its glowing plunge through Earth's atmosphere, the remaining unvaporized chunk is named a *meteorite*.

MeV One million electron volts (10^6 eV); a large energy unit commonly encountered in the study of nuclear reactions.

micro- (symbol: μ) A prefix in the SI system meaning divided by one million; for example, a micrometer (μm) is a millionth of a meter (10^{-6} meter). Scientists and engineers often use the term as a prefix to indicate that something is very small, as in *micromachine* or *micrometeoroid*.

microgravity (common symbol: μg) Because its inertial trajectory compensates for the force of gravity, an Earth-orbiting spacecraft travels in a state of continual free fall. In this state, all objects inside the spacecraft appear "weightless"—as if they were in a zero-gravity environment. However, the venting of gases, the minuscule drag exerted by Earth's residual atmosphere (at low orbital altitudes), and crew motions tend to create nearly imperceptible forces on the people and objects inside the orbiting vehicle. These tiny forces are collectively designated "microgravity." In a microgravity environment, astronauts and their equipment are almost, but not entirely, weightless.

micron (symbol: μm) A unit of length in the SI system equal to one-millionth (10^{-6}) of a meter. Also called a *micrometer*.

military satellite (MILSAT) A satellite used primarily for military or defense purposes, for example, intelligence gathering, missile surveillance, or secure communications.

milli- (symbol: m) The SI system prefix meaning multiplied by one-thousandth (10^{-3}). For example, a millimeter (mm) is 0.001 meter; a millisecond (ms) is 0.001 second; and a millivolt (mV) is 0.001 volt.

mini- A common contraction in the aerospace literature for "miniature"; for example, MINISAT means miniature satellite.

missile In general, any object thrown, dropped, fired, launched, or otherwise projected with the purpose of striking a target. In aerospace usage, short for *ballistic missile* or *guided missile*. "Missile" should not be used loosely as a synonym for rocket or launch vehicle.

missile silo A hardened protective container, usually buried in the ground, that maintains land-based long-range ballistic missiles under conditions that permit rapid launching.

mission 1. In scientific (civilian) aerospace operations, the performance of a set of investigations or operations in space to achieve program goals; for example, NASA's *Galileo* mission to Jupiter. 2. In military aerospace operations, the dispatching of one or more aircraft, spacecraft, or aerospace vehicles to accomplish a particular task, for example, an enemy-missile-site "search and destroy" mission.

mock-up In aerospace engineering, a full-sized replica or dummy of something, such as a spacecraft. Mock-ups are often made of some substitute material (such as wood) and sometimes incorporate actual functioning pieces of equipment, such as engines or power supplies. Engineers use mock-ups to study construction procedures, to examine equipment interfaces, and to train personnel.

modulation In aerospace operations (telemetry), the process of modifying a radiofrequency (RF) signal by shifting its phase, frequency, or amplitude to carry information. The respective processes are designated *phase modulation* (PM), *frequency modulation* (FM), and *amplitude modulation* (AM).

module 1. A self-contained unit of a launch vehicle or spacecraft that serves as a building block for the overall structure. Aerospace engineers often refer to a module by its primary function, for example, "command module" or "service module." 2. In engineering, a one-package assembly of functionally related parts, usually a "plug-in" unit arranged to function as a system or subsystem; a "black box." 3. With respect to human spaceflight, a pressurized, crewed facility/laboratory suitable for conducting science, applications research, and technology demonstrations; for example, the *Spacelab* module in NASA's Space Transportation System.

monopropellant A liquid-rocket propellant consisting of a single chemical substance (such as hydrazine) that decomposes exothermally and produces a heated exhaust jet without the use of a second chemical substance. Aerospace engineers frequently use a monopropellant in attitude-control systems on spacecraft and aerospace vehicles. *See also* hydrazine.

multistage rocket A vehicle that has two or more rocket units, each firing after the one behind it has exhausted its propellant. This type of rocket vehicle then discards (or jettisons) each exhausted stage in sequence. Sometimes called a *multiple-stage rocket* or a *step rocket*.

mutual assured destruction (MAD) The Cold War strategic situation in which either superpower (i.e., the United States or the former Soviet Union) could inflict massive nuclear destruction on the other, no matter which side attacked first. *See also* Cold War.

nadir 1. The direction from a spacecraft directly down toward the center of a planet; the opposite of zenith. 2. That point on the celestial sphere directly beneath an observer and directly opposite the zenith.

nano- (symbol: n) A prefix in the SI system meaning multiplied by 10^{-9}; for example, a nanometer is 10^{-9} meter, a very small distance.

nearside In astronomy, the side of the Moon that always faces Earth.

nebula (plural: nebulae) A cloud of interstellar gas or dust. Astronomers observe a nebula either as a dark hole against a brighter background (a *dark nebula*) or as a luminous patch of light (a *bright nebula*).

newton (symbol: N) The unit of force in the SI system, honoring Sir Isaac Newton (1642–1727). A newton is the amount of force that gives a one-kilogram mass an acceleration of one meter per second per second.

Newton's law of gravitation The important and universal physical law proposed by the brilliant English scientist Sir Isaac Newton (1642–1727) in 1687. This law states that every particle of matter in the universe attracts every other particle. The force of gravitational attraction (F_G) acts along the line joining the two particles, is proportional to the product of the particles' masses (m_1 and m_2), and is inversely proportional to the square of the distance (r) between the particles. Expressed in the form of an equation, $F_G = [Gm_1m_2]/r^2$, where G is the universal gravitational constant with a value of approximately $6.6732\ (\pm\ 0.003) \times 10^{-11}$ $N \cdot m^2/kg^2$ (in SI units).

Newton's laws of motion The three postulates that form the basis of rigid-body mechanics. Sir Isaac Newton (1642–1727) formulated these laws in about 1685 as he was studying the motion of the planets around the Sun. Newton's first law is the law of conservation of momentum. It states that a body continues in a state of uniform motion (or rest) unless it is acted upon by an external force. The second law states that the rate of change of momentum of a body is proportional to the force acting upon the body and occurs in the direction of the applied force. The third law is the action-reaction principle. It states that for every force acting upon a body, there is a corresponding force of the same magnitude exerted by the body in the opposite direction. This particular law is the basic principle by which every rocket operates.

nominal A term commonly used in aerospace operations to indicate that a system is performing within prescribed or acceptable limits; for example, the rate of propellant consumption by a liquid-fueled rocket engine is "nominal" during a launch ascent, or the spacecraft is on a "nominal" trajectory to Venus.

nose cone The cone-shaped leading edge of a rocket vehicle. Aerospace engineers carefully design a nose cone to protect and contain a warhead or other payload, such as a satellite, scientific instruments, biological specimens, or auxiliary equipment. For example, the outer surface and structure of a nose cone are built to withstand the high temperatures arising from aerodynamic heating and any vibrations (buffeting) due to high-speed flight through the atmosphere.

nozzle A rocket engine's nozzle is a flow device that promotes the efficient expansion of the hot gases from the combustion chamber. As these gases leave the nozzle at high velocity, a propulsive (forward) thrust also occurs in accordance with Newton's third law of motion (the action-reaction principle).

nuclear-electric propulsion (NEP) A space-deployed propulsion system that uses a nuclear reactor to produce the electricity needed to operate the vehicle's electric propulsion engine(s). *See also* electric propulsion.

nuclear rocket A rocket vehicle that derives its propulsive thrust from nuclear energy sources. There are two general classes of nuclear rockets: the *nuclear-thermal rocket* and the *nuclear-electric propulsion (NEP) system*. The nuclear-thermal rocket uses a nuclear reactor to heat a working fluid (generally hydrogen) to extremely high temperatures before expelling it through a thrust-producing nozzle. The nuclear-electric propulsion system uses a nuclear reactor to generate electric power that then supports an electric propulsion system.

one g In physics and engineering, a term describing the downward acceleration of gravity at Earth's surface; at sea level, one g corresponds to an acceleration of approximately 9.8 meters per second per second (m/s^2).

Oort cloud A large population or "cloud" of comets thought to orbit the Sun at a distance of between 50,000 and 80,000 astronomical units (i.e., out to the limits of the Sun's gravitational attraction). First hypothesized by the Dutch astronomer Jan Hendrik Oort (1900–1992) in 1950.

orbit 1. In astronomy and aerospace operations, the path followed by a satellite around an astronomical body, such as Earth or Mars. When an object moves around a primary body under the influence of gravitational force alone, the closed path forms an elliptical (or circular) orbit. For example, the planets have the Sun as their primary body and follow elliptical (but nearly circular) orbits. When a satellite makes a complete trip around its primary body, it completes a *revolution*, and the time required is the *period of revolution* or *orbital period*. 2. In physics, the region occupied by an electron as it moves around the nucleus of an atom.

orbital elements The set of six parameters (e.g., apogee, perigee, and inclination) that specify the size, shape, and orientation of a Keplerian orbit. Aerospace mission controllers use orbital elements to define the precise position of an Earth-orbiting satellite at a particular time. Also called *Keplerian elements*.

orbital velocity The average velocity at which a satellite, spacecraft, or natural body travels around its primary.

order of magnitude 1. A factor of 10. 2. A value expressed to the nearest power of 10; for example, a cluster containing 9,450 stars has approximately 10,000 stars in an order-of-magnitude estimate.

oxidizer In aerospace engineering, a substance whose main function is to supply oxygen or other oxidizing materials for deflagration (burning) of a rocket engine's solid propellant or combustion of its liquid fuel.

parking orbit In aerospace operations, a temporary, but relatively stable, orbit around a celestial body that a spacecraft uses for repair activities, rendezvous and transfer of components (e.g., the rendezvous and docking of a lander spacecraft with its mother ship), or simply to wait ("parked in space") for conditions to become favorable for it to depart from that orbit to another orbit or trajectory.

passive In general, a system that contains no power sources to augment output power or signal, such as a passive electrical network or a passive reflector. The term is usually applied to a device that draws all its power from the input signal. In aerospace operations, a dormant satellite that does not transmit a signal. This type of silent human-made space object might be a dormant (but functional) replacement satellite, a retired or decommissioned satellite, a satellite that has failed prematurely and is now nothing more than a hunk of "space junk," or possibly even a dangerous, target-stalking space mine.

payload Originally, the revenue-producing portion of an aircraft's load, such as passengers, cargo, and mail. By universally used extension, the term *payload* now applies to anything that a rocket, aerospace vehicle, or spacecraft carries over and above what is necessary for the operation of the vehicle during flight.

peri- A prefix meaning near, as in perigee.

perigee In general, the point at which a satellite's orbit is the closest to the primary (central body); the minimum altitude attained by an Earth-orbiting object. *See also* apogee.

perihelion The place in an elliptical orbit around the Sun that is nearest to the center of the Sun. *See also* aphelion.

photon An elementary bundle or packet of electromagnetic radiation, such as a photon of light. Photons have no mass and travel at the speed of light. From quantum theory, the energy (E) of the photon is equal to the product of the frequency (ν) of the electromagnetic radiation and Planck's constant (h): $E = h\,\nu$, where h is equal to 6.626×10^{-34} joule/s and ν is the frequency (hertz).

photon engine A reaction engine that produces thrust by emitting photons, such as light rays. The thrust from such a proposed rocket engine is actually very modest. However, if a continuous portable power source became available, the photon engine could theoretically thrust continuously in the vacuum of outer space, and the photon rocket might eventually achieve speeds appropriate for interstellar flight.

pitch The rotation or oscillation of an aircraft, missile, or aerospace vehicle about its lateral axis.

pitchover In aerospace operations, the programmed turn from the vertical that a rocket or launch vehicle under power takes as it describes an arc and points in a direction other than vertical.

planet A nonluminous celestial body that orbits around the Sun or some other star. The word comes from the ancient Greek word *planetes* (wanderers), since early astronomers identified the planets as the points of light that appeared to wander relative to the fixed stars. There are nine such large objects, or "major planets," in the solar system, and numerous "minor planets," or asteroids. The distinction between a planet and its large satellite may not always be precise. For example, our Moon is nearly the size of the planet Mercury and is very large in comparison to its parent planet, Earth. Therefore, some planetary scientists treat the Earth and the Moon as a double-planet system. There is also a current astronomical debate whether Pluto should really be considered one of the major planets because of its tiny size. For now, astronomers treat icy Pluto as a planet. Along with its large moon, Charon, Pluto also belongs to a double-planet system at the frigid extremes of the solar system.

planetary albedo The fraction of incoming solar radiation (sunlight) that a planet's surface (and atmosphere, if any) reflects back to space. Earth has a variable planetary albedo of approximately 30 percent (0.30). Polar ice sheets and clouds in the atmosphere contribute significantly to how much incoming solar radiation our planet reflects back to space at any given time.

plasma An electrically neutral gaseous mixture of positive and negative ions.

plasma engine An electric rocket engine that uses electromagnetically accelerated plasma as its reaction mass.

polar orbit In aerospace operations, an orbit around a planet that passes over or near its poles; an orbit with an inclination of about 90 degrees.

posigrade rocket An auxiliary rocket that fires in the direction in which the vehicle is pointed. In aerospace operations, small posigrade rockets help separate two stages of a multistage launch vehicle or a payload from its final propulsion stage. A posigrade firing adds to the space vehicle's speed, while a *retrograde* firing slows it down.

pressurized habitable environment Any enclosure or module deployed in outer space or on the surface of a planetary body in which an astronaut may perform activities in a "shirtsleeve" environment.

primary body In astronomy and aerospace operations, the celestial body around which a satellite, moon, or space vehicle orbits, from which it is escaping, or toward which it is falling. For example, Earth is the Moon's primary body, while the Sun is Earth's primary body.

probe 1. Generally, any device inserted into an environment for the purpose of obtaining information about the environment. 2. In aerospace operations, an instrumented spacecraft or vehicle that moves through the upper atmosphere, travels in outer space, or lands on another celestial body to obtain scientific information about that particular environment; for example, a deep-space probe, a lunar probe, or an atmospheric probe.

propellant Fundamentally, any material, such as a fuel, an oxidizer, an additive, a catalyst, or any compound mixture of these, carried in a rocket vehicle that releases energy during combustion. The hot combustion gases escape through a nozzle and provide reactive thrust to the vehicle. Propellants commonly occur in either liquid or solid form. Modern launch vehicles use three types of liquid propellants: petroleum based, cryogenic (very cold), and hypergolic (self-igniting upon contact)

propellant mass fraction (symbol: ζ) In aerospace operations and rocketry, the ratio of the propellant mass (m_p) to the total initial mass (m_o) of the launch vehicle before operation, including propellant load, payload, structure, and so on.

propulsion system The collection of rocket engines, propellant tanks, fluid lines, and all associated equipment necessary to provide the propulsive force for a launch vehicle or space vehicle. Propulsion systems involving solid-propellant chemical rockets are generally less complicated than those involving liquid-propellant chemical rockets. High-thrust propulsion systems are needed by launch vehicles to take a payload from the surface of Earth into orbit (generally low Earth orbit). Compact, solid-propellant chemical rockets generally make up the upper-stage propulsion systems used to send payloads from low Earth orbit to geosynchronous Earth orbit or to place a spacecraft on an interplanetary trajectory. Aerospace engineers have also considered using nuclear propulsion systems (thermal or electric), solar-electric propulsion systems, and even very exotic propulsion systems (based on nuclear-fusion reactions or matter-antimatter-annihilation reactions).

prototype In engineering, a production model suitable for complete evaluation of a device's design and performance. 2. The first of a series of similar devices. 3. A spacecraft or aerospace vehicle that has passed qualification tests, releasing the design for the fabrication of complete flight units. 4. A physical standard to which replicas are compared.

pulsar A subclass of rapidly spinning neutron stars; a stellar radio source that emits radio waves in a pulsating rhythm.

purge To rid a line or tank of residual fluid; in aerospace operations, to remove residual fuel or oxidizer from the tanks or lines of a liquid-propellant rocket after a test firing or simulated test firing.

pyrophoric fuel A fuel that ignites spontaneously in air. *See also* hypergolic fuel.

Q In aerospace engineering, the symbol commonly used for dynamic pressure; for example, after liftoff, the launch vehicle encountered maximum q (that is, maximum dynamic pressure) 45 seconds into the flight.

quasar A mysterious, very distant object with a high redshift (i.e., traveling away from Earth at great speed). These objects appear almost like stars, but are far more distant than any individual star astronomers can now observe. Astrophysicists speculate that quasars might be the very luminous centers of active distant galaxies. When radio astronomers first identified these unusual objects in 1963, they called them "quasi-stellar radio sources," or quasars, for short. Quasars emit tremendous quantities of energy from very small volumes. Some of the most distant quasars yet observed are so far away that they are receding at more than 90 percent of the speed of light.

radar astronomy The use of radar to study objects in our solar system, such as the Moon, the planets, asteroids, and planetary ring systems. A powerful radar telescope, like the Arecibo Observatory in Puerto Rico, can transmit a radar pulse through the "opaque" Venusian clouds (about 80 kilometers thick). Astronomers then analyze the return signal to obtain detailed information for the preparation of high-resolution surface maps. Radar astronomers precisely measure distances to celestial objects, estimate rotation rates, and also develop unique maps of surface features. Orbiter spacecraft with imaging radar systems, like NASA's *Magellan* spacecraft, create detailed surface maps even when the celestial object's physical surface is obscured from view by thick layers of clouds.

radiator In aerospace engineering, a device that rejects waste heat from a space vehicle or satellite to outer space by radiant-heat-transfer processes. The radiator plays an important role in the thermal control of a spacecraft. Its design depends on both the surface (operating) temperature and the amount of heat (thermal energy) to be rejected. From heat-transfer theory, the Stefan-Boltzmann law determines the amount of waste heat that can be radiated to space by a given surface area. The amount of heat transferred by radiation is proportional to the fourth power of the absolute temperature of the radiating surface. Higher heat-rejection temperatures correspond to smaller radiator areas and lower radiator masses.

radioactivity In physics and nuclear engineering, the spontaneous decay or disintegration of an unstable atomic nucleus. The emission of ionizing nuclear radiation, such as alpha particles, beta particles, and gamma rays, usually accompanies this radioactive decay process. The radioactivity, often shortened to "activity," of natural and human-made (artificial) radioisotopes decreases exponentially with time, governed by the fundamental relationship $N = N_0 e^{-\lambda t}$, where N is the number of radionuclides (of a particular radioisotope) at time t, N_0 is the number of radionuclides of that particular radioisotope at the start, λ is the decay constant of the particular radioisotope, and t is time. The decay constant (λ) is related to the half-life (T_2) of a radioisotope by the equation $\lambda = (\ln 2)/T_2 = 0.69315/T_2$. Half-lives vary widely for different radioisotopes and range in value from as short as 10^{-8} seconds to as long as 10^{10} years or more. The longer the half-life of the radioisotope, the more slowly it undergoes radioactive decay or the less "radioactive" a substance is.

radio astronomy The branch of astronomy that collects and evaluates radio signals from extraterrestrial sources. Radio astronomy started in the 1930s when an American radio engineer, Karl Jansky (1905–1950), detected the first extraterrestrial radio signals. Astronomers now use giant, sensitive radio telescopes to explore the radio-frequency portion of the universe in great detail. Often they discover unusual extraterrestrial radio sources. One of the strangest of these cosmic radio sources is the pulsar, a collapsed giant star that emits pulsating radio signals as it spins. When scientists detected the first pulsar in 1967, the event caused a great deal of excitement in the technical community. Because of the regularity of the radio signal, scientists initially thought that they had just detected the first radio signal from an intelligent interstellar civilization. Another unusual extraterrestrial radio source is the quasar (quasi-stellar radio source). Quasars are the most distant objects yet observed in the universe.

radioisotope An unstable (radioactive) isotope of an element that spontaneously decays or disintegrates. When a radioisotope decays, it emits nuclear radiation (such as an alpha particle, beta particle, or gamma ray). There are more than 1,300 natural and artificial (human-made) radioisotopes.

radioisotope thermoelectric generator (RTG) A versatile, compact space nuclear power system. The operating principle of the RTG is quite simple. The device converts the heat (thermal energy) deposited by the absorption of alpha particles from a radioisotope source (generally plutonium 238) directly into electricity. Aerospace engineers use RTGs to provide spacecraft electric power on missions where long life, high reliability, operation independent of the distance or orientation to the Sun, and operation in severe environments (e.g., lunar night, Martian dust storms) are important design criteria. For example, all NASA spacecraft that explored the outer regions of the solar system used the RTG for electric power.

radio telescope A large, parabolic (dish-shaped), metallic antenna that collects radio-wave signals from extraterrestrial objects or from distant spacecraft and focuses these very weak signals onto a sensitive radio-frequency (RF) receiver for identification and analysis.

reaction engine An engine that develops a forward thrust by ejecting a substance in the opposite direction. Usually, the reaction engine ejects a high-velocity stream of hot gases created by combusting or heating a propellant within the engine, but more exotic reaction engines could eject photons or nuclear radiation. The reaction engine operates in accordance with Newton's third law of motion (the action-reaction principle). Both rocket engines and jet engines are reaction engines.

real time Time in which reporting on or recording events is simultaneous with the events; essentially, "as it happens." For example, in aerospace operations, real-time data are those available in usable form at the same time the event occurs.

reconnaissance satellite A military satellite that orbits Earth and performs intelligence gathering against enemy nations and potential adversaries. In the 1960s, the United States developed and flew its first generation of photoreconnaissance satellites, called the Corona, Argon, and Lanyard systems. Also called a spy satellite.

redline In engineering, a term indicating a critical value for a parameter or a condition. If the specified parameter exceeds its redline value, then a physical threat exists to the integrity of the system, the performance of a vehicle, or the success of a mission.

redshift In astronomy, the apparent increase in the wavelength of a light source caused by its receding motion. The Doppler shift of the visible spectra of distant galaxies toward red light (i.e., longer wavelength) indicates that these galaxies are receding. The greater redshift observed in more distant galaxies has been interpreted that the universe is expanding. Some of the most distant quasars are significantly redshifted and are receding at more than 90 percent of the speed of light. *See also* blueshift; Doppler shift; quasar.

redundancy (of design) In engineering, the existence of more than one means for accomplishing a given task, where all means must fail before there is an overall failure of the system. *Parallel redundancy* applies to a system where two or more means are on-line (i.e., working) at the same time to accomplish the critical task. Any one of these means is capable of handling the important task by itself if necessary, should one or all of the other means suffer failure. *Standby redundancy* applies to a system where there is an alternative means of accomplishing a critical task that is turned on (or activated) when a malfunction sensor detects an impending failure of the primary means of accomplishing the task. Through redundant design principles, the failure of an individual unit or means of accomplishing a critical task will not cause failure of the entire system or an abort of the mission.

reentry In aerospace operations, the return of an object, originally launched from Earth, back into the sensible atmosphere; the action involved in this event. The major types of reentry are ballistic, gliding, and skip. To perform a safe, controlled reentry, a spacecraft or aerospace vehicle must be able to carefully dissipate its kinetic and potential energies. For an aerospace vehicle like NASA's space shuttle, a successful reentry ends with a safe ("soft") landing on the surface of Earth and requires a very precisely designed and maintained flight trajectory. When a derelict satellite or piece of space debris undergoes a random or uncontrolled reentry, it usually burns up in the atmosphere due to excessive aerodynamic heating, although at times, large natural or human-made objects experience uncontrolled reentry and survive the fiery plunge through the atmosphere to impact on Earth. Also called *entry* in the aerospace literature.

reentry vehicle (RV) Generally, the part of a space vehicle designed to reenter Earth's atmosphere in the terminal portion of its trajectory; specifically, the part of a ballistic missile (or postboost vehicle) that carries the nuclear warhead to its target. The military reentry vehicle is designed to enter Earth's atmosphere in the terminal portion of its trajectory and proceed to its assigned target. Aerospace engineers carefully design the military RV so it can survive rapid heating during high-velocity flight through the atmosphere and protect its warhead until the nuclear weapon detonates at the target.

regenerative cooling In aerospace engineering, a common approach to cooling large liquid-propellant rocket engines and nozzles, particularly those engines that must operate for an appreciable period of time. In this technique, one of the liquid propellants (say, the liquid oxygen) first flows through specially designed cooling passages in the thrust chamber and nozzle walls. The flowing liquid propellant cools the walls while recovering (regenerating) some of the otherwise wasted heat. The prewarmed propellant then enters the combustion chamber.

remote sensing The sensing of an object, event, or phenomenon without having the sensor in direct contact with the object being studied. Practically all branches of modern ground-based and space-based astronomy depend on remote sensing instruments. These instruments use many different portions of the electromagnetic spectrum, not just the visible portion we see with our eyes. Telltale electromagnetic radiations in a variety of wavelengths carry important information about the object to the sensor across the vacuum. Planetary scientists also use remote sensing to study Earth in detail from space or to study other objects in the solar system, generally using flyby and orbiter spacecraft.

rendezvous In aerospace operations, the close approach of two or more spacecraft in the same orbit so that docking can take place. The co-orbiting objects meet at a preplanned location and time and carefully come together (rendezvous) with essentially zero relative velocity. Aerospace mission planners use the rendezvous operation during the construction, servicing, or resupply of a space station, or when the space shuttle performs in-orbit repair/servicing of a satellite. NASA also applies the term to space missions, such as the *Near Earth Asteroid Rendezvous* (NEAR) mission, in which a scientific spacecraft so maneuvers as to fly alongside a target celestial body (such as a comet or asteroid) at essentially zero relative velocity.

retrograde rocket An auxiliary rocket that fires in the direction opposite to that in which the vehicle is traveling (pointed); in aerospace operations, small retrograde rockets produce a retarding thrust that opposes the vehicle's forward motion. Also called a retrorocket. *See also* posigrade rocket.

reusable launch vehicle (RLV) A space launch vehicle that includes simple, fully reusable designs that support airline-type operations; primarily achieved through the use of advanced technology and innovative operational techniques.

robotics The science and technology of designing, building, and programming robots. A robot is simply a machine that does routine (mechanical) tasks on human command. In aerospace operations, space robots are "smart machines" with manipulators that can be programmed to do a variety of human labor tasks automatically. Robots can be operated at a distance in real time by their human controllers (*teleoperation*), or they can function with varying degrees of autonomy.

rocket A completely self-contained projectile, pyrotechnic device, or flying vehicle propelled by a reaction (rocket) engine. A rocket carries all of its propellant and can function in the vacuum of outer space. In the early part of the twentieth century, the three founders of astronautics, Konstantin Tsiolkovsky, Robert Goddard, and Hermann Oberth, each independently recognized that the rocket (especially the large liquid-propellant rocket) represented the key to space travel. Rockets obey Newton's third law of motion, which states that "for every action there is an equal and opposite reaction." Aerospace engineers often classify rockets by the energy source the reaction engine uses to accelerate the ejected matter that creates the vehicle's thrust, as, for example, chemical rocket, nuclear rocket, and electric rocket. They further divide chemical rockets into two general subclasses: solid-propellant rockets and liquid-propellant rockets.

roll The rotation or oscillation of an aircraft, missile, or aerospace vehicle about its longitudinal (lengthwise) axis.

rumble With respect to a liquid-propellant rocket engine, a form of combustion instability, characterized by a low-pitched, low-frequency rumbling noise.

satellite A secondary (smaller) body in orbit around a primary (larger) body. Planet Earth is a natural satellite of the Sun, while the Moon is a natural satellite of Earth. Aerospace engineers and scientists call human-made spacecraft placed in orbit around Earth "artificial satellites," or more commonly just "satellites."

satellite power system (SPS) A proposed very large space structure that takes advantage of the continuous availability of sunlight to provide useful energy to a terrestrial power grid. After being constructed and assembled in space (possibly from lunar materials), each SPS unit would operate in geosynchronous orbit (i.e., in a fixed position above Earth's equator). There, the SPS would collect raw sunlight and convert and transmit the harvested solar energy as either microwave or laser-beam energy to special receiving stations on Earth's surface.

screaming With respect to a liquid-propellant rocket engine, a relatively high-frequency form of combustion instability, characterized by a high-pitched noise.

scrub In aerospace operations, to cancel or postpone a rocket firing, either before or during the countdown.

search for extraterrestrial intelligence (SETI) A modern attempt to answer this important philosophical question: Are we alone in the Universe? The major objective of contemporary SETI programs (mostly conducted by private foundations and public contributions) is to detect coherent radio-frequency (microwave) signals generated by intelligent extraterrestrial civilizations. While technical in approach, the efforts are also considered highly speculative.

sensible atmosphere That portion of a planet's atmosphere that offers resistance to a body passing through it.

sensor The portion of an instrument that detects or measures some type of physically observable phenomenon. Sensors can be in direct contact with the object being observed or at a distance (remotely sensed). Engineers use direct-contact sensors to measure a variety of physical and mechanical properties of an object, including vibration, temperature, and pressure. In remote sensing, the noncontact sensor often detects characteristic electromagnetic radiation or nuclear particles. The sensor responds to such input phenomena by converting them into an internal electronic signal. Another part of the instrument then amplifies, digitizes (quantifies), and displays (or records) the sensor's signal. A passive remote sensor uses characteristic emissions from the target as its input signal, while an active remote sensor places a burst of energy (such as electromagnetic radiation or nuclear particles) on the target and then uses any returned signal as its input.

shake-and-bake test In aerospace engineering, a series of prelaunch tests performed on a completed, near-flight-ready spacecraft to simulate the launch vibrations and thermal environment (e.g., temperature extremes) it will experience during the mission. A typical test plan often involves the use of enormous speakers to quite literally blast the spacecraft with acoustic vibrations similar to those encountered during launch. Aerospace engineers put the entire spacecraft in a large environmental chamber and expose it to the high and low temperature extremes and thermal cycles it will encounter during the mission. The spacecraft either emerges from this test series ready for flight or else requires some redesign to overcome deficiencies that appeared during the test.

sloshing The back-and-forth movement of a liquid-rocket propellant in its tank(s). This movement often creates stability and control problems for the rocket vehicle. Aerospace engineers use antislosh baffles in propellant tanks to avoid or reduce the problem.

soft landing The process of landing a spacecraft on the surface of a planet without damaging any portion of the craft, except possibly an expendable landing-gear structure. Aerospace engineers successfully designed the Surveyor and Viking Lander robot spacecraft for soft landings on the Moon and Mars, respectively.

solar Of or pertaining to the Sun or caused by the Sun.

solar cell A direct-energy-conversion (DEC) device that turns sunlight directly into electricity. Aerospace engineers extensively use solar cells (often in combination with rechargeable storage batteries) to provide electric power for spacecraft. Also called photovoltaic cell.

solar constant A relatively stable physical quantity that describes the total amount of the Sun's radiant energy (in all wavelengths) crossing perpendicular to a unit area at the top of Earth's atmosphere. At one astronomical unit from the Sun, the solar constant is about 1,371 ± 5 watts per square meter. The spectral distribution of the Sun's radiant energy approximates the radiant emissions of a blackbody radiator with an effective temperature of 5,800 degrees Kelvin. Therefore, most of the Sun's radiant energy lies in the visible portion of the electromagnetic spectrum, with a peak value near 0.45 micrometer (μm).

solar-electric propulsion (SEP) A low-thrust propulsion system in which a solar-thermal conversion system or a solar-photovoltaic conversion system provides the electric power needed to operate the system's electric rocket engines.

solar system In astronomy, any star and its gravitationally bound collection of nonluminous objects, such as planets, asteroids, and comets; specifically, our own solar system, consisting of the Sun and all the objects bound to it by gravitational attraction. These celestial objects include the nine major planets with more than 60 known gravitationally bound moons of their own, more than 2,000 minor planets, and a very large number of comets. Except for the comets, all the other celestial objects orbit around the Sun in the same direction. Astronomers often divide eight of the major planets into two general categories: (1) the terrestrial or Earthlike planets, consisting of Mercury, Venus, Earth, and Mars; and (2) the outer or Jovian planets, consisting of the gaseous giants Jupiter, Saturn, Uranus, and Neptune. Scientists treat tiny Pluto as a special "frozen snowball." As a group, the terrestrial planets are dense, solid bodies with relatively shallow or negligible atmospheres. In contrast, the Jovian planets contain modest-sized rock cores surrounded by concentric layers of frozen hydrogen, liquid hydrogen, and gaseous hydrogen, respectively. Their frigid outer atmospheres also contain helium, methane, and ammonia.

solar wind The variable stream of plasma (i.e., electrons, protons, alpha particles, and other atomic nuclei) that flows continuously outward from the Sun into interplanetary space.

solid-propellant rocket engine A rocket propelled by a chemical mixture of fuel and oxidizer that are in solid form. Sometimes simply called a *solid rocket*.

sounding rocket A solid-propellant rocket used to carry scientific instruments on parabolic trajectories into the upper regions of Earth's sensible atmosphere (i.e., beyond the reach of high-altitude aircraft and scientific balloons) and into near-Earth space. Generally, it has two major components: the solid-rocket motor and the payload. Payloads often include the scientific instrument or experiment package, the nose cone, a telemetry system, an attitude-control system, a radar-tracking beacon, the firing de-spin module, and the recovery section. Many of the sounding-rocket payloads are recovered for refurbishment and reuse.

space commerce The business or commercial portion of space operations and activities. Currently recognized areas of space commerce include (1) space transportation, (2) satellite communications, (3) satellite-based geopositioning and navigational services, (4) satellite remote sensing (including support for geographic information systems), (5) materials research and processing in space, and (6) space-based industrial facilities.

spacecraft A platform that can function, move, and operate in outer space or on a planetary surface. Aerospace engineers design all types of spacecraft. For example, spacecraft can be human-occupied or uncrewed (robotic) platforms. They can operate in orbit about Earth or while on an interplanetary trajectory to another celestial body. Some spacecraft travel through space and orbit another planet, while others descend to the planet's surface to make a hard (collision-impact) or soft (survivable) landing. To the aerospace engineer, a spacecraft is a well-designed space platform that supports its payload (human crew, instruments, deployable probes, and so on) by providing the mechanical structure, thermal control, wiring, attitude control, computer/command functions, data handling, and power necessary for a successful space mission. Aerospace engineers will often custom-design a spacecraft to meet the demanding requirements of a particular space mission. Space-mission planners sometimes categorize scientific spacecraft as flyby spacecraft, orbiter spacecraft, atmospheric-probe spacecraft, atmospheric-balloon packages, lander spacecraft, surface-penetrator spacecraft, and surface-over spacecraft.

spacecraft clock The timing component within a spacecraft's command and data-handling system. This clock is a very important device because it chronicles the passing time during the life of the spacecraft and regulates nearly all onboard activity.

space settlement A very large, human-made habitat in space within which from 1,000 to 10,000 people would live, work, and play while supporting space commercial activities, such as the operation of a large space manufacturing complex or the construction of satellite power systems.

spaceship An interplanetary spacecraft that carries a human crew.

space sickness The space-age form of motion sickness whose symptoms include nausea, vomiting, and general malaise. This temporary condition usually lasts no more than a day or so. About 50 percent of astronauts or cosmonauts experience space sickness when they initially encounter the microgravity ("weightless") environment of an orbiting spacecraft after a launch. At present, medications can treat the discomfort, but cannot prevent the onset of the condition. Also called *space adaptation syndrome*.

space station An orbiting facility designed to support long-term human habitation in space.

space suit The flexible, outer-garment-like structure (including visored helmet) that protects an astronaut in the hostile environment of space or on the surface of an alien world. The well-designed space suit provides portable life-support functions, supports communications, and accommodates some level of movement and flexibility, so the astronaut can perform useful tasks while in outer space.

space vehicle The general term describing a crewed or robotic vehicle capable of traveling through outer space. An *aerospace vehicle* can operate both in outer space and in Earth's atmosphere.

specific impulse (symbol: I_{sp}) An index of performance for rocket propellants. Aerospace engineers define specific impulse as the thrust (or thrust force) produced by propellant combustion divided by the propellant mass flow rate; that is, the specific impulse (I_{sp}) = thrust/mass flow rate. Unit-system confusion can cause a problem in understanding the value of specific impulse. In the SI system, thrust is expressed in newtons and mass flow rate in kilograms per second. Therefore, the specific impulse ultimately acquires the units of meters per second (m/s).

spy satellite Popular term for a military reconnaissance satellite. *See also* reconnaissance satellite.

stage 1. In aerospace engineering and rocketry, the part (element) of the missile or launch-vehicle system that separates and falls away from the vehicle at burnout or rocket engine cutoff. In multistage rockets, the stages are numbered chronologically in the order of burning (i.e., first stage, second stage, third stage, and so on). When the first stage stops burning, it separates from the rest of the vehicle and falls away. Then the second-stage rocket ignites, fires until burnout, and also separates and falls away from the remaining vehicle. This process continues up to the last stage of the vehicle, the stage that contains the payload. 2. In thermodynamics, a step or process through which a working fluid passes, especially in compression or expansion. 3. In mechanical engineering, a set of rotor blades and stator vanes in a turbine or in an axial-flow compressor.

stationkeeping In aerospace operations, the sequence of maneuvers that keeps a space vehicle or spacecraft in a desired, predetermined orbit or trajectory.

Sun-synchronous orbit A very useful polar orbit that allows a satellite's sensor to maintain a fixed relation to the Sun during each local data collection. This feature is especially important for meteorological and Earth-observation satellites. Each day, a satellite in a Sun-synchronous orbit will fly over a particular area at the same local time. Aerospace workers characterize a particular Sun-synchronous orbit by the time when the satellite crosses Earth's equator. These equator crossings (called "nodes") occur at the same local time each day, with the descending crossing occurring 12 hours (local time) from the ascending crossing. Spacecraft operators use the terms AM and PM polar orbiters to describe those satellites with morning and afternoon equator crossings, respectively.

supernova The catastrophic explosion of certain dying stars. During this end of the stellar life process, the star collapses and explodes, manufacturing (by nuclear transmutation) heavy elements that it throws out into space. In this spectacular explosion, the brightness of the exploding star increases several million times in a matter of days, and it outshines all other objects in its galaxy.

supersonic Of or pertaining to speed in excess of the speed of sound in a particular fluid medium.

sustainer engine A rocket engine that maintains the velocity of a launch vehicle once the vehicle has achieved its intended (programmed) ascent velocity by means of more powerful booster engines (usually jettisoned). Aerospace engineers also use a sustainer engine to provide the modest amount of thrust needed to maintain the speed of a low-altitude spacecraft as it dips into the upper regions of a planet's sensible atmosphere.

synchronous satellite An equatorial west-to-east satellite orbiting Earth at an altitude of approximately 35,900 kilometers. At this altitude, the satellite makes one revolution in 24 hours and remains synchronous with Earth's rotation. *See also* geostationary earth orbit.

telemetry The engineering science of making measurements at one point and transmitting the data to a distant location for evaluation and use. In aerospace operations, the process of transmitting data on a spacecraft's communications downlink. These telemetered data can include scientific data collected by the spacecraft's instruments and spacecraft-subsystem "state-of-health" data.

teleoperation The overall process by which a human controller (usually in a safe and comfortable environment) operates a versatile robot system that is at a distant, often hazardous, location. High-resolution vision and tactile sensors on the robot, reliable telecommunication links, and computer-generated virtual-reality displays at the control station enable the human worker to experience "telepresence"—that is, while operating the distant robot, he or she feels almost physically present in the remote, dangerous work location.

terminator In planetary science and astronomy, the distinctive boundary line separating the illuminated (i.e., sunlit) and dark portions of a nonluminous celestial body, like the Moon.

terrestrial Of or relating to planet Earth.

thermal radiation In physics and engineering, the electromagnetic radiation emitted by any object as a consequence of its temperature. Thermal radiation ranges in wavelength from the longest infrared wavelengths to the shortest ultraviolet wavelengths and includes the optical (or visible) portion of the electromagnetic spectrum.

thermocouple A device that converts thermal energy directly into electricity. In its basic form, the thermocouple consists of two dissimilar metallic conductors, joined at both ends. This configuration creates a closed loop in which an electric current will flow when there is a difference in temperature between the two junctions. The amount of current flow depends on the temperature difference between the measurement (hot) and reference (cold) junction and the physical characteristics of the two different metals. Engineers use various combinations of thermocouple materials to create thermometers that operate over specific temperature ranges.

thrust (symbol: *T*) The forward force provided by a reaction motor.

tracking In aerospace operations, the process of following the movement of a satellite, rocket, or aerospace vehicle; usually performed with optical, infrared, radar, or radio systems.

trajectory The three-dimensional path traced by any object or body moving because of an externally applied force. In aerospace operations (orbital mechanics), the term sometimes means flight path or orbit; more precisely, the term *orbit* describes a closed path, and the term *trajectory* generally refers to an open path (i.e., one that is not closed).

transfer orbit In aerospace operations, an elliptical interplanetary trajectory tangent to the orbits of both the departure planet and the target planet or moon. *See also* Hohmann transfer orbit.

translunar Of or pertaining to the region of outer space beyond the Moon's orbit around Earth.

T-time In aerospace operations, any specific time (minus or plus) that uses "launch time" or "zero" at the end of a countdown as its reference. Aerospace workers use T-time to refer to times and events during a live-fire countdown sequence. For example, the phrase "T minus 30 seconds and counting" refers to the point in the launch sequence that occurs 30 seconds before launch controllers ignite the vehicle's rocket engines. T-plus time describes times and events after rocket-engine ignition.

ullage The amount that a container, such as a propellant tank, lacks of being full.

ultraviolet (UV) radiation The region of the electromagnetic spectrum between visible (violet) light and X-rays. The UV-wavelength range extends from 400 nanometers (just past violet light) down to about 10 nanometers (the extreme ultraviolet cutoff and the beginning of X-rays). *See also* electromagnetic radiation.

umbilical An electrical or fluid (prelaunch) servicing line between the ground or a launch tower and an upright rocket vehicle. Also called the umbilical cord.

unmanned vehicle A space vehicle without a human crew; *unpersoned* or *uncrewed* are more appropriate terms.

upper stage In aerospace engineering, the second, third, or later stage of a multistage rocket vehicle. Solid-propellant or liquid-propellant rocket engines are used as upper stages with expendable launch vehicles. However, for safety, NASA's space shuttle carries and deploys only solid-propellant upper-stage vehicles. Getting into low Earth orbit (LEO) is usually only part of the effort necessary to position a spacecraft at its mission location. Once the spacecraft is lifted into LEO, it often uses an attached upper-stage rocket vehicle to reach its final destination. The upper-stage rocket provides the extra thrust or "kick" to move spacecraft into a higher-altitude orbit around Earth or onto an interplanetary trajectory.

vernier engine A small-thrust rocket engine, used primarily to make fine adjustments in a rocket vehicle's velocity, attitude, or trajectory after the main rocket engines have shut down.

virtual reality (VR) A computer-generated artificial reality. Computer engineers use equipment like a data glove, headphones, and a head-mounted stereoscopic display to project a person into the three-dimensional "virtual" world created by the computer. The virtual world is a computerized description (i.e., the database) of the physical world scene or event under study. For example, it can be an interesting, but currently inaccessible, place, such as the surface of Mars, realistically created from thousands of digitized images sent back by robot space probes. The virtual world might even be quite abstract, like a model of the astrophysical processes occurring inside a black hole.

visible radiation The region of the electromagnetic spectrum to which the human eye is sensitive. The wavelength range of visible radiation spans roughly 0.4 to 0.7 micrometers (or 4,000 to 7,000 angstroms). Beyond this narrow wavelength region, short-wavelength visible (violet) light becomes ultraviolet radiation and long-wavelength visible (red) light becomes infrared radiation. *See also* electromagnetic radiation.

warhead The portion of a missile or rocket that contains either a nuclear or thermonuclear weapon system, a high-explosive system, chemical or biological agents, or harmful materials intended to inflict damage upon the enemy.

watt (symbol: W) The SI unit of power (i.e., work per unit time). One watt represents one joule (J) of energy per second. In electrical engineering, one watt corresponds to the product of one ampere (A) times one volt (V). This unit honors James Watt (1736–1819), the Scottish engineer who developed the steam engine.

wavelength (symbol: λ) In physics and engineering, the mean distance between maxima (or minima) of a periodic pattern; specifically, the least distance between particles moving in the same phase of oscillation in a wave disturbance. Scientists measure wavelength along the direction of propagation of the wave, usually from the midpoint of a crest (or trough) to the midpoint of the next crest or trough. They relate wavelength (λ) to frequency (ν) and phase speed (c) (i.e., here c is the speed of propagation of the wave disturbance) by the simple formula: $\lambda = c/\nu$. The *wave number* is the reciprocal of the wavelength.

weapons of mass destruction (WMD) In defense and arms-control usage, weapons that can cause a high order of destruction and kill many people. These types of weapons include nuclear, chemical, biological, and radiological. The term generally excludes the means of transporting or propelling the weapon where such means is a separable and divisible part of the weapon (e.g., a guided missile).

weather satellite An Earth-orbiting spacecraft that carries a variety of special environmental sensors to observe and measure a wide range of atmospheric properties and processes. For example, imaging instruments provide detailed pictures of clouds and cloud motions, as well as measurements of sea-surface temperature. Sounders collect data in several infrared or microwave spectral bands. Meteorologists process these data to generate profiles of temperature and moisture as a function of altitude. Weather satellites support sophisticated weather-warning and forecasting activities. There are two basic types of weather satellites: the geostationary and the polar-orbiting weather satellites. Also called meteorological satellite or environmental satellite.

weightlessness The condition of free fall or zero g, in which objects inside an Earth-orbiting, unaccelerated spacecraft appear "weightless," even though the objects and the spacecraft are still under the influence of Earth's gravity; the condition in which no acceleration, whether of gravity or other force, can be detected by an observer within the system in question. *See also* microgravity.

window A gap in a linear continuum. In remote sensing, an *atmospheric window* is the range of wavelengths in the electromagnetic spectrum to which the atmosphere is transparent. In aerospace operations, a *launch window* is the time during which conditions are favorable for launching an aerospace vehicle or spacecraft on a specific mission.

X-ray astronomy The most advanced of the three high-energy astrophysics disciplines: X-ray, gamma-ray, and cosmic-ray astronomy. Very energetic and violent processes throughout the universe emit characteristic X-rays. These X-ray emissions carry detailed information about the temperature, density, age, and other physical conditions of celestial objects that produced them. Modern space-based X-ray observatories provide important data that help scientists investigate supernova remnants, pulsars, black-hole candidates, active galaxies, and even energetic solar flares from our parent star, the Sun.

yaw The rotation or oscillation of an aircraft, missile, or aerospace vehicle about its vertical axis. This rotation causes the longitudinal axis of the vehicle to deviate from the flight line or heading in its horizontal plane.

zenith The point on the celestial sphere vertically overhead. Compare with *nadir*, the point 180° from the zenith.

zero g The condition of continuous free fall and apparent weightlessness experienced by passengers and objects in an orbiting spacecraft. *See also* microgravity.

Chapter 9

Associations

This chapter presents a selected collection of interesting organizations from around the world that are involved in developing, applying, or promoting space technology. Some of the organizations described here are major, government-sponsored agencies whose raison d'être is the timely development and application of space technology within either the defense, scientific, commercial, or public-services sector. Other, more modest-sized organizations support and encourage the great appeal that space exploration has on human imagination and the deeply rooted human urge to explore the unknown. Still other entries represent societies within the aerospace industry or important international nongovernmental organizations (NGOs) that provide forums for professional dialogue at the national and international levels. The collection is not meant to be comprehensive. Inclusion is based on two factors: the importance of the agency in space technology and the quality of information it offers the average researcher. Consequently, the Chinese and Russian agencies are not on the list. Although they are important, the information they offer is not reliable and Internet contact is problematic.

The Aerospace Corporation

(Mailing Address)
P.O. Box 92957
Los Angeles, CA 90009-2957 USA
(Street Address)
2350 E. El Segundo Blvd.
El Segundo, CA 90245-4691 USA

1-310-336-5000
1-310-336-7055 (Fax)
http://www.aero.org

The Aerospace Corporation, a private, nonprofit corporation created in 1960, is responsible for the architecture, development, and orbit operation of national security space systems and missions and is the systems engineer for the Space and Missile Systems Center (SMC) of the U.S. Air Force and the National Reconnaissance Office. It also performs work for other government agencies, international organizations, and foreign governments when this is deemed in the national (U.S. government's) interest. Its technical involvement in a project has historically helped reduce the risk of launch failure and increased in-orbit satellite endurance.

Aerospace Education Foundation

1501 Lee Highway
Arlington, VA 22209 USA
1-800-291-8480
1-703-247-5853 (Fax)
(http://www.aef.org/

The Aerospace Education Foundation (AEF), founded in 1956, is a nonprofit organization dedicated to helping the United States maintain primacy in aerospace technology through education. The organization offers scholarships and sponsors symposia, roundtables, workshops, and other programs designed to educate the public about the importance of space technology to the national defense.

American Institute of Aeronautics and Astronautics National Headquarters

1801 Alexander Bell Drive, Suite 500
Reston, VA 20191 USA
1-703-264-7500
1-800-639-2422
1-703-264-7657 (Fax)
http://www.aiaa.org/

The American Institute of Aeronautics and Astronautics (AIAA) is a major global organization, society, and voice of advocacy serving the aerospace profession. It was formed in 1963 through a merger of the American Rocket Society (ARS) and the Institute of Aerospace Sciences (IAS). The primary purpose of the AIAA is to advance the arts, sciences, and technology of aeronautics and astronautics and to foster and promote professional behavior by those individuals engaged in such pursuits. Although the AIAA was founded and formed in the United States, it is now a worldwide organization with nearly 30,000 individual professional members, more than 50 corporate members, and an active international outreach program. AIAA represents the United States on the International Astronautical Federation (IAF) and the International Council on Aeronautical Sciences (ICAS). AIAA and its predecessor organizations have published more than 350 books and 250,000 technical papers. Current publications include six journals, two magazines, more than 40 standards, electronic-format information products, and a Web site.

The Astronauts Memorial Foundation
The Center for Space Education
Mail Code AMF
State Road 405, Building M6-306
Kennedy Space Center, FL 32899 USA
1-321-452-2887
1-321-452-6244 (Fax)
http://www.amfcse.org/

The Astronauts Memorial Foundation (AMF) is a private, not-for-profit organization founded in the wake of the space shuttle *Challenger* explosion on January 28, 1986. It is dedicated to memorializing 23 very special Americans who lost their lives while supporting national goals in space exploration. AMF receives half of the proceeds from the sale of a special *Challenger* automotive license plate, issued by the state of Florida to honor the seven crew members who died in the disaster. Through these funds, corporate donations, and individual contributions, AMF operates the Center for Space Education at the Kennedy Space Center as a living memorial to fallen American astronauts. The sale of *Challenger* license plates funded the design and construction of this building (located adjacent to the Kennedy Space Center Visitor Complex, described in chapter 10). NASA personnel conduct many space-education programs in this center, and its popular "Exploration Station" hosts more than 1,000 school field trips each year. Qualified schoolteachers are invited to visit NASA's Education Resource Center, located within the AMF building. The center is also home to the AMF Educational Technology Institute, a national organization that provides training, demonstration, and developmental programs for educators in partnership with corporations, government agencies, educational institutes, and national associations.

British Interplanetary Society
27/29 South Lambeth Road
London SW8 1SZ
England, The United Kingdom
+44 (0)20-7735-3160
+44 (0) 20-7820-1504 (Fax)
http://www.bis-spaceflight.com/homepage.htm

The British Interplanetary Society (BIS) was founded in 1933 and is the world's longest-established organization devoted exclusively to supporting and promoting astronautics and the exploration of space. The space visionary Arthur C. Clarke was a founding member and guiding spirit of this highly respected organization. The BIS has its headquarters in London, the United Kingdom, where it maintains charitable status. The society is financially independent and obtains most of its income from the annual fees paid by its worldwide membership. The society produces two popular and influential publications: *Spaceflight* and the *Journal of the British Interplanetary Society (JBIS)*. *Spaceflight* is a monthly magazine that covers all aspects of space technology and exploration, astronomy, and international space developments. First published in 1956 (a year before *Sputnik 1* and the birth of the space age), this magazine continues to serve as an authoritative reference for both the aerospace professional and those who simply have a general interest in space. First published in 1934, the *Journal of the British Interplanetary Society (JBIS)* remains a scientific space journal that focuses on interesting, yet far-reaching, ideas associated with space travel. In fact, this leading-edge, bimonthly journal maintains an enviable record of consistently being the first publication to describe some important, but "futuristic," aspect of space technology that eventually becomes commonplace.

The British National Space Centre
151 Buckingham Palace Road
London SW1W 9SS
England, The United Kingdom
+44 (0) 20-7215-5000 (switchboard)
+44 (0) 20-7215-0807 (general inquiries)
+44 (0) 20-7215-0936 (Fax)
http://www.bnsc.gov.uk/

The British National Space Center (BNSC) serves as the major information pathway and advocacy source for space-industry and science activities within the United Kingdom (UK). This organization's publications and extensive Web site provide current information about UK space policy, ongoing space research projects, and the latest activities involving the commercial space industry within the United Kingdom. Recent speeches by the UK's space minister and space news items are also available.

Canadian Space Agency (CSA)
6767 route de l' Aéroport
Saint-Hubert, Quebec J3Y 8Y9
Canada
1-450-926-4800
http://www.space.gc.ca/

The Canadian Space Agency (CSA) serves as the leader of the Canadian space program. The Canadian Space Agency operates under a legislated mandate to "promote the peaceful use and development of space, to advance the knowledge of space through science and to ensure that space science and technology provide social and economic benefits for Canadians." The agency is positioning Canada to pursue five specific areas deemed of critical importance to Canada: Earth and environment, space science, human presence in space, satellite communications, and generic space technologies (including excellence in space robotics). Through the efforts of CSA, the Canadian space program is now a critical element within the Canadian government's overall strategy for science and technology development in Canada.

Cape Canaveral Council of Technical Societies (CCTS)

(Mailing Address)
P.O. Box 245
Cape Canaveral, FL 32920-0245 USA
(Mailing/Office Address)
1980 N. Atlantic Avenue, Suite 401
Cocoa Beach, FL 32931 USA
http://www.canaveralcts.org/

Canaveral Council of Technical Societies (CCTS), founded in 1960, is a voluntary, not-for-profit association of engineering, technical, and scientific societies. CTTS provides professional support to the members of a wide variety of aerospace societies who live and work along Florida's Space Coast. This region of Florida includes Cape Canaveral Air Force Station and NASA's Kennedy Space Center. Member organizations of CCTS represent a mix of Space Coast chapters of national professional societies and groups with local space-science and educational interests. CCTS was founded in 1960 at the start of the American space program. In addition to monthly meetings, one of the major activities of CCTS is to sponsor and host the annual Space Congress™, an interdisciplinary conference held in the Cape Canaveral area that provides a well-attended forum for space professionals to meet and share their knowledge of technical areas and issues facing both the American space program and the global space industry.

Center for Earth and Planetary Studies (CEPS)

National Air and Space Museum
Smithsonian Institution
Washington, DC 20560-0315 USA
1-202-357-1457
1-202-786-2566 (Fax)
http://www.nasm.edu/ceps/

The Center for Earth and Planetary Studies (CEPS) is a scientific research unit within the Collections and Research Department of the National Air and Space Museum of the Smithsonian Institution (see chapter 10). CEPS performs original research and outreach activities on topics involving planetary sciences, terrestrial geophysics, and the remote sensing of environmental change using data from Earth-orbiting satellites and crewed and uncrewed space missions. For example, CEPS serves as a repository for an extensive collection of Earth images acquired during NASA space shuttle missions. CEPS staff members participate in the development and presentation of public programs, including workshops and special events at the National Air and Space Museum, as well as outreach activities in the community. CEPS also houses a NASA-supported Regional Planetary Image Facility (RPIF) that serves as a reference library for scientific researchers who need professional access to NASA's extensive collection of planetary-mission imagery data.

Centre National d'Etudes Spatiales (CNES)
(Mailing Address/CNES Headquarters)
2 place Maurice Quentin
75 039 Paris Cedex 01
France
01-44-76-75-00 (Domestic)
+33-1-44-76-75-00 (International)
01-44-76-76-76 (Fax)
+33-1-44-76-75-00 (Fax)
http://www.cnes.fr/ (French language)
http://www.cnes.fr/WEB_UK/index.htm (English language)

The Centre National d'Etudes Spatiales (CNES) is the public body in France responsible for all sectors of space technology and activity. The government of France created CNES as the French space agency in 1961. Today, it has three major roles as a state institution: first, to ensure the space-technology independence of France and Europe; second, to develop the use of space in all sectors that can possibly benefit from satellite technology; and third, to prepare for the future by exploring those innovative technical concepts that form the basis of future space systems. (See chapter 10 for additional information about the Centre Spatial Guyanais [Guiana Space Center].)

European Space Agency
(Mailing Address/Paris Headquarters)
8, 10 rue Mario-Nikis
75738 Paris Cedex 15
France

(Mailing Address/Washington, D.C., Liaison Office)
955 L'Enfant Plaza SW, Suite 7800
Washington, DC 20024 USA
(Paris Headquarters, Public Relations)
33-1-53-69-71-55
(Washington, D.C., Liaison Office)
1-202-488-4158
(Fax/Paris Headquarters, Public Relations)
33-1-53-69-76-90
(Fax/Washington, D.C., Liaison Office)
1-202-488-4930
http://www.esa.int/

The European Space Agency (ESA) is an international organization whose task is to provide for and promote, for exclusively peaceful purposes, cooperation among European states in space research and technology and their applications. ESA has 15 member states: Austria, Belgium, Denmark, Finland, France, Germany, Ireland, Italy, the Netherlands, Norway, Portugal, Spain, Sweden, Switzerland, and the United Kingdom. Canada is a cooperating state. ESA's activities include research in space science, planetary exploration, Earth observation, telecommunications, space segment technologies (such as space-station components and modules), and space transportation systems. (See chapter 3 for the history of the agency.)

Federation of American Scientists (FAS)
307 Massachusetts Avenue NE
Washington, DC 20002 USA
1-202-546-3300
1-202-675-1010 (Fax)
http://www.fas.org/

The Federation of American Scientists (FAS) is a privately funded, nonprofit policy organization whose Board of Sponsors includes 51 of America's Nobel laureates in the sciences. FAS conducts analysis and advocacy on science, technology, and public policy. FAS was founded in 1945 as the Foundation of Atomic Scientists by members of the Manhattan Project who produced the first nuclear bombs. Their purpose was to create a nongovernmental organization of American scientists capable of addressing the implications and dangers of the nuclear age. The current FAS Space Policy Project promotes American national security and international stability by providing the public and decision makers with information and analysis on civil and military space issues, policies, and programs. This ongoing project is dedicated to increasing international cooperation in space as a means of improving global cooperation to solve problems on Earth.

Numerous FAS publications and the organization's Web site provide a great deal of interesting, but sometimes controversial, information about many of today's most important issues involving the application and use of space technology.

Indian Space Research Organization (ISRO)

Director, Publications and Public Relations
New BEL Road
Bangalore 560 094
India
+91-80-341-5275
+91-80-341-2253 (Fax)
http://www.isro.org/

The Indian Space Research Organization (ISRO) under the Department of Space (DOS) executes the national space program for the government of India. The primary emphasis of the ISRO program is the application of space technology to solve the problems of human beings and society. The government of India established the Space Commission and the DOS in 1972. The main objectives of the Indian space program include the development of satellites (especially communications and remote sensing), launch vehicles, and sounding rockets.

International Astronautical Federation (IAF)

Executive Director
3–5 rue Mario-Nikis
75015 Paris
France
+33-(0)-1-45-67-42-60
+33-(0)-1-42-73-21-20 (Fax)
http://www.iafastro.com/

The International Astronautical Federation (IAF) is an organization that serves the global space community and provides a technical forum for professionals from many different nations. Along with specialized conferences and symposia, the IAF sponsors the annual International Astronautical Congress—a major international meeting that represents a premier space-industry event for its many participants from all over the world.

National Aeronautics and Space Administration (NASA)

Headquarters Information Center
Washington, DC 20546-0001 USA
1-202-358-0000
1-202-358-3251 (Fax)
http://www.nasa.gov/

The National Aeronautics and Space Administration (NASA) is the civilian space agency of the U.S. government and was created in 1958 by an act of Congress. NASA's overall mission is to plan, direct, and conduct American civilian (including scientific) aeronautical and space activities for peaceful purposes. NASA's overall program is composed of five strategic enterprises: (1) to pioneer in aeronautics and space transportation technologies; (2) to conduct research to support human exploration of space and to take advantage of space as a scientific laboratory; (3) to use space to provide information about Earth's environment; (4) to facilitate the human exploration of space; and (5) to explore the universe. The Information Center at NASA Headquarters is complemented in its task of providing a wide variety of interesting data about NASA and its mission to government contractors and officials, educators, and members of the national and international public by Public Affairs Offices at each of the major NASA centers.

National Reconnaissance Office (NRO)
Office of Corporate Communications
14675 Lee Road
Chantilly, VA 20151-1715 USA
1-703-808-1198
1-703-808-1171 (Fax)
http://www.nro.gov/

The National Reconnaissance Office (NRO) is the organization that designs, builds, and operates U.S. reconnaissance satellites. NRO products, provided to government customers like the Central Intelligence Agency (CIA) and the Department of Defense (DOD), can warn of potential trouble spots around the world, help plan military operations, and monitor the environment. The NRO is an agency within the Department of Defense and part of the U.S. intelligence community. The NRO is responsible for unique and innovative technology, large-scale systems engineering, development and acquisition, and operation of space systems.

National Space Development Agency of Japan (NASDA)
Public Relations Office
28F World Trade Center Bldg.
2-4-1, Hamamatsu-cho, Minato-ku
Tokyo, 105-8060
Japan
+81-3-3438-6111
+81-3-5402-6513 (Fax)
http://www.nasda.go.jp/index_e.html

The National Space Development Agency of Japan (NASDA) is the government agency, founded in October 1969, that serves as the nucleus for the development of space and the promotion of the peaceful use of space by Japan. NASDA is responsible for the development of satellites (including space experiments and space-station modules), launch vehicles, and their supporting operational facilities and equipment. The Tanegashima Space Center is NASDA's largest facility (see chapter 10).

National Space Society

600 Pennsylvania Avenue SE, Suite 201
Washington, DC 20003 USA
1-202-543-1900
1-202-546-4189 (Fax)
http://www.nss.org/

The National Space Society (NSS) is a grassroots organization that relies on its nearly 100 chapters worldwide to help promote the vision of people living and working in thriving communities beyond Earth. NSS members support change in technical, social, economic, and political conditions as a means of advancing the day when human beings will permanently live and work in space. Society members receive the bimonthly magazine titled *Ad Astra* (To the stars).

Office for Outer Space Affairs (OOSA)

(Mailing Address)
United Nations Office at Vienna
Vienna International Centre
P.O. Box 500
A-1400 Wien, Austria
+43-1-260-60-4950
+43-1-260-60-5830 (Fax)
http://www.oosa.unvienna.org/

The Office for Outer Space Affairs (OOSA) of the United Nations has the dual objective of supporting intergovernmental discussions on various technical and legal aspects of space activity and of assisting developing countries in using space technology. OOSA follows legal, scientific, and technical developments relating to space activities, technology, and applications to provide technical information and advice to UN member states, international organizations, and other United Nations offices. In addition, the Office for Outer Space Affairs is the secretariat for the Legal Subcommittee of the United Nations Committee on the Peaceful Uses of Outer Space (COPUOS)—the primary international forum for the development of laws and principles governing outer space.

Office of the Associate Administrator for Commercial Space Transportation
FAA/AST
U.S. Department of Transportation
800 Independence Avenue SW, Room 331
Washington, DC 20591 USA
1-202-267-8308
(Customer Service)
1-202-267-5473
(Fax/Customer Service)
http://ast.faa.gov/

The Office of the Associate Administrator for Commercial Space Transportation (AST) of the U.S. government resides in the Federal Aviation Administration (FAA) in the Department of Transportation. Under current federal regulations, this office is given the responsibility for licensing commercial (American) space launches and launch-site operations. This office also encourages and promotes commercial space activities by the private sector.

The Planetary Society
65 North Catalina Avenue
Pasadena, CA 91106-2301 USA
1-626-793-5100
1-626-793-5528 (Fax)
http://www.planetary.org/

The Planetary Society is a nonprofit organization founded in 1980 by Carl Sagan and other scientists that encourages all spacefaring nations to explore other worlds. This organization, which claims more than 100,000 members from countries all around the world, serves as a major space advocacy group, provides public information, and supports educational activities focusing on the exploration of the solar system and the search for extraterrestrial life (SETI). Members of the society receive the bimonthly magazine the *Planetary Report*.

Satellite Industry Association.
225 Reinekers Lane, Suite 600
Alexandria, VA 22314 USA
1-703-549-8697
1-703-549-9188 (Fax)
http://www.sia.org/

The Satellite Industry Association (SIA) is a trade organization, formed in 1998, that represents U.S. space and communications companies in the global commercial satellite marketplace. SIA's executive member companies (such as Boeing Commercial Space Company, COMSAT Corporation, Hughes Communications, and TRW are the leading satellite service providers, satellite manufacturers, launch-services companies, and ground-equipment suppliers in America.

United States Space Foundation
2860 South Circle Drive, Suite 2301
Colorado Springs, CO 80906-4184 USA
1-719-576-8000
1-719-576-8801 (Fax)
http://www.spacefoundation.org/

The United States Space Foundation (USSF) was organized in 1983 as a nonprofit affiliate of the National Aeronautics and Space Administration (NASA) to promote national awareness and support for American space endeavors. Today, the United States Space Foundation has a twofold mission: first, to provide and support educational excellence through the excitement of space, and second, to enthusiastically advocate civil, commercial, and national-security-related space activities. Each spring, this organization conducts its National Space Symposium in Colorado Springs, Colorado—an annual event that has become a major gathering of space leaders.

United States Strategic Command
Public Affairs
901 SAC Blvd., Suite 1A1
Offutt AFB, NE 68113-6020
1-402-294-5961
http://www.stratcom.af.mil/

The United States Strategic Command (USSTRATCOM) is a unified command formed within the Department of Defense on October 1, 2002, by the merger of the U.S. Space Command with USSTRATCOM. The new organization, headquartered at Offutt Air Force Base in Nebraska, serves as the command and control center for all U.S. strategic (nuclear) forces. The command also controls American military space operations through three subsidiary components: Army Space Command (ARSPACE) in Arlington, Virginia; Naval Space Command (NAVSPACE), in Dahlgren, Virginia; and Space Air Force (SPACEAF), at Vandenberg AFB, California.

Universities Space Research Association
American City Building, Suite 212
10227 Wincopin Circle
Columbia, MD 21044-3498 USA
1-410-730-2656
1-410-730-3496 (Fax)
http://www.usra.edu/

The Universities Space Research Association (USRA) is a private, nonprofit corporation operating under the auspices of the National Academy of Sciences. Founded in 1969, USRA provides a mechanism through which member universities can cooperate effectively with one another, with the government, and with other organizations to further space science and technology and to promote education in these areas. The organization's mission is accomplished through a variety of affiliated institutes, centers, divisions, and programs. The great majority of USRA's activities are funded by grants and contracts from the National Aeronautics and Space Administration (NASA).

Young Astronaut Council

5200 27th Street NW
Washington, DC 20015 USA
1-202-682-1984
1-202-244-4800 (Fax)
http://www.yac.org/yac/

The Young Astronaut Council (YAC), formed by the White House in 1984, is a major youth-oriented aerospace organization that develops and promotes space-related educational activities. Using space as the framework, YAC has developed a variety of curricula spanning kindergarten through ninth (grade) that integrate Earth, space life, and physical sciences as well as other disciplines. The Young Astronaut Council works with counterpart programs in foreign countries under an international umbrella organization called Young Astronauts International (YAI). Some of the foreign nations represented within YAI include Australia, Belarus, Bulgaria, China, Japan, Kazakhstan, Korea, Russia, and the Ukraine.

Chapter 10

Demonstration Sites

This chapter provides a selective international listing of facilities, technical exhibits, and space museums at which a person can learn about the astronomical discoveries that eventually gave rise to the space age and experience space-technology developments—past, present, or planned. Several of the facilities listed (like the National Air and Space Museum in Washington, D.C.) have become major "space tourism" attractions that host millions of guests each year. Other demonstration sites included here are more modest in size and scope or perhaps physically quite remote (like the Guiana Space Center), but are worthy of mention because they represent a special space-technology-related experience. As with planning any type of successful travel, it is wise to inquire ahead (preferably by telephone or via the Internet) to make sure that the particular site you wish to visit will actually be accessible during the time period desired. This is especially important for facilities like the U.S. Air Force Space and Missile Museum at Cape Canaveral that are an integral part of an operational launch site or an active space-technology development center. Web sites provide a great deal of useful information about the facility, including hours of operation, admission prices, location, and driving directions. They are a good first-stop source for planning.

Many of the fine science museums and planetariums around the world provide their guests with some type of space-technology experience (on a permanent or temporary basis) as part of their overall general science or astronomy programs. However, only a representative number of such "partial" space-experience facilities are mentioned here, with preference being given instead to facilities and sites that are primarily space technology related. As part of its extensive education and public outreach programs, the National Aeronautics and Space Administration (NASA), through its various field centers, sponsors excellent traveling space-technology exhibits. Several of these traveling exhibits are included in this section because (with some advance planning) they represent a special opportunity to bring an exciting space-technology demonstration directly to you no matter where you live in the United States. Finally, as described in chapter 11, the Internet also provides a continuously expanding opportunity to take "virtual tours" of interesting space-technology demonstration sites from the comfort of your home or school.

Adler Planetarium and Astronomy Museum

1300 South Lakeshore Drive
Chicago, IL 60606 USA
1-312-922-7827
http://www.adlerplanetarium.org

Founded in 1930, the Adler Planetarium had the first planetarium theater in America. It contains a special collection of more than 2,000 historic astronomical, navigational, and mathematical instruments that provide historical context for the space age. The astronomy museum currently features an exhibit describing the works of Copernicus, Hevelius, and other historic Polish astronomers who made significant contributions to our understanding of the universe at the beginning of the scientific revolution. The Adler Education Department provides information about astronomy for a variety of academic users, including lesson plans and classroom activities for teachers. There is a charge for admission.

NASA Ames Research Center (ARC)

Attn: Visitors Center
Moffett Field, CA 94035 USA
1-650-604-6247
http://www.arc.nasa.gov/

NASA's Ames Research Center (ARC) is the agency's primary center for astrobiology, information technology, and aeronautics. The Ames visitor facility, located in a former hypervelocity-flight-test facility used to evaluate early space-capsule designs, is intended to educate the general public about the research and technological developments taking place at the facility. The facility also offers a special "Aerospace Encounter" feature, created especially for fourth-, fifth-, and sixth-grade students to help stimulate their imaginations and increase their enthusiasm for science, mathematics, and technology. There is no charge for admission. Individual (drop-in) visitors are always welcome during normal business hours at the NASA Ames Research Center, but it is advisable to make advance arrangements for visits by large (class-size) groups, especially those groups wishing to enjoy the "Aerospace Encounter" fun classroomlike experience.

Centre National d'Etudes Spatiales (CNES)—Centre Spatial Guyanais
Relations Publiques
BP 726
97387 Kourou
French Guiana
+05-94-33-43-47
+05-94-33-45-55 (Fax)
http://www.cnes.fr/WEB_UK/1 (English-language site for CNES)
http://www.cnes.fr/WEB_UK/enjeux/cnes/establissements/lcentres_csg.htm (Guiana Space Center)

The Guiana Space Center (Centre Spatial Guyanais [CSG]) serves as the launch and rocket-test base for the French Space Agency (CNES) and the European Space Agency (ESA). The location of the Guiana Space Center on the northeast coast of South America, near the equator (5 degrees north latitude), is ideal for launching payloads into geostationary orbit. Created in 1964, this complex became operational in April 1968. Today, it supports the Ariane family of launch vehicles. Visitors are welcome to enjoy the site's space museum (at no charge on certain days), as well as to view an *Ariane 4* launch (by advance reservation) or to take a guided tour of the launch complex (for a modest fee by advance reservation). However, there are safety-imposed age restrictions for children under 16 for viewing Ariane launches. Visitors are strongly encouraged to direct any questions about the space museum or requests for tour reservations and for invitations to view an *Ariane 4* launch directly to the Guiana Space Center's Public Relations Office, preferably well in advance of any planned travel.

The English-language Web-site for CNES leads directly to the Web site for the Guiana Space Center (under "Establishments—CSG"). The CSG Web site provides a great deal of useful information about the launch complex, hours of operation of the space museum, and points of contact for tours and launch invitations.

Cheyenne Mountain Operations Center
Public Affairs and Presentations
1 NORAD Road, Suite 101-213
Cheyenne Mountain AFS, CO 80914-6066
1-719-474-2238
http://www.cheyennemountain.af.mil/cmoc/

Cheyenne Mountain Operations Center (CMOC) has been the center of military space defense activities since the beginning of the Cold War. It is the central collection and coordination center for a worldwide system of satellites, radars, and sensors that provide early warning for any air, missile, or space threat against North America. It is also one of the few joint and binational military organizations in the world, composed of more than 200 men and women from the U.S. Army, Navy, Marine Corps, Air Force, and the Canadian forces. Housed deep inside a mountain about 600 meters underground, the facility supports the North American Air Defense Command (NORAD) and the United States Strategic Command (USSTRATCOM), as well as elements of Space Air Force (SPACEAF). CMOC serves as the command center for NORAD.

The Cheyenne Mountain Operations Center does not offer tours inside the mountain (an active, 24-hour-a-day military center) for the general public. However, in lieu of going inside Cheyenne Mountain, CMOC Public Affairs and Presentations offers a general public presentation about the history of Cheyenne Mountain, the facilities inside the mountain, and the missions of CMOC for visitors who wish to explore a special aspect of military space technology. This presentation is available to the general public at no charge, but children must be at least 12 years old to participate. The one-hour presentation is conducted every Thursday in the James E. Hill Technical Support Facility (TSF) Building 101 at Cheyenne Mountain Air Force Station. Because of limited seating in the visitor center, reservations must be made in advance by contacting the Cheyenne Mountain Public Affairs and Presentations Office.

Coca-Cola Space Science Center

701 Front Avenue
Columbus, GA 31901 USA
1-706-649-1470
1-706-649-1478 (Fax)
http: //www.ccssc.org/

The Coca-Cola Space Science Center in Columbus, Georgia, is operated by Columbus State University and serves as a regional resource for teachers and students by providing unique, on-site learning experiences. The center includes a Challenger Learning Center, in which students use simulator programs to enact missions to the moon, rendezvous with a comet, or repair a satellite, as well as space-technology exhibits, including a full-sized replica of the first 16 meters of the space shuttle and an Apollo capsule replica. The center also houses the Omnisphere Planetarium and the Mead Observatory, which has regular public observing sessions. There is an admission charge for the Omnisphere Planetarium.

Edmonton Space and Science Centre

11211 142 Street
Edmonton, Alberta, Canada T5M 4A1

1-780-452-9100
1-780-455-5882 (Fax)
http://www.planet.eon.net/~essc/

The Edmonton Space and Science Centre is a popular regional space- and science-education center. Its mission is to inspire and motivate people to learn about and contribute to the science and technology advances that make up modern life. The facility includes an IMAX theater, a star theater, science classrooms, exhibits, and an observatory. There is a charge for admission to the theaters, but no charge to visit the observatory.

Euro Space Center
Rue Devant les Hetres, 1
B-6890 Transinne
Belgium
+32-61-65-64-65
+32-61-65-64-61 (Fax)
http://www.ping.be/eurospace/envisit.htm

The Euro Space Center, located in the Ardennes Forest of Belgium, provides a unique space-technology environment for visitors who wish to experience astronaut training and spaceflight firsthand. The facility has a collection of astronaut-training devices and spaceflight simulators that allow participants to experience various aspects of spaceflight and orbital operations. It also houses displays and mock-ups of the *Ariane 4* and *Ariane 5* launch vehicles, as well as NASA's space shuttle and the European Space Agency's *Artemis* communications satellite (with a 25-meter wingspan). There is a fee for using these facilities.

NASA Glenn Research Center
21000 Brookpark Road
Cleveland, OH 44135 USA
1-216-433-2000
1-216-433-8143 (Fax)
http://www.grc.nasa.gov/

The NASA Glenn Research Center develops propulsion, power, and communications technologies for the agency. Its visitor center provides guests with an interactive space-technology learning experience that features the science of flight and propulsion, as well as space power and communications systems. The facility has eight galleries that contain a wide variety of exhibits, including the Apollo spacecraft used on the *Skylab 3* mission and the achievements of astronaut John Glenn, the first American to orbit Earth in a spacecraft. There is also an interactive gallery that describes the role played by the NASA Glenn Research Center in supporting pioneering commercial space communications programs—efforts that have helped bring about the current information revolution. The center is home to Star Station One®, a dynamic exhibit that connects the public with activities and results being accomplished by the *International Space Station* (*ISS*). The facility also offers a variety of lectures, group programs, and tours, including those of the research facilities, but these must be arranged in advance by contacting personnel at the visitor center or the Community and Media Relations Office (CMRO). The CMRO at the NASA Glenn Research Center also manages traveling programs to bring the excitement of space technology to locations throughout the Great Lakes region (Ohio, Indiana, Illinois, Michigan, Minnesota, and Wisconsin). There is no charge for admission to the NASA Glenn Visitor Center.

Goddard Space Flight Center
Code 130, Public Affairs Office
Greenbelt, MD 20771
1-301-286-8955
1-301-286-1707 (Fax)
http://www.gsfc.nasa.gov/

The NASA Goddard Space Flight Center is home to the largest collection of scientists and engineers dedicated to exploring Earth from space. This technical focus provides an exciting heritage for the numerous displays and presentations found in the visitor center. Visitors can tour the Spacecraft Operations Facility, which provides communications links for the space shuttle and the *International Space Station* and controls operations for several space research missions. There is no charge for admission, but all visitors to Goddard must enter via the main gate and obtain visitor and vehicle passes. Foreign visitors must make arrangements in advance through the Goddard International Office (1-301-286-8300).

NASA Johnson Space Center
Lunar Sample Curator
Houston, TX 77058 USA
1-281-483-6187
1-281-483-5347 (Fax)
http://curator.jsc.nasa.gov/

Through a variety of interesting outreach programs, the National Aeronautics and Space Administration (NASA) shares the legacy of the Apollo Project by bringing Moon rocks and soil samples directly to qualified people and institutions in a variety of locations on Earth. Decades after the last human being walked on the Moon (1972) in the twentieth century, examining lunar materials up close provides a uniquely exciting space-technology experience for an entire new generation of future space travelers. In one public "Meet the Moon" program, NASA provides for a limited number of lunar rock samples to be used for short-term or long-term displays at museums, planetariums, expositions, or professional events that are open to the public. The Public Affairs Office at the Johnson Space Center (JSC), Houston, Texas, handles requests for lunar samples under this display program. NASA has also created a special educational-disk program. Each "hand-on" disk contains lunar samples embedded in rugged acrylic. Supported by companion learning materials, these disks have proven very suitable for classroom use by qualified teachers in upper-primary- and secondary-school programs. Requests for use of a lunar-sample disk should be made through the regional NASA Teacher Resource Center appropriate for the school's location. (See chapter 11 for a listing of the NASA Teacher Resource Centers.) For college-level students, NASA has created collections of thin samples of representative lunar rocks on rectangular glass slides. Each set of 12 slides is accompanied by a sample disk (as mentioned earlier) and supporting technical materials. Professors may request use of a thin-section collection by sending a letter on institutional stationery to the Lunar Sample Curator at the Johnson Space Center. The Web site is an excellent starting point for all questions or inquiries concerning lunar materials for research, education, or public viewing.

Johnson Space Center Traveling Exhibits Program

NASA Johnson Space Center
Exhibits Manager
HA/Technology Transfer and Commercialization Office
2101 NASA Road One
Houston, TX 77058-3696 USA
1-281-483-5111 (JSC Public Affairs Office)
1-281-483-4876 (Fax/Exhibit Requests)
http://www.jsc.nasa.gov/pao/exhibits

As a free service of NASA and the U.S. government, the Johnson Space Center (JSC), along with other NASA centers, operates an active traveling exhibits program whereby space-technology exhibits, displays, spacecraft models, space suits, and various space artifacts are made available to organizations and institutions on a short-term (1–29 days) or long-term (30–90 days) basis. For example, a new *International Space Station* Mobile Exhibit travels the country in two 16-meter (48-foot) trailers to provide a contemporary space-technology experience in a large number of communities. The JSC Traveling Exhibits Program primarily serves the NASA Region 4 states of Colorado, Kansas, Nebraska, New Mexico, North Dakota, Oklahoma, South Dakota, and Texas, with other NASA centers being responsible for servicing the remainder of the United States, the District of Columbia, Puerto Rico, and the Virgin Islands with similar traveling exhibits. However, the JSC program can respond to a request from other states or countries when the normally responsible NASA center is unable to satisfy the needs of the requestor. Under this mobile program, NASA exhibits are loaned to an organization for educational displays that are open to the public. The JSC Traveling Exhibits Program includes trailers, panels, photography, models, and space suits. The panels program, for example, includes such themed displays as space food, the space station, the commercial use of space, Project Apollo, and exploring the universe. The JSC Traveling Exhibits Program and the companion mobile exhibit programs sponsored by other NASA centers provide an effective way of bringing a space-technology experience to a community facility (like a science museum) or a special event (like a technical meeting or convention). However, requests should be made well in advance so the necessary administrative details and equipment scheduling can take place. The Web site provides a great deal of well-illustrated information about the many different, high-quality traveling exhibits available under the JSC program, the rules governing that loan program, and the points of contact at JSC. The Public Affairs Office at other NASA centers (listed in chapter 11) can assist in identifying the traveling space-technology exhibits available for their respective regions of the United States.

Kansas Cosmosphere and Space Center

1100 North Plum
Hutchinson, KS 67501 USA
1-316-662-2305
1-800-397-0330 (Fax)
http://www.cosmo.org/

The Kansas Cosmosphere and Space Center in Hutchinson, Kansas, is home to the Hall of Space Museum, a facility whose exhibits and artifacts chronicle the space programs of the United States and the former Soviet Union during the great space race of the Cold War. For example, visitors can see a full-scale Apollo-Soyuz Test Project spacecraft, the *Apollo 13* command module, German V-1 and V-2 rockets, a Mercury Redstone rocket, and a Gemini Titan rocket. There is an SR-71 Blackbird spy plane on display in the lobby entrance to the museum, as well as a full-scale replica of the space shuttle orbiter. The facility also hosts a Star Station One® educational display that provides an entertaining and educational view of activities on the *International Space Station* (*ISS*). The complex hosts an OMNIMAX7 theater and a planetarium. There is a charge for admission to these facilities.

Kennedy Space Center Visitor Complex

Delaware North Parks Services of Spaceport, Inc.
Mail Code DNPS
Kennedy Space Center, FL 32899 USA
1-321-449-4444
http://www.kennedyspacecenter.com
http://www-pao.ksc.nasa.gov/

The Kennedy Space Center Visitor Complex on Florida's east central coast is a major space tourism attraction that hosts million of guests each year and provides a truly unique space-technology experience. Visitors should anticipate spending a full day exploring the Visitor Complex, enjoying its Rocket Garden (a large outdoor collection of rockets on display), viewing a space-themed IMAX movie, examining space-history exhibits, walking through a full-sized model of the space shuttle, and taking one of several possible bus tours around the sprawling Kennedy Space Center and adjacent Cape Canaveral Air Force Station, both of which are active, working space launch complexes. Three guided bus tours are especially popular: "Cape Canaveral: Then and Now," which visits the historic launch pads of the Mercury, Gemini, and Apollo Project era; "NASA Up Close," which provides a detailed look at NASA's space shuttle program and Launch Complex 39; and (by advance reservation) "KSC Wildlife," a guided tour of the Merritt Island Wildlife Refuge. Delaware North Parks Services of Spaceport operates the Visitor Complex for NASA without the use of public tax dollars. Consequently, there is a charge for admission to the Visitor Complex and for the bus tours. Depending on NASA's launch schedule, it is also possible for visitors in Florida to witness a space shuttle launch. Details are available on the second Web site and at the phone number listed. The second Web site also links to additional visitor information from the Public Affairs Office of the NASA Kennedy Space Center.

Kirkpatrick Science and Air Space Museum

Omniplex
2100 NE 52nd Street
Oklahoma City, OK 73111 USA
1-405-602-6664
1-800-532-7652
1-405-602-3768 (Fax)
http://www.omniplex.org/

The Kirkpatrick Science and Air Space Museum offers visitors a diverse collection of interactive and historic exhibits on space exploration. Visitors can view memorabilia from the Apollo program as well as full-scale models of the Mercury, Gemini, and Apollo space capsules and the lunar excursion and command and service modules. Exhibits also honor aviation and space pioneers from Oklahoma. There is a charge for admission. The Web site provides useful information about the facility and its exhibits, operating hours, admission prices, location, and travel directions.

H.R. MacMillan Space Centre
1100 Chestnut Street, Vanier Park
Vancouver, BC V6J 3J9
Canada
1-604-738-7827
1-604-736-5665 (Fax)
http://pacific-space-centre.bc.ca/

The H.R. MacMillan Space Centre in Vancouver, British Columbia, is a nonprofit community resource that brings the wonder of space to Earth while providing each visitor with a chance to experience a personal sense of ongoing discovery. Through innovative programming, exhibits, and activities, the center pursues a goal of inspiring sustained interest in the fields of Earth science, space science, and astronomy. The center features state-of-the-art interactive exhibits, shows, and demonstrations, including the Virtual Voyages® full-motion simulator and GroundStation Canada, which has extensive interactive programs about the *International Space Station* as well as exhibits on the principles of rocketry and the science of simulation. The Cosmic Courtyard enables visitors to try to land the space shuttle and plan a mission to Mars. The H.R. MacMillan Planetarium remains the cornerstone of this Pacific space center, presenting multimedia shows on space and astronomy. As a community resource, this delightful facility operates both public and fee-based programs.

National Air and Space Museum (NASM)
Smithsonian Institution
Washington, DC 20560 USA
1-202-357-2700
1-202-633-8982 (Fax)
http://www.nasm.edu/

The National Air and Space Museum (NASM) of the Smithsonian Institution maintains the largest collection of historic aircraft and spacecraft in the world. It is an exceptionally exciting space-technology site as well as a highly respected center for research into the history, science, and technology of aviation and spaceflight. Located on the National Mall in Washington, D.C., the museum offers its millions of annual visitors hundreds of professionally displayed artifacts, including the *Apollo 11* command module and a lunar rock sample that guests can touch. The Langley IMAX7

Theater and the Albert Einstein Planetarium offer exciting shows related to aviation, space exploration, and astronomy. General admission to the museum is free, but there is a charge for admission to the Langley IMAX Theater and the Einstein Planetarium. The staff at the museum continues to develop new exhibits that examine the impact of air and space technology on science and technology. Here is a very small sampling of the exhibits and displays that await the NASM visitor: the spacecraft used by the primates Able and Baker, an Apollo Lunar Roving Vehicle, a *Corona KH-4B* reconnaissance satellite camera and film-return capsule, John Glenn's *Friendship 7* Mercury spacecraft, a Jupiter launch vehicle and nose cone, a Minuteman III ICBM, the *Pioneer 10* spacecraft and its interstellar message plaque, rockets developed by Sir William Congreve and Robert Goddard, a Russian SS-20 ICBM, a huge collection of space suits, *Sputnik 1*, a V-2 rocket, a Voyager spacecraft, and much more (see the complete listing of exhibits on the NASM Web site).

Neil Armstrong Air and Space Museum

500 South Apollo Drive
Wapakoneta, OH 45895 USA
1-419-738-8811
1-419-738-3361 (Fax)
http://www.ohiohistory.org/phases/armstron/

The Neil Armstrong Air and Space Museum in Wapakoneta, Ohio (Armstrong's hometown), honors the *Apollo 11* astronaut who was the first human being to step on the surface of the Moon. This museum features the *Gemini 8* spacecraft flown by astronauts Neil Armstrong and David Scott in 1966, Armstrong's Gemini and Apollo space suits, a Moon rock brought back by the *Apollo 11* mission, and other interesting aviation and space artifacts that celebrate the contribution of Ohio to the development of human flight. There is a charge for admission. The Ohio History Society Web site provides useful information about the Neil Armstrong Air and Space Museum, which does not maintain a separate, dedicated Web site.

Rose Center for Earth and Space

American Museum of Natural History
Central Park West at 79th Street
New York, NY 10024 USA
1-212-769-5100
http://www.amnh.org/rose/

The Rose Center for Earth and Space at the American Museum of Natural History (AMNH) in New York City provides visitors with an exhilarating multimedia space and astronomy experience. For example, visitors can explore the Cullman Hall of the Universe, an interactive exhibit that is divided into four zones, each illuminating the processes that led to the creation of the planets, stars, galaxies, and the universe, respectively. The Heilbrunn Cosmic Pathway takes guests on a journey through 13 billion years of cosmic evolution. The Scales of the Universe exhibit uses the powers of 10 to illustrate the relative scale of objects in the cosmos. The Gottesman Hall of the Planet Earth allows visitors to explore the dynamic processes that produce global change on Earth, while the Space Theater at the new Hayden Planetarium provides interesting space and astronomy programs. There is a charge for admission. The American Museum of Natural History is one of the world's premiere museums, scientific institutions, and cultural and educational resources. Since its founding in 1869, the museum has advanced a mission to discover, interpret, and disseminate knowledge about human cultures, the natural world, and the universe through a broad program of field exploration, scientific research, innovative exhibitions, and pioneering educational programs.

The New Mexico Museum of Space History
P.O. Box 5430
Alamogordo, NM 88311-5430
1-505-437-2840
1-877-333-6589
1-505-434-2245 (Fax)
http://www.spacefame.org

The New Mexico Museum of Space History in Alamogordo, New Mexico, is a division of the New Mexico Office of Cultural Affairs and contains a space museum, planetarium, IMAX7 theater, and the International Space Hall of Fame. The center helps guests appreciate the developments in modern rocketry and space technology, with a special emphasis on the role that New Mexico (White Sands Missile Range) has played in the American space program. Visitors can trace the story of rocket development and view American and Soviet space capsules and satellites, including a rare replica of a Sputnik. The facility has a distinctive outdoor collection of rockets and space hardware. There is a charge for admission to the theater and planetarium.

Space Center Houston
1601 NASA Road 1
Houston, TX 77058 USA
1-281-244-2100
1-281-283-7724 (Fax)
http://www.spacecenter.org/

The Space Center Houston is the official visitor center for NASA's Johnson Space Center. This major space tourism and experience facility is owned and operated by the Manned Space Flight Education Foundation. The mission of the center is to help celebrate and commemorate the accomplishments of NASA and the American human spaceflight program. Visitors can view numerous historic spacecraft and artifacts, such as the *Apollo 17* command capsule, an Apollo lunar excursion module (LEM), a Gemini capsule, and a Mercury capsule. They can also enjoy a variety of hands-on interactions with astronaut-training devices, shuttle simulators, and space-station-equipment mock-ups. The center does not receive any federal funds, and there is a charge for admission.

Tanegashima Space Center

Mazu, Kukinaga, Minamitane-machi
Kumage-gun, Kagoshima 891-37
Japan
+81-9972-6-2111
+81-9972-4-4004 (Fax)
http://www.nasda.go.jp/Home/Facilities/e/tnsc_e.html (Tanegashima Space Center home page)
http://spaceboy.nasda.go.jp/gallery/gallery-e/t_tour_e.html (virtual tour of Tanegashima Center)

The Tanegashima Space Center is the Japanese rocket-launching base and the largest facility within the National Space Development Agency (NASDA) of Japan. The space complex contains all the facilities and support equipment to perform prelaunch, launch, and postlaunch (i.e., tracking) operations for a variety of Japanese rockets, including the H-I and H-II launch vehicles. Except for launch and rocket-engine-test days, visitors are welcome to tour the entire launch complex. Visitors can tour the Takesaki Range Control Center (RCC), the Launch Simulation Center, and the Space Museum. The Space Museum is housed in the Space Development Exhibition Hall that welcomes the general public. There is no charge for admission. When the complex is closed due to a rocket launch, members of the general public are invited to observe the launch from designated observation points. Reservations are not required to visit these launch-observation points, but they are often crowded. The Web sites provide a great deal of well-illustrated information about the center, its facilities, and the Space Museum, operating hours, general location, travel directions, and an effective virtual tour of the launch complex.

United States Air Force Space and Missile Museum

Public Affairs (Community Relations Office)

45th Space Wing
Patrick AFB, FL 32925 USA
1-321-494-1110 (Main base)
1-321-853-9171 (Museum site)
http://www.patrick.af.mil/

The United States Air Force Space and Missile Museum at Cape Canaveral Air Force Station, Florida, preserves both the hardware and the spirit of America's earliest adventures into space. The predominantly outdoor museum displays numerous missiles, rockets, and related pieces of space-system equipment. The museum was originally opened at the historic Space Launch Complex 26 from which the United States placed its first satellite (*Explorer 1*) into Earth orbit on January 31, 1958. The museum grounds also include historic Space Launch Complex 5/6, from which astronaut Alan Shepard and then astronaut Gus Grissom took off on their suborbital flights. The early blockhouse that served both of these launch sites is preserved and contains much of the original launch-support equipment. Adjacent to the blockhouse is an exhibit hall that features a number of displays describing the numerous contributions the U.S. Air Force has made to the development of space technology. Visitors can also walk through an outdoor "rocket garden" on the museum grounds that contains one of the largest collections of rocket vehicles in the world, including many rare winged missiles like the Navaho and the Bull Goose.

Unfortunately, this richly historic and interesting space-technology site is located deep within an active military launch complex, so people who do not have an access badge to Cape Canaveral Air Force Station cannot normally visit the museum on their own. Visitors, however, can enjoy the museum by taking the special ("Cape Canaveral") escorted bus tour that originates at the Kennedy Space Center Visitor Complex. While admission to the Air Force Space and Missile Museum is free, there is a charge for the bus tour, as mentioned in a previous entry of this chapter. Special arrangements can also be made for escorted groups to tour this historic museum facility by contacting Public Affairs (Community Relations Office) at Patrick Air Force Base, Florida. The Web site provides useful and historic information about the museum (http://www.patrick.af.mil/museum.htm), as well as important points of contact at Patrick AFB (e.g., History Office and Community Relations Office).

U.S. Space and Rocket Center
c/o Guest Relations
U.S. Space Camp
P.O. Box 070015

Huntsville, AL 35807-7015 USA
1-256-721-7160 (Media Relations)
1-800-637-7223 (U.S. Space and Rocket Center/U.S. Camp)
http://www.spacecamp.com/

The U.S. Space and Rocket Center (USSRC) in Huntsville, Alabama, is a major space tourism destination, providing visitors with an extensive collection of rockets and space artifacts and the opportunity to interact with many "hands-on" displays and training simulators. Here they can "fly" a space shuttle or view various kinds of space food. The center has an IMAX7 theater that features space-themed shows. The museum and space-experience complex serve as the official NASA Visitor Center for the Marshall Space Flight Center (MSFC). However, USSRC receives no direct federal funding for operational support, and there is a charge for admission.

Since 1982, the U.S. Space and Rocket Center has also been the home of U.S. Space Camp7, an organization promoting youthful space training experiences and activities as a means of encouraging young people to pursue studies in mathematics, science, and technology. U.S. Space Camp is a legal entity of the U.S. Space and Rocket Center, established under the laws of the state of Alabama. In addition to the U.S. Space Camp at the U.S. Space and Rocket Center in Huntsville, Alabama, U.S. Space Camp operations now take place in Titusville, Florida (see the Astronaut Hall of Fame entry in this chapter), in Mountain View, California, and at several international locations. There are tuition and other expenses associated with each of the various U.S. Space Camp space-experience and astronaut-training activity programs. The Web site also provides information about the U.S. Space and Rocket Center (http://www.spacecamp.com/museum).

Virginia Air and Space Center
600 Settlers Landing Road
Hampton, VA 23669-4044 USA
1-757-727-0900 (Phone)
1-757-727-0898 (Fax)
http://www.vasc.org/

The Virginia Air and Space Center (VASC) serves as the official visitor center for the NASA Langley Research Center, which specializes in aeronautics research. The mission of this center is to preserve and interpret national achievements in air and space exploration and development and to stimulate visitor interest in the sciences by providing them entertaining experiences with innovative educational programs, interactive exhibits, and multimedia shows. Exhibits enable visitors to explore humankind's relationship with Mars, launch a rocket, and simulate landing the space shuttle. Some of the interesting space artifacts on display include the *Apollo 12* command module, a Moon rock, and a Martian meteorite. A giant-screen IMAX7 theater offers guests a variety of air and space shows, and the center has developed a number of space-related youth camp-in educational programs. There is a charge for admission to the exhibits and to the IMAX theater shows.

NASA Wallops Flight Facility

NASA Visitor Center
Bldg I-17
Wallops Island, VA 23337 USA
1-757-824-2298
1-757-824-1776 (Fax)
http://www.wff.nasa.gov/

NASA's Wallops Island Flight Facility conducts sounding-rocket flights and suborbital space-probe launches. Its visitor center offers guests exhibits describing current and future NASA projects, scale models of space probes and satellites, a Moon rock from the *Apollo 17* mission, and full-scale rockets. There are also interactive computer displays such as "Fly Your Own Sounding Rocket" and "Can You Find Your Way on Earth from Space?" that let visitors explore space technology on a personal basis. There is no charge for admission.

Chapter 11

Sources of Information

This chapter describes additional sources of information about space technology. The list of more traditional sources (such as selected books, publications, and educational resource centers) is complemented by a special collection of cyberspace sources. The information revolution and the exponential growth of the Internet have produced an explosion in electronically distributed materials. Unfortunately, unlike a professionally managed library or a well-stocked bookstore within which you can confidently locate desired reference materials, the Internet is a vast digitally formatted information reservoir that is overflowing with both high-quality, technically accurate materials and inaccurate, highly questionable interpretations of history, technology, or the established scientific method. To help you make the most efficient use of your travels through cyberspace in pursuit of information about outer space, this chapter provides a selected list of Internet addresses (i.e., Web sites) that can conveniently serve as your starting point whenever you seek additional source materials about a particular aspect of space technology. Many of the Web sites suggested here contain links to other interesting Internet locations. With some care and reasoning, you should be able to rapidly branch out and customize any space-technology information search. With the contents of this book and especially this chapter as a guide, you can effectively harness the power of the modern global information network.

The following key words and phrases should prove quite useful in starting your customized Internet searches: astrobiology, astronaut, astrophysics, communications satellite, cosmonaut, Earth-observing satellite, Earth-system science, exobiology, global change, launch site, launch vehicle, military satellite, navigation satellite, planetary science, remote sensing, rocket propulsion, satellite power system (SPS), search for extraterrestrial intelligence (SETI), space agencies, space exploration, space station, space suit, space technology, surveillance satellite, and terraforming. Also, as found within this book, the proper names of space-technology pioneers (such as Robert Goddard), spacecraft, launch vehicles, projects, and programs (such as the Voyager spacecraft), and major solar-system bodies (such as the planet Mars) will prove helpful in initiating other specialized information searches on the Internet.

SELECTED BOOKS

Angelo, Joseph A., Jr. *The Dictionary of Space Technology*. 2nd ed. New York: Facts on File, 1999.

Angelo, Joseph A., Jr. *Encyclopedia of Space Exploration*. New York: Facts on File, 2000.

Brown, Robert A., ed. *Endeavour Views the Earth*. New York: Cambridge University Press, 1996.

Burrows, William E., and Walter Cronkite. *The Infinite Journey: Eyewitness Accounts of NASA and the Age of Space*. Discovery Book, 2000.

Cole, Michael D. *International Space Station: A Space Mission*. Springfield, NJ: Enslow Publishers, 1999.

Ginsberg, Irving W., and Joseph A. Angelo, Jr., eds. *Earth Observations and Global Change Decision Making, 1989: A National Partnership*. Malabar, Fla.: Krieger Publishing, 1990.

Heppenheimer, Thomas A. *Countdown: A History of Space Flight*. New York: Wiley, 1997.

Kluger, Jeffrey. *Journey beyond Selene: Remarkable Expeditions Past Our Moon and to the Ends of the Solar System*. New York: Simon & Schuster, 1999.

Kraemer, Robert S. *Beyond the Moon: A Golden Age of Planetary Exploration, 1971–1978* Smithsonian History of Aviation and Spaceflight Series. Washington, D.C.: Smithsonian Institution Press, 2000.

Lewis, John S. *Rain of Iron and Ice: The Very Real Threat of Comet and Asteroid Bombardment*. Reading, Mass.: Addison-Wesley, 1996.

Logsdon, John M. *Together in Orbit: The Origins of International Participation in the Space Station*. NASA History Division, Monographs in Aerospace History 11. Washington, D.C.: Office of Policy and Plans, November 1998.

Neal, Valerie, Cathleen S. Lewis, and Frank H. Winter. *Spaceflight: A Smithsonian Guide*. New York: Macmillan, 1995.

Pebbles, Curtis L. *The Corona Project: America's First Spy Satellites*. Annapolis, Md.: Naval Institute Press, 1997.

SELECTED PERIODICALS

Ad Astra (literally, To the stars). Bimonthly publication of the National Space Society. http://www.nss.org/adastra/

Aerospace Power Journal. Scholarly, professional journal from the Air University of the U.S. Air Force. http://www.airpower.maxwell.af.mil/airchronicles /apje.html

Air and Space Magazine. Informative publication of the Smithsonian National Air and Space Museum. http://www.airspacemag.com.

Airman. The popular monthly magazine of the U.S. Air Force. http://www.af.mil/news/airman.

Astronomy. Popular monthly commercial publication that deals with space exploration and astronomy. http://www.astronomy.com/.

Journal of the British Interplanetary Society (JBIS). The scientific space-travel journal published bimonthly by the British Interplanetary Society. http://www.bis-spaceflight.com/public B. htm

Planetary Report. The bimonthly space-exploration magazine of the Planetary Society. http://www.planetary.org/.

Spaceflight. the magazine of astronautics and outer space published bimonthly by the British Interplanetary Society. http://www.bis-spaceflight.com/publicA.htm

Space News. A weekly newspaper that deals with all aspects of space technology and exploration. http://www.spacenews.com/.

NASA Educator Resource Center Network

Through its Educator Resource Center Network (ERCN), the National Aeronautics and Space Administration (NASA) provides expertise and facilities to help educators access and use science, mathematics, and technology instructional products aligned with national standards and appropriate state frameworks and based on NASA's unique mission and results. Educator Resource Centers (ERCs) are located on or near NASA Field Centers, as well as at planetariums, museums, colleges, universities, and other nonprofit organizations around the United States. The Educator Resource Centers at NASA Field Centers are identified here along with appropriate contact information (including Web-site address when relevant), as well as the regions of the United States each NASA ERC primarily serves. Primary and secondary schoolteachers are especially encouraged to take advantage of the great variety of space-technology-related videocassettes, slides, computer software, printed materials, lesson plans, and historical NASA materials that can be found at each NASA ERC. These centers also serve as a gateway to NASA's lunar and meteorite materials loan programs.

NASA Ames Research Center
Educator Resource Center
Mail Stop 253-2
Moffett Field, CA 94035-1000 USA
1-650-604-3574
1-650-604-3445 (Fax)
http://amesnews.arc.nasa.gov/erc/erchome.html

This ERC serves educators in the following states: Alaska, northern California, Hawaii, Idaho, Montana, Nevada, Oregon, Utah, Washington, and Wyoming.

NASA Educator Resource Center
NASA Dryden Flight Research Center
45108 North Third Street East
Lancaster, CA 93535 USA
1-661-948-7347
1-661-948-7068 (Fax)
http://www.dfrc.nasa.gov/trc/ERC/

This ERC serves educators in southern California and Arizona.

NASA JPL Educator Resource Center
Village at Indian Hills Mall
1460 East Holt Avenue, Suite 20
Pomona, CA 91767 USA
1-909-397-4420
1-909-397-4470 (Fax)
http://learn.jpl.nasa.gov/resource/resources-index.html

This ERC serves educators in California.

NASA John H. Glenn Research Center
NASA Educator Resource Center
21000 Brookpark Road,
Cleveland, OH 44135 USA
1-216-433-2017
1-216-433-3601 (Fax)
http://www.grc.nasa.gov/WWW/PAO/html/edteachr.htm

This ERC serves educators in the following states: Illinois, Indiana, Michigan, Minnesota, Ohio, and Wisconsin.

NASA Goddard Space Flight Center
Educator Resource Laboratory
Mail Code 130.3

Greenbelt, MD 20771 USA
1-301-286-8570
1-301-286-1781 (Fax)
http://www.gsfc.nasa.gov/vc/erc.htm

This ERC serves educators in the following locations: Connecticut, Delaware, District of Columbia, Maine, Maryland, Massachusetts, New Hampshire, New Jersey, New York, Pennsylvania, Rhode Island, and Vermont.

GSFC/Wallops Flight Facility
Visitor Center
Building J-17
Wallops Island, VA 23337 USA
1-757-824-2298
1-757-824-1776 (Fax)
http://www.wff.nasa.gov/~WVC/ERC.htm

This location serves educators in the Eastern Shores region of Virginia and Maryland.

NASA Educator Resource Center
NASA Johnson Space Center
Space Center Houston
1601 NASA Road One
Houston, Texas 77058 USA
1-281-244-2129
1-281-483-9638 (Fax)
http://www.spacecenter.org/educator;_resource.html

This ERC serves educators in the following states: Colorado, Kansas, Nebraska, New Mexico, North Dakota, Oklahoma, South Dakota, and Texas.

NASA Kennedy Space Center
Educator Resource Center
Mail Code ERC
J.F. Kennedy Space Center, FL 32899 USA
1-321-867-4090
1-321-867-7242 (Fax)
http://www-pao.ksc.nasa.gov/kscpao/educate/teacher.htm#educate

This ERC serves educators in the following locations: Florida, Georgia, Puerto Rico, and the Virgin Islands.

Educator Resource Center for NASA Langley Research Center
Virginia Air and Space Center
600 Settlers Landing Road

Hampton, VA 23669-4033 USA
1-757-727-0900, ext 757
1-757-727-0898 (Fax)
http://www.vasc.org/erc/

This ERC serves educators in the following states: Kentucky, North Carolina, South Carolina, Virginia, and West Virginia.

NASA Educator Resource
Center for NASA Marshall
Space Flight Center
U.S. Space and Rocket Center
One Tranquility Base
Huntsville, AL 35807
1-256-544-5812
1-256-544-5820 (Fax)
http://erc.msfc.nasa.gov

When functioning as the NASA MSFC ERC, this location serves educators in the following states: Alabama, Arkansas, Iowa, Louisiana, Missouri, and Tennessee.

NASA Stennis Space Center
Educator Resource Center
Building 1200
Stennis Space Center, MS 39529-6000 USA
1-228-688-3338
1-800-237-1821
1-228-688-2824 (Fax)
http://education.ssc.nasa.gov/erc/erc.htm

This ERC serves educators in the state of Mississippi.

CYBERSPACE SOURCES: A COLLECTION OF SELECTED SPACE-TECHNOLOGY-RELATED INTERNET SITES

Agencies and Organizations of the U.S. Government

National Aeronautics and Space Administration (NASA)

Headquarters Washington, D.C. (main site): http://www.nasa.gov

Selected NASA Centers

Ames Research Center, Mountain View, CA: http://www.arc.nasa.gov

Dryden Flight Research Center, Edwards, CA: http://www.dfrc.nasa.gov

Glenn Research Center, Lewis Field, OH: http://www.grc.nasa.gov

Goddard Space Flight Center, Greenbelt, MD: http://www.gsfc.nasa.gov

Jet Propulsion Laboratory, Pasadena, CA: http://www.jpl.nasa.gov

Johnson Space Center, Houston, TX: http://www.jsc.nasa.gov

Kennedy Space Center, FL: http://www.ksc.nasa.gov

Langley Research Center, Hampton, VA: http://www.larc.nasa.gov

Marshall Space Flight Center, Huntsville, AL: http://www.msfc.nasa.gov

Stennis Space Center, MS: http://www.ssc.nasa.gov

Wallops Flight Facility, Wallops Island, VA: http://www.wff.nasa.gov

White Sands Test Facility, White Sands, NM: http://www.wstf.nasa.gov

Selected Space Missions

Cassini Mission (Saturn): http://www.saturn.jpl.nasa.gov/cassini/index.shtml

Galileo Mission (Jupiter): http://www.jpl.nasa.gov/galileo

Ulysses Mission (Sun's polar regions): http://ulysses.jpl.nasa.gov

Voyager (deep space/interstellar): http://voyager.jpl.nasa.gov/

Exploration of Mars (numerous missions): http://mars.jpl.nasa.gov

National Space Science Data Center (NSSDC) [numerous space missions including planetary]: http://nssdc.gsfc.nasa.gov/planetary

Military Space

Aerospace Corporation (supports U.S. Air Force): http://www.aero.org/

45th Space Wing, Patrick Air Force Base, FL (includes links to Cape Canaveral history): https://www.patrick.af.mil

National Reconnaissance Office (NRO): http://www.nro.gov/

U.S. Strategic Command (USSTRATCOM) [includes links to Army, Navy, and Air Force Space Commands and other military sites]: http://www.stratcom.af.mil/

Other U.S. Government Agencies and Organizations

Commercial Space Transportation Office (FAA/DOT) [includes information about launch sites throughout the United States and the world]: http://ast.faa.gov/

National Oceanic and Atmospheric Administration (NOAA) [environmental/weather satellites]: http://www.noaa.gov

Smithsonian National Air and Space Museum (NASM): http://www.nasm.edu

Selected Foreign Space Agencies and Organizations

Argentinian Space Agency: http://www.conae.gov.ar

Brazilian Space Agency: http://www.inpe.br

British National Space Centre (BNSC): http://www.bnsc.gov.uk

Canadian Space Agency (CSA): http://www.space.gc.ca

CNES (French space agency): http://www.cnes.fr/

European Space Agency (ESA): http://www.esa.int/

German Space Agency (DLR): http://www.dlr.de/

Indian Space Research Organization (ISRO): http://www.isro.org/

International Astronautical Federation (IAF): http://www.iafastro.com/

Italian Space Agency (ASI): http://www.asi.it

National Space Development Agency of Japan (NASDA): http://www.nasda.go.jp/index_e.html

Office for Outer Space Affairs (UN): http://www.oosa.unvienna.org/

Swedish Space Agency: http://www.ssc.se

Russian space program (detailed background from FAS) http:www.fas.org/spp/civil/russia/rsa.htm

Selected Space Societies and Advocacy Groups

Aerospace Education Foundation (AEF): http://www.aef.org

American Institute of Aeronautics and Astronautics (AIAA): http://www.aiaa.org

British Interplanetary Society (BIS): http://bis-spaceflight.com/homepage.htm

Challenger Center for Space Science: http://www.challenger.org

National Space Society (NSS): http://www.nss.org

Planetary Society: http://www.planetary.org

U.S. Space Foundation: http://www.spacefoundation.org/

Young Astronaut Council: http://www.yac.org/yac/

Other Interesting Space-Related Educational Sites

NASA Space Educators' Handbook: http://vesuvius.jsc.nasa.gov/er/seh/seh.html

NASA Space Resource Links for Education: http://spacelink.msfc.nasa.gov/index.html

Observing Earth from Space: http://earthobservatory.nasa.gov

Teaching Earth Science: http://www.earth.nasa.gov/education/index.html

Tours of Solar System (UCSB): http://www.deepspace.ucsb.edu/ia/nineplanets/overview.html

(JPL/Caltech): http://pds.jpl.nasa.gov/planets

Index

Page numbers in *italics* indicate illustrations.

About the Author

JOSEPH A. ANGELO, JR., a retired U.S. Air Force officer (lieutenant colonel), is currently a consulting futurist and technical writer. He is an Adjunct Professor in the College of Engineering at Florida Tech and served with distinction as a commissioner for the governor of Florida's Commission on Space in 1987–1988. Dr. Angelo is the author of seven other books on space technology and space exploration.